Advances in Experimental Medicine and Biology

Volume 1258

Advances in Experimental Medicine and Biology provides a platform for scientific contributions in the main disciplines of the biomedicine and the life sciences. This series publishes thematic volumes on contemporary research in the areas of microbiology, immunology, neurosciences, biochemistry, biomedical engineering, genetics, physiology, and cancer research. Covering emerging topics and techniques in basic and clinical science, it brings together clinicians and researchers from various fields.

Advances in Experimental Medicine and Biology has been publishing exceptional works in the field for over 40 years, and is indexed in SCOPUS, Medline (PubMed), Journal Citation Reports/Science Edition, Science Citation Index Expanded (SciSearch, Web of Science), EMBASE, BIOSIS, Reaxys, EMBiology, the Chemical Abstracts Service (CAS), and Pathway Studio.

2019 Impact Factor: 2.450 5 Year Impact Factor: 2.324

More information about this series at http://www.springer.com/series/5584

Eugenie S. Kleinerman • Richard Gorlick
Editors

Current Advances in the Science of Osteosarcoma

Research Perspectives: Tumor Biology, Organ Microenvironment, Potential New Therapeutic Targets, and Canine Models

Second Edition

Springer

Editors
Eugenie S. Kleinerman
Department of Pediatric Research
Department of Cancer Biology
University of Texas MD Anderson
Cancer Center
Houston, TX, USA

Richard Gorlick
Division of Pediatrics
University of Texas MD Anderson
Cancer Center
Houston, TX, USA

ISSN 0065-2598 ISSN 2214-8019 (electronic)
Advances in Experimental Medicine and Biology
ISBN 978-3-030-43084-9 ISBN 978-3-030-43085-6 (eBook)
https://doi.org/10.1007/978-3-030-43085-6

This Springer imprint is published by the registered company Springer Nature Switzerland AG
The registered company address is: Gewerbestrasse 11, 6330 Cham, Switzerland

To my father, Jerome I. Kleinerman, M.D., who was Chair of Pathology at Mt. Sinai Medical School in New York City, and later MetroHealth Medical Center, Case Western Reserve University School of Medicine. He inspired me to love the discipline of medicine and laboratory research and the importance of education. During his distinguished career as a pulmonary pathologist, his research integrated clinical medicine and experimental models of lung disease in an effort to improve the health of the people with occupational and chronic obstructive lung disease and lung cancer. He was my first role model, a man of ethics and conviction. His memory and words of wisdom continue to guide and inspire me.

This book is also dedicated to my mother, Seretta Miller Kleinerman, who preached and fought for equal opportunity for women in the workplace – a woman way ahead of her time. She pounded into my head that career and family were not mutually exclusive. Yet, she also stressed the importance of maintaining a lady-like decorum, insisting on perfect manners and gracious behavior.

My sisters and I often said we would write a book entitled "Seretta Says."

Finally, I thank my husband, Dr. Leonard Zwelling, for supporting, encouraging, and believing in me. He gave me the strength I needed to push through the insecurities that stem from being a professional woman and mother.

Eugenie S. Kleinerman, M.D.

Preface

Osteosarcoma: The State of Affairs Dictates a Change in Clinical Practice and Clinical Trial Design

We have made many new discoveries with regard to osteosarcoma biology and uncovered potential new targets for therapy. The challenge for us moving forward is: Can we apply these discoveries and alter clinical research practices to achieve success?

Osteosarcoma continues to claim the lives of too many children, adolescents, and young adults. Being both a rare and a pediatric cancer, the resources allocated to finding a cure and improving outcomes have been and will continue to be sparse. This is why, as we move forward, we must be judicious and strategic in the selection of which new agents we incorporate into our clinical treatment regimens and the clinical trial design constructed to assess the activity of these new agents. Experience and multiple clinical trials have defined an accepted three-drug chemotherapy regimen that results in a 65–70% overall survival at 5 years. However, clinical trial after clinical trial adding additional chemotherapeutic agents to this three-drug backbone failed to have an impact with no improvement in outcome since 1987. This is an unacceptable statistic. We need to recognize that we have achieved what we can with combination chemotherapy and move on.

The era of "targeted therapy" based on genomics and proteomics of the tumor cells has emerged. Genomic analysis of tumor tissue has identified potential targets for other solid tumors. However, the genetic signatures from individual osteosarcoma patient samples and even different metastatic tumor nodules in the same patient are not consistent, showing diverse genetic mutations and alterations. Furthermore, tumor cells do not grow in isolation. In my opinion, this approach will fail therapeutically unless we also understand (a) the interactions between the osteosarcoma cells and the lung microenvironment (the most common site of metastases), (b) which molecular pathways are altered epigenetically that permit bone cells to grow in the lung, and (c) how the osteosarcoma cells circumvent the immune response. We also need to understand how the osteosarcoma cells adapt to the lung microenvironment.

Recognizing the success of using chemotherapy to treat newly diagnosed osteosarcoma patients but also admitting that we have reached a plateau using this approach dictates that we must incorporate non-chemotherapy agents into our current three-drug regimen to improve patient outcomes. Such new agents can include those that target the dysregulated pathways that have been identified in the tumor cells, the tumor microenvironment, and the immune response.

How best to combine the new agent with chemotherapy and how to interdigitate it into the treatment schema based on our knowledge of the agent's target and whether chemotherapy can help or interfere must be a primary focus. These two books (the first focused on clinical practice and novel therapeutic discoveries and the second on laboratory research that will hopefully inspire new therapeutic ideas) have been compiled to bring the latest findings in regard to these three areas. National and international authorities have summarized the historical perspectives and their own clinical, translational, and laboratory research in an effort to provide a single resource to serve as the starting point for discussions as we move forward in designing novel therapeutic strategies. We cannot continue to merely add one new agent and measure success by evaluating response in the setting of bulky, visible relapsed disease. This has been our approach for the last 40 years. While it was successful in identifying the active chemotherapy agents, it is not appropriate for assessing the activity of immunotherapies, agents that target the tumor microenvironment or even agents that target specific pathways. In addition, we cannot continue to assess therapy activity by tumor shrinkage. Agents that activate an immune response resulting in immune cell infiltration into the tumor may be interpreted as tumor progression if response is judged by radiographic measurements. Without histologic evaluation, we cannot decipher whether an enlarged mass is a growing tumor or the result of immune cell infiltration, dead amorphous tissue, and edema. We must incorporate histologic evaluation and biologic measures that confirm that the target of the chosen agent is being affected. Proper resources must be devoted, and carefully designed clinical trials must be implemented. It is imperative that we use the discoveries made by the authors in this book to design our clinical trials, keeping in mind the biology of both the tumor and the organ microenvironment. If we do not implement such changes in our clinical research practice, we will continue to struggle and fail.

In this spirit, I express my gratitude to all of my distinguished colleagues for their willingness to contribute to this book. Without their assistance and their expertise, this project would not have been possible. It is my hope that the information in this book will provide inspiration, data, and the rationale needed to change the way we practice clinical research and design our clinical trials for patients with newly diagnosed and relapsed osteosarcoma.

Houston, TX, USA Eugenie S. Kleinerman

Contents

Contributors

Rebekah Belayneh Department of Orthopaedic Surgery, University of Pittsburgh Medical Center, Pittsburgh, PA, USA

Musculoskeletal Oncology Lab, University of Pittsburgh, Pittsburgh, PA, USA

Christopher D. Collier Department of Orthopaedics, University of Chicago Pritzker School of Medicine, Chicago, IL, USA

Timothy P. Cripe Nationwide Children's Hospital, The Ohio State University, Columbus, OH, USA

David S. Geller Department of Orthopedic Surgery, Montefiore Medical Center, Albert Einstein College of Medicine, Bronx, NY, USA

Patrick J. Getty Department of Orthopaedics, Case Western Reserve University, Cleveland, OH, USA

Seidman Cancer Center, University Hospitals – Cleveland Medical Center, Cleveland, OH, USA

Nancy Gordon Department of Pediatrics-Research, The University of Texas MD Anderson Cancer Center, Children's Cancer Hospital, Houston, TX, USA

Edward M. Greenfield Department of Orthopaedic Surgery, Indiana University School of Medicine and Indiana University Melvin and Bren Simon Cancer Center, Indianapolis, IN, USA

Bang H. Hoang Department of Orthopedic Surgery, Montefiore Medical Center, Albert Einstein College of Medicine, Bronx, NY, USA

Weidong Jin Department of Pediatrics, Section of Hematology/Oncology, Texas Children's Cancer Center, Baylor College of Medicine, Houston, TX, USA

E. Markel Justin West Virginia University School of Medicine, Morgantown, WV, USA

Eugenie S. Kleinerman Department of Pediatrics-Research, The University of Texas MD Anderson Cancer Center, Children's Cancer Hospital, Houston, TX, USA

Nadya Koshkina Department of Pediatrics-Research, The University of Texas MD Anderson Cancer Center, Children's Cancer Hospital, Houston, TX, USA

Brock A. Lindsey West Virginia University School of Medicine, Morgantown, WV, USA

J. Andrew Livingston Department of Sarcoma Medical Oncology, The University of Texas MD Anderson Cancer Center, Houston, TX, USA

Department of Pediatrics, The University of Texas MD Anderson Cancer Center, Houston, TX, USA

David M. Loeb Departments of Pediatrics and Developmental and Molecular Biology, Children's Hospital at Montefiore, Albert Einstein College of Medicine, Bronx, NY, USA

Linchao Lu Department of Pediatrics, Section of Hematology/Oncology, Texas Children's Cancer Center, Baylor College of Medicine, Houston, TX, USA

Nicola J. Mason Department of Clinical Sciences and Advanced Medicine, School of Veterinary Medicine, University of Pennsylvania, Philadelphia, PA, USA

Department of Pathobiology, School of Veterinary Medicine, University of Pennsylvania, Philadelphia, PA, USA

Parker Institute for Cancer Immunotherapy, University of Pennsylvania, Philadelphia, PA, USA

Grace Nehme Department of Pediatrics-Research, The University of Texas MD Anderson Cancer Center, Children's Cancer Hospital, Houston, TX, USA

Hillary G. Pratt West Virginia University School of Medicine, Morgantown, WV, USA

James Brandon Reinecke Department of Pediatrics, Nationwide Children's Hospital and The Ohio State University, Columbus, OH, USA

Ryan David Roberts Division of Hematology, Oncology, and Blood and Marrow Transplant, Department of Pediatrics, Center for Childhood Cancer and Blood Diseases, Abigail Wexner Research Institute at Nationwide Children's Hospital and The James Comprehensive Cancer Center at The Ohio State University, Columbus, OH, USA

Courtney Schott Department of Pediatrics, Division of Hematology and Oncology, University of California San Francisco, San Francisco, CA, USA

Avanthi Tayi Shah Department of Pediatrics, Division of Hematology and Oncology, University of California San Francisco, San Francisco, CA, USA

Amit Singla Department of Orthopedic Surgery, Montefiore Medical Center, Albert Einstein College of Medicine, Bronx, NY, USA

E. Alejandro Sweet-Cordero Department of Pediatrics, Division of Hematology and Oncology, University of California San Francisco, San Francisco, CA, USA

Jichuan Wang Department of Orthopedic Surgery, Montefiore Medical Center, Albert Einstein College of Medicine, Bronx, NY, USA

Musculoskeletal Tumor Center, Beijing Key Laboratory for Musculoskeletal Tumors, Peking University People's Hospital, Beijing, China

Lisa L. Wang Department of Pediatrics, Section of Hematology/Oncology, Texas Children's Cancer Center, Baylor College of Medicine, Houston, TX, USA

Mary Frances Wedekind National Institutes of Health, Bethesda, MD, USA

Kurt Weiss Department of Orthopaedic Surgery, University of Pittsburgh Medical Center, Pittsburgh, PA, USA

Musculoskeletal Oncology Lab, University of Pittsburgh, Pittsburgh, PA, USA

Kerri Wolf-Dennen Department of Pediatrics-Research, The University of Texas MD Anderson Cancer Center, Children's Cancer Hospital, Houston, TX, USA

Chia-Chin Wu Department of Genomic Medicine, The University of Texas MD Anderson Cancer Center, Houston, TX, USA

Rui Yang Department of Orthopedic Surgery, Montefiore Medical Center, Albert Einstein College of Medicine, Bronx, NY, USA

Yuanzheng Yang Department of Pediatrics-Research, The University of Texas MD Anderson Cancer Center, Children's Cancer Hospital, Houston, TX, USA

Genomic Complexity of Osteosarcoma and Its Implication for Preclinical and Clinical Targeted Therapies

1

Courtney Schott, Avanthi Tayi Shah, and E. Alejandro Sweet-Cordero

Abstract

Osteosarcoma is a genomically complex disease characterized by few recurrent single-nucleotide mutations or in-frame fusions. In contrast, structural alterations, including copy number changes, chromothripsis, kataegis, loss of heterozygosity (LOH), and other large-scale genomic alterations, are frequent and widespread across the osteosarcoma genome. These observed structural alterations lead to activation of oncogenes and loss of tumor suppressors which together contribute to oncogenesis. To date, few targeted therapies for osteosarcoma have been identified. It is likely that effectiveness of targeted therapies will vary greatly in subsets of tumors with distinct key driver events. Model systems which can recapitulate the genetic heterogeneity of this disease are needed to test this hypothesis. One possible approach is to use patient-derived xenograft (PDX) models characterized with regards to their similarity to the human tumor samples from which they were derived. Here we review evidence pointing to the genomic complexity of osteosarcoma and how this is reflected in available model systems. We also review the current state of preclinical testing for targeted therapies using these models.

Keywords

Patient-derived xenografts · Targeted therapy · Combination therapy · MYC · CCNE1 · AKT · PTEN · VEGFR · CDK4

General Introduction

The genome of osteosarcoma is highly complex, and tumors are extraordinarily heterogeneous between patients. This heterogeneity may be one reason why it has been difficult to successfully identify targeted therapies for this disease. Osteosarcomas are characterized primarily by structural rearrangements, aneuploidy, and copy number alterations, whereas recurrent point mutations are few. Whether the observed structural alterations are truly targetable vulnerabilities remains to be fully explored, and it is almost certain that combination approaches need to be developed, since single agents are unlikely to lead to significant tumor regression. Given that osteosarcoma is a rare disease, the development of new therapies will depend heavily on preclinical testing in appropriate models. In this chapter,

C. Schott · A. T. Shah · E. A. Sweet-Cordero (✉)
Department of Pediatrics, Division of Hematology and Oncology, University of California San Francisco, San Francisco, CA, USA
e-mail: courtney.schott@ucsf.edu; avanthi.shah@ucsf.edu; alejandro.sweet-cordero@ucsf.edu

E. S. Kleinerman, R. Gorlick (eds.), *Current Advances in the Science of Osteosarcoma*, Advances in Experimental Medicine and Biology 1258, https://doi.org/10.1007/978-3-030-43085-6_1

we review the current understanding of the complex genome of osteosarcoma and discuss how this complexity is reflected in available cell lines and patient-derived xenografts (PDXs). We also review efforts to evaluate possible targeted therapies in these models and briefly discuss the impact of this testing on clinical trial development. We focus here only on human model systems and defer discussion of mouse models and the canine patient to other chapters.

Genomics of Human Osteosarcoma

Osteosarcoma has long been known to be characterized by a complex karyotype. An early cytogenetic analysis of 73 samples from 53 patients identified widespread aneuploidy, recurrent numerical abnormalities, and recurrent chromosomal breakpoints [1]. The authors concluded that the majority of osteosarcomas were characterized by complex chromosomal abnormalities with marked cell-to-cell heterogeneity. This study was one of the first to determine that chromosomal breakpoints were nonrandom, with increased frequency occurring in defined regions of chromosomes 1, 2, 3, 4, 11, 12, 14, 16, 17, 19, and 22. Loss of whole chromosomes was also noted to be more common than whole chromosome gains, highlighting the importance of loss of heterozygosity (LOH) in this disease. In particular, frequent LOH of chromosome 17 was observed, a region which contains the tumor suppressor TP53. This early pioneering work noted a distinct cytogenetic profile in parosteal osteosarcoma, which they observed to be characterized by the presence of ring chromosomes. This provided an early indication that there may be molecular subtypes of osteosarcoma with unique features. Applying spectral karyotyping (SKY) to refine these prior cytogenetic studies, Bayani and colleagues analyzed 14 primary tumors and 4 established osteosarcoma cell lines [2]. They quantified a high rate of structural rearrangements, with an average of 38.5 breakpoints per tumor with those involving chromosomes 8 and 20 being disproportionately prevalent. Gains in 8q23–24 and 17p11–13 were also frequently observed. In addition to these and other studies evaluating the landscape of alterations in osteosarcoma [3, 4], a large body of literature has highlighted specific alterations in tumor suppressors and oncogenes including CDK4 [5], p53 and MDM2 [6], MYC [7, 8], Rb [9], and others. Other comprehensive molecular profiling studies of osteosarcoma have demonstrated that copy number amplification and overexpression of genes in chromosome 8 and chromosome 17 strongly correlate with osteosarcoma progression and relapse [10, 11]. Summarizing the work of several laboratories, Martin et al. concluded that conventional osteosarcoma is characterized by losses of portions of chromosomes 3q, 6q, 9, 10, 13, 17p, and 18q and gains of portions of chromosomes 1p, 1q, 6p, 8q, and 17p [9]. 6p gains commonly involve RUNX2, VEGFA, E2F3, PIM1, and CCND3, all of which could potentially play oncogenic roles in this disease. 8q contains MYC, which has been reported to be frequently amplified in osteosarcoma by many groups [12, 13]. 8q also contains RECQL4. While germ line loss of RECQL4 causes Rothmund-Thomson syndrome and predisposes patients to osteosarcoma development, in sporadic osteosarcoma tumors, RECQL4 has been reported to be upregulated at the level of protein expression or amplified at the gene level [14].

Cheng et al. evaluated copy number changes in a large panel of 117 osteosarcomas using Affymetrix SNP arrays [15]. They noted frequent gains in chromosomes 8, 12, 21, and X and frequent deletions in chromosomes 2, 10, and 13. These authors also correlated copy number alterations with therapy response in sarcoma cell lines. For example, they noted a correlation between IGFIR copy number gain and sensitivity to clofarabine. Similarly, Smida et al. profiled 45 osteosarcomas using Affymetrix SNP arrays and identified frequent alterations in 6p21 (including VEGFA, CCND2, and RUNX2), 8q24 (including MYC), and 12q14 (including CDK4) as well as loss of 10q21.1. They noted that a high LOH score (greater than 1500 loci with LOH) was predictive of a poor response to chemotherapy and a higher risk of recurrence. This was one of the first papers to correlate specific alterations with poor event-free survival [16]. In a more recent study, Smida et al. evaluated a larger cohort of 160 pretreatment osteosarcoma samples using high-density arrays. They described frequent loss of tumor suppressors WWOX (31%), DLG2 (27%), and LSAMP (8%), in addition to Rb and Trp53. They also described the frequency of "chromothripsis-like pattern" (CTLP) and noted that the presence of CTLP in osteosarcoma was associated with a worse outcome [17]. Overall, it should be emphasized that loss of the tumor suppressors Trp53 (17p13.1) and Rb (13q14.2) is very common in sporadic osteosarcoma and germ line deletions are strongly associated with increased incidence [18, 19]. In addition to loss of function of Rb itself, osteosarcomas often exhibit deregulation of this pathway through other means, such as loss of the tumor suppressor CDKN2A/p16/INK4A or amplification of cyclin-dependent kinase 4 (CDK4).

The widespread availability of next-generation sequencing (NGS) has allowed further refinement of our understanding of the genomics of osteosarcoma. Two initial studies using NGS confirmed the prediction that these tumors are characterized by recurrent structural rearrangements. A very interesting observation was that Trp53 is more commonly altered by structural rearrangements in intron 1, rather than by deep deletion or mutation [20]. DLG2, ATRX, and RB alterations were also frequently noted. Another study identified frequent alterations in the PI3K pathway and suggested that this may represent a targetable opportunity for a subset of patients [21]. More recently, Sayles and Breese et al. also identified a number of subsets of osteosarcoma patients with potentially targetable alterations using whole-genome sequencing [22]. Alterations in MYC, CDK4, VEGFA, AKT or PTEN, and CCNE1 were frequently observed across a cohort of 63 tumors profiled using whole-genome sequencing. These and other studies suggest that some subsets of osteosarcoma may contain gains or losses in genes that represent potential therapeutic vulnerabilities. However, further identification and validation of such therapy biomarkers will likely require a combination of preclinical and clinical studies.

A critical issue that remains to be fully explored is the prognostic and therapeutic value of specific genomic alterations in osteosarcoma. To address this, Suehara et al. evaluated 66 patients with osteosarcoma using a clinical grade panel sequencing assay and found a fraction of patients with targetable or potentially targetable alterations [23]. Twenty-one percent of patients had a genomic alteration suggestive of an actionable alteration including CDK3, MDM2, BRCA2, PDGFRA, and VEGFR. In another study, Kovac et al. sequenced the exomes of 31 tumors and identified 14 genes as the main drivers in osteosarcoma. They also suggested that a large percentage of osteosarcomas have genome instability signatures characteristic of BRCA1/2-deficient tumors [24]. Whether this will translate into a therapeutic opportunity as it has for BRAC1/2-deficient breast and ovarian cancers remains to be explored.

To date, relatively few studies have systematically evaluated the evolution of osteosarcoma using matched samples; thus, we know little regarding how this disease progresses over time in individual patients, which is a significant gap in knowledge with regard to the development of targeted therapies. Negri and colleagues used whole-genome and whole-exome sequencing to deeply characterize a set of 13 primary and metastatic matched pairs [25]. High conservation of copy number in the matched pairs was seen, suggesting that perhaps the concept of osteosarcoma as a "genomically unstable" cancer needs to be further refined. Alternatively, it is possible that many of the genomic events seen in osteosarcoma are early events that remain stable over time. These authors also identified a recurrent amplification in the gene KDR in metastatic osteosarcoma. In another recent study, Brady and colleagues performed a deep analysis of tumor evolution in four osteosarcoma patients. They described a pattern of "branching evolution" shaped by treatment with cisplatin [26]. In general, these studies point to the need for analysis of more longitudinal samples, so we can begin to understand the evolutionary dynamics of osteosarcoma and how they may define therapeutic vulnerabilities at relapse.

Overall, a great deal of knowledge has been gained regarding the genomic alterations present in osteosarcoma. Here we have emphasized studies evaluating structural alterations in the genome of osteosarcoma. However, many studies have also pointed to an important role for epigenetic mechanisms including DNA methylation. For example, promoter methylation may be an important mechanism of silencing the p16INK4A locus in osteosarcoma [27]. Studies using osteosarcoma models for preclinical therapeutic development will need to take the above summarized complexity of this disease into account. It is our belief that because of this complexity, osteosarcoma should be considered to include subsets of tumors with different drivers and thus will require an evaluation of specific genomic and epigenomic alterations to define appropriate subgroups for targeted therapy. Below we discuss the current understanding of the genomics of available preclinical models followed by a summary of some therapeutic studies with an emphasis on those that have addressed genomics while evaluating response.

Cell Line Genomics

Cell lines are the most widely utilized model system to study osteosarcoma due to their ease of use and wide availability. Several authors have attempted to evaluate the genomic characteristics of available cell lines to correlate the observed alterations with those found in patient samples. Ottaviano et al. used multiplex ligation-dependent probe amplification (MLPA) to screen for loss of 38 tumor suppressors across 19 osteosarcoma cell lines, including HOS, Saos2, U2OS, and other commonly used cell lines [28]. Loss of CDKN2A (42%) and TP53 (47%) was the most frequently observed alteration. Notably, these percentages are higher than what has been reported when evaluating tumors in vivo [22]. Lorenz et al. performed whole-genome sequencing of osteosarcoma cell lines including IOR/ OS15, IOR/ OS18, MG63, and ZK-58. They found that inactivating genomic rearrangements, most commonly involving TP53, were frequent in both osteosarcoma patient samples (10/25) and cell lines (7/11). Interestingly, they also reported that osteosarcoma cell lines had numerous deletions, tandem duplications, inversions, and interchromosomal translocations at frequencies similar to human tumor samples [29]. They suggest that involvement of nonhomologous end-joining (NHEJ) and microhomology-mediated end-joining (MMEJ) DNA repair contribute to the generation of structural alterations in osteosarcoma. Additionally, RB1 rearrangements were similarly common between cell lines (3/11) and patient samples (5/25). Lastly, the presence of multiple fusion transcripts in osteosarcoma was also reported, including a previously undescribed fusion of *PMP22-ELOVL5*.

Cell Line Models of Metastatic Disease

Most osteosarcoma patients who do not survive die from metastatic disease; therefore, there has been a strong emphasis on developing metastatic cell line models. Several studies have evaluated the tumorigenicity and metastatic capacity of established osteosarcoma cell lines [30–33]. Some, but not all, commerical cell lines are capable of forming tumors after subcutaneous implantation [31]. Formation of spontaneous metastasis from those subcutaneous xenografts is less common [31, 34, 35]; however, formation of metastatic nodules after intravenous injection does occur with many cell lines [30, 35–39] (Table 1.1). Notably, there is discordance with respect to the metastatic capacity of certain cell lines between published studies, perhaps reflective of genetic drift in sublines expanded in different laboratories (Table 1.1).

To help overcome the variation in metastatic capacity observed in commercially available cell lines, in vivo passaging or transformation of established cell lines has been used to create metastatic derivatives (e.g., MG63.2 [71], MG63.3 [30], 143B, SaOS2-LM2 to LM7 [37, 38],

Table 1.1 Genomic characteristics of cell line models for osteosarcoma

Cell line model (derivation)				Genomic characteristics	Tumorigenicity	Metastatic capacity
MG63 [40, 41]				MTAP-BCN2 fusion transcript (inversion ch9) [42]; fusion transcript with TP53 [42]; P53 arrangement intron 1 [43]; homozygous deletion CDKN2A [42, 44]; p53 wildtype [44]	*Yes*: SQ [45]; OT [46, 47] *No*: SQ [48]	*No*: IV [47, 49]; SQ/IM [48]; OT [47, 50]; *Yes*: OT [46]
	MG63.2 [46]				*Yes*: OT [46, 47]	*Yes*: IV [47]; OT [46, 47]
	MG63.3 [47]				*Yes*: OT [47]	*Yes*: IV [47]; OT [47]
U20S [51]				PMP22-ELOVL5 fusion [42]; P53 [44, 52]; single gene deletion P53 [43]; hemizygous deletion CDKN2A MLPA normal aCGH [44]	*Yes*: SQ [45, 48, 53]; OT [47]	*Yes*: IV [53] *No*: IV [49]; SQ [48, 53];IM [48]
Saos2				Fusion transcript involving TP53, RB1 transcript truncated after exon 20 [42]; P53 deletion [43, 44, 54–56]; normal CDKN2A [44]	*Yes*: SQ [45, 57]; OT [47, 50, 57–60] *No*: SQ [48]	*Yes*: OT [47, 58–60] *No*: IV [37, 47, 49]; SQ/IM [48]; OT [50]
	SaOS-LM2 [37]					*No*: IV [37, 61]
	SaOS-LM3–6 [37]					Yes: IV [37, 61]
	SaOS-LM7 [61]				*Yes*: OT [47, 50]	*Yes*: IV [47, 61];OT [47, 50]
TE85					*Yes*: OT [47] *No*: OT [50]	*No*: IV [47]; OT [47]
	HOS [62]			TP53 mutant, Rb wildtype [63]; homozygous CDKN2A deletion, TP53 p.Arg156Pro (c.467G > C) [44]	*Yes*: SQ [45]; OT [47] *No*: SQ [48]	*No*: IV [47, 49]; SQ/IM [48]; OT [47]
		MNNG/HOS [64]		KRAS [64]	*Yes*: OT [47, 65]; SQ [45, 48]; IM [48]	*Yes*: IV [47]; OT [47, 65] *No*: SQ/IM [48]
			KHOS			
		143B			*Yes*: SQ [45, 48]; IM [48]; OT [47, 50, 65, 66]	*Yes*: IV [47, 66]; SQ/IM [48]; OT [47, 50, 65, 66]
		KRIB [67]			*Yes*: OT [47, 68]	*Yes*: IV [47]; OT [47, 68]
G292					*Yes*: SQ [45] *No*: OT [50]	*No*: IV [49]; SQ/IM [48]
SJSA (OSA)				MDM2 amplification [42]; hemizygous deletion CDKN2A, P53 wildtype [44]	*Yes*: SQ [45, 48]	*No*: SQ/IM [48]
OHS [69]				Gain CKN2A aCGH, P53 mutant p.Glu286Lys [44]	*Yes*: SQ [45, 48, 70]; OT [70]	*No*: SQ/IM [48]

KRIB [67] etc.). These metastatic derivatives can then be compared to their cell line of origin to investigate possible mechanisms of metastatic progression [30, 72, 73]. For example, Muff et al. showed that metastatic derivatives of established osteosarcoma cell lines have increased Hedgehog and WNT signaling pathways compared to their cell line of origin [73].

Orthotopic implantation is becoming increasingly common, despite the increased technical challenges [30, 33, 39, 71, 72, 74–85]. In general, osteosarcoma cell line xenografts more consistently form tumors at orthotopic sites than at subcutaneous or intramuscular sites (Table 1.1). To facilitate metastatic progression in orthotopic models, limb amputation is frequently performed as it allows for more time for metastasis development before the mice need to be sacrificed [30, 33, 39, 76]. Amputations are typically performed when the orthotopic tumor reaches 1–2 cm in diameter, and the mice are sacrificed 2–8 weeks later or when metastasis-associated morbidity is observed [30, 33, 39, 76]. A thorough investigation by Berlin et al. using KRIB cells demonstrated that a minimum of 2 weeks post orthotopic implantation is required for formation of pulmonary nodules [84]. Amputation models are arguably the ones that most accurately recapitulate the natural disease.

The phenotypic and genomic stability of continuously passaged cells has been evaluated. One commercial osteosarcoma cell line, SaOS2, exhibited increased proliferation and matrix mineralization with passage number, although many other phenotypic properties were stable over at least 100 passages and no significant changes in the expression of various growth factors were observed [86]. The use of cell lines to model osteosarcoma has certain pitfalls and limitations. Cross contamination with other cell lines and mycoplasma infection are frequent problems and must be considered [87, 88]. Additionally, when comparing cell line gene expression profiles to primary tumor samples of the same cancer type, not all cell lines have gene expression profiles that correlate with samples from their presumed tumor of origin [89]. Finally, cancer cell line xenografts lack complexity in both cellular heterogeneity and with regard to their tumor microenvironment [90]. The need for a more representative and reliable model to study osteosarcoma has led to increased utilization of the PDXs, which may more faithfully recapitulate features of the primary disease, and their derivative cell lines.

An increasingly utilized approach to model metastasis is the use of the ex vivo pulmonary metastasis assay (PuMA) [91]. Recently, this assay was used to carry out a screen for over 100 potential therapies for metastatic osteosarcoma, with CDK12 inhibition emerging as a strong candidate [92]. In another powerful approach, Morrow et al. recently mapped putative enhancer elements in matched human osteosarcoma tumors and in metastatic/nonmetastatic cell line pairs and identified metastasis-associated variant enhancer loci. This led them to identify a potential role for individual genes as key metastatic drivers [93].

PDX Models

To produce a PDX model, tumor tissue, as a fragment or after digestion into a single-cell suspension, is transplanted directly into an immunodeficient mouse [90]. PDX establishment was initially made possible by the introduction of nude mice and, more recently, the more immunodeficient NSG mouse [94]. In 1982, Ishii et al. reported that 24 of 30 patient-derived osteosarcoma samples established subcutaneous tumors in mice and they were able to continuously passage and maintain 2 of these PDXs in nude mice for 3 years [95]. Bauer et al. described establishment of six models in nude mice and noted that one line became polyploid after extended passage [34]. Meyer et al. described establishing eight transplantable osteosarcoma lines in mice made immunodeficient by whole-body irradiation [96].

Reported success rates vary widely for the initial establishment of osteosarcoma PDXs in mice (20–100%) [34, 95, 97–101]. Success rates can vary with tumor type, tumor stage (biopsy of a primary tumor vs. metastatic tumor), mouse strain, and implantation site [90]. PDX maintenance requires continuous passaging through immunodeficient mice. Published osteosarcoma PDX models report widely variable implantation sites, including bone (femur [98], intratibial [102]), lung [103], muscle [104], subrenal capsule [22], and subcutis [34, 79, 101, 105–109]. Overall osteosarcoma PDXs maintain similar gross and histologic features to their primary tumor of origin [22, 96, 97, 102, 109–111], with preservation of the patient's tumor transcriptome [112] and tumor microenvironment [113]; however, some reports have described divergence [114]. Using microarray-based comparative genomic hybridization, Kresse et al. analyzed nine matched osteosarcoma patient/xenograft pairs over multiple xenograft passages. They concluded that most alterations were maintained, although some drift was identified [108]. Other authors have also found close similarity of their PDX models to the primary tumor and stable characteristics of PDX models in serial passage [109].

Several authors have reported the development of orthotopic osteosarcoma models (PDOX). This model is ideal because the transplanted tissue grows at a site that is most similar to its natural environment [90]. The orthotopic location could, in theory, more closely recapitulate the biology of osteosarcoma, including the metastatic phenotype [115]. Many researchers have developed osteosarcoma PDOX models [98, 102, 104, 110, 111, 115–120], including some with amputation protocols that facilitate the development of metastasis [84, 104, 116]. Goldstein et al. showed that metastatic spread was not observed after subcutaneous implantation of two osteosarcoma PDXs, whereas after orthotopic implantation, pulmonary metastases developed with both PDXs [104].

Cell lines can also be derived from PDXs, and several groups have used this approach [22, 101, 121, 122]. An important consideration is whether the PDX-derived cell lines are obtained from pretreatment biopsies, localized recurrence, or distant metastasis. For example, some lines obtained from pretreated tumor type have been reported to be cisplatin-resistant and can therefore be useful in the evaluation of combination therapies for relapsed disease such as trabectedin and temozolomide [115].

In perhaps the most comprehensive analysis to date, Stewart et al. evaluated 31 osteosarcoma patient-derived orthotopic models, compared them to their tumor of origin, and characterized the alterations that occur over time. This work also evaluated a large number of models for other pediatric solid tumors. Extensive evaluation determined there was preservation of clonal complexity between the primary tumors and the corresponding PDX. Furthermore, evolution of the PDX over subsequent passages was also evaluated. Notably, among all the tumors examined, osteosarcomas had the best preservation of the primary tumor's clonal complexity [98]. These results suggest that the use of osteosarcoma PDX models matched to the primary tumor may be a particularly fruitful strategy for identification of biomarker-driven therapeutic opportunities.

Preclinical Models for Drug Testing

Many authors have tested individual targeted therapies using cell lines. We will not make a comprehensive effort to evaluate all of these due to space limitations. Importantly, very few studies have systematically evaluated a large number of drugs in screens using cell lines. A recent notable effort utilized short-term cultures from PDX models, representing one possible approach [110]. Perhaps the most comprehensive effort to utilize osteosarcoma PDX models for preclinical evaluation of targeted agents has been done through the Pediatric Preclinical Testing Consortium (PPTC), formerly known as the Pediatric Preclinical Testing Program (PPTP). The PPTC is a National Cancer Institute (NCI)-sponsored initiative for the investigation, consideration, and prioritization of drugs for early-phase pediatric clinical trials. This multi-institutional consortium uses cell line and xenograft tumor panels to evaluate the antitumor activity of agents in osteosarcoma, as well as other sarcomas, renal tumors, neuroblastoma, CNS tumors, and hematologic malignancies [123].

The PPTC has evaluated a significant number of potential therapeutics for osteosarcoma. Here we highlight a few of these rather than attempting a comprehensive review. Glycoprotein NMB (GPNMB), also known as osteoactivin, is a transmembrane glycoprotein that is expressed in many nonmalignant cells, including osteoblasts, and is overexpressed in many malignancies, including osteosarcoma, making it a good candidate for targeting [124]. Glembatumumab vedotin is an antibody-drug conjugate that combines an anti-GPNMB antibody with the antimitotic agent vedotin [125]. Upon binding of the antibody to GPNMB, the drug is internalized and active glembatumumab vedotin is released, causing cell-cycle arrest and death. Because transmembrane expression of GPNMB is required for binding and downstream cytotoxicity of glembatumumab vedotin, the PPTC tested it in a subset of osteosarcoma xenografts that were known to express GPNMB. It yielded high activity in some xenografts and response seemed to be related to GPNMB expression [126].

The insulin-like growth factor-1 receptor (IGF1R) is a transmembrane receptor tyrosine kinase. Upon binding of insulin-like growth factor-1 or -2, IGF1R becomes autophosphorylated and activates multiple downstream signaling pathways that regulate cell growth and development, including the PI3K/Akt pathway [127]. The IGF1R pathway has been implicated in the pathogenesis of both pediatric and adult sarcomas [128, 129]. The PPTC evaluated anti-IGF1R monotherapies, including the monoclonal antibodies robatumumab and cixutumumab, and the small molecule inhibitor BMS-754807. Robatumumab demonstrated in vitro and in vivo activity [130]. These results were reproduced by other investigators, who also showed that combining robatumumab with cisplatin or cyclophosphamide further enhanced its activity [131]. Cixutumumab only demonstrated in vivo activity in osteosarcoma [132]. Furthermore, cixutumumab in combination with rapamycin resulted in increased antitumor activity, compared to either agent alone [133]. BMS-754807 also demonstrated in vitro and in vivo activity [134]; however, this drug is no longer in development for pediatric use. All three anti-IGF1R agents demonstrated varying degrees of antitumor activity; however, IGF1R copy number (assessed by PCR, FISH, and dot blot analysis), IGF1R mRNA expression (determined by RT-PCR), and IGF1R surface antibody expression (measured by flow cytometry) were not correlated with response to therapy [135].

The vascular endothelial growth factor (VEGF) pathway plays a critical role in angiogenesis, and activation of this pathway has been described in a variety of inflammatory disease as well as cancer [136]. Several tyrosine kinase inhibitors with activity against the VEGF receptor (VEGFR) family have been investigated by the PPTC. For example, cediranib, a highly potent VEGFR inhibitor, demonstrated promising antitumor activity [137]. Subsequently, phase I clinical studies with this receptor tyrosine kinase (RTK) inhibitor have been performed in a variety of advanced pediatric solid tumors, including osteosarcoma [138]. Sunitinib and sorafenib, which are multi-RTK inhibitors with high activity for VEGFRs, also demonstrated tumor growth inhibition [139, 140]. In a phase I trial of sunitinib, one osteosarcoma patient demonstrated stable disease, but no objective responses were observed in any tumor type [141]. Although a phase II Children's Oncology Group (COG) trial of sorafenib in refractory pediatric solid tumors did not demonstrate objective responses [142], a phase II trial by the Italian Sarcoma Group did demonstrate activity, when administered as a combination therapy, in relapsed and unresectable high-grade osteosarcoma [143]. Another multi-RTK, regorafenib, which targets VEGFR1–3, as well as BRAF, FGFR1, KIT, PDGFR-β, RAF-1, and RET, demonstrated modest tumor inhibition in osteosarcoma xenograft models [144]. The SARC024 phase II trial of regorafenib demonstrated improved progression-free survival in progressive, metastatic osteosarcoma patients [145].

SRC is overexpressed in osteosarcoma and results in increased cell proliferation and decreased apoptosis [146]. Dasatinib, a multi-RTK inhibitor that also has activity against the Src family of kinases, demonstrated intermediate tumor growth inhibition in osteosarcoma models when tested by the PPTC [147]. Interestingly, in another study, dasatinib altered metastatic potential of osteosarcoma in vitro, but not in vivo [148]. Phase I testing of dasatinib monotherapy in pediatric solid tumor patients demonstrated poor activity [149]; however, investigations of dasatinib combination therapies in both pediatric and adult advanced solid tumors are ongoing (NCT00788125, NCT03041701, NCT02389309).

mTOR activation has also been implicated in osteosarcoma tumorigenesis. The mTOR inhibitor rapamycin demonstrated intermediate to high activity in osteosarcoma xenografts as a single agent [150] and in combination with cytotoxic chemotherapy [151]. However, a phase I trial of temsirolimus, another mTOR inhibitor, did not result in significant antitumor activity in osteosarcoma [152]. Furthermore, phase II testing of rapamycin in combination with cyclophosphamide failed to meet its primary endpoint [153]. In contrast, phase II and III studies of ridaforolimus, another mTOR inhibitor, have had more promising results [154, 155]. Treatment with dual mTORC1 and mTORC2 inhibitors may help circumvent some of the resistance that develops after mTOR inhibition; however, the dual inhibitor AZD8055 demonstrated poor activity in vivo [156].

PI3K and Akt activation are thought to contribute to tumorigenesis in osteosarcoma [157]. MK-2206, a highly selective, non-ATP-competitive, pan-Akt inhibitor, and GSK690693, an ATP-competitive, pan-Akt inhibitor, both demonstrated prolonged event-free survival, but did not result in any objective responses in osteosarcoma xenografts [158, 159]. As tumor cells often respond to inhibition of this pathway by activating feedback loops, a large challenge to therapeutic targeting of the PI3K/Akt pathway is the development of resistance. Several other groups have investigated targeting of the PI3K and Akt pathways with other agents, and these results will be discussed below.

The MAPK/ERK pathway is activated by many different growth factor receptors, including IGF1R, VEGFR, and PDGFR, which are often overexpressed in osteosarcoma, and thus it has been considered a promising candidate for inhibition [160]. Unfortunately, selumetinib, a potent MEK1/2 (a.k.a MAP2K1/2) inhibitor, had poor in vitro and in vivo antitumor activity, as a single-agent therapy, despite demonstrating significant pathway inhibition [161]. Though MEK inhibitors hold much potential, clinical application of these inhibitors can prove challenging due to the complex signaling cascades and cross talk between pathways. To this end, clinical trials have investigated combination therapies incorporating MEK inhibitors for refractory sarcoma patients, but did not demonstrate significant antitumor activity [162, 163].

As mentioned above, cell-cycle aberrations occur frequently in osteosarcoma, with amplifications in CDK4, CDK6, CCND, and CCNE and deletions in CDKN2A/B. Dinaciclib, a potent inhibitor of CDK1, CDK2, CDK5, and CDK9, was tested by the PPTC, and stable disease was the best response that was seen [164]. In contrast, when tested in PDXs harboring copy number changes in cell-cycle checkpoint genes [22], dinaciclib and palbociclib, a CDK4/6 inhibitor, demonstrated significant inhibition of tumor growth. This suggests that CDK inhibition may be efficacious in a subset of osteosarcoma patients who harbor genomic aberrations in cell-cycle genes. Several cell-cycle inhibitors are available, and these results support further preclinical testing in a genomically informed manner.

The Aurora kinase family members are key regulators of mitosis and cell-cycle progression, and amplifications in Aurora kinase A and B have been described in osteosarcoma [2]. Testing of the Aurora kinase A inhibitor alisertib by the PPTC resulted in high antitumor activity in one osteosarcoma xenograft and intermediate activity in the remaining five xenografts [165]. However, phase II testing of alisertib in relapsed and refractory solid tumors demonstrated very poor response rates (<5%) [166]. The serine/threonine kinase Chk1 directs the DNA damage response (DDR) and cell-cycle checkpoint response. Prexasertib, a CHK1 inhibitor, demonstrated poor single-agent activity when tested by the

PPTC; however, when combined with irinotecan, a topoisomerase I inhibitor, it showed prolonged event-free survival in osteosarcoma xenograft studies [167].

As described above, inactivating structural alterations in TP53 occur in the majority of osteosarcoma tumors. In a subset of osteosarcoma, inactivation can occur through amplification of MDM2. Serdemetan is an MDM2 antagonist that reactivates p53 and results in apoptosis. It resulted in tumor growth inhibition when tested by the PPTC. Unfortunately, a phase I trial in adults with advanced solid tumors did not demonstrate significant clinical activity [168].

Finally, several other targeted agents have been tested by the PPTC that failed to demonstrate significant antitumor activity. These include navitoclax [169], a potent Bcl-2 inhibitor; lapatinib [170], an EGFR inhibitor; RG7112 [171], an MDM2 inhibitor; pevonedistat [172], a NEDD8 inhibitor; RO4929097 [173], a gamma-secretase inhibitor that targets the NOTCH pathway; and seclidemstat, an LSD1 inhibitor [174].

Other PDX Studies

Several other investigators have evaluated targeted agents in in vivo osteosarcoma models. The Italian Sarcoma Group investigated inhibition of the mTOR pathway and showed that sorafenib monotherapy resulted in decreased mTORC1 signaling but resulted in mTORC2 activation as an escape mechanism. However, when sorafenib was combined with everolimus, another mTOR inhibitor, it caused increased antitumor activity and complete inhibition of the mTOR pathway in osteosarcoma xenograft models [175]. As mentioned above, a nonrandomized phase II trial investigating the utility of this combination in patients with unresectable, progressive, high-grade osteosarcoma was subsequently activated and showed promise. Although the combination therapy demonstrated activity, it failed to achieve the prespecified outcome of a 6-month progression-free survival of >50% [176].

One group investigated PI3K inhibition in osteosarcoma using two different agents, NVP-BEZ235, a dual PI3K/mTOR inhibitor, and BYL719, a PI3K inhibitor that specifically targets the alpha isoform. Both agents slowed osteosarcoma tumor growth in allograft and xenograft mouse models; however, neither drug induced in vivo tumor shrinkage [177, 178]. Interestingly, administration of ifosfamide in combination with BYL719 resulted in a synergistic effect [178]. Combining NVP-BEZ235 with cisplatin also enhances its antitumor effects in osteosarcoma xenografts [179]. Furthermore, in other in vivo sarcoma models, vincristine and NVP-BEZ235 combination therapy decreased metastasis and slowed tumor growth [180]. These studies highlight the importance of considering combination therapy to help potentiate an agent's efficacy. There are several ongoing phase I/II clinical trials investigating these two agents in adult cancers as monotherapies or in combination with other agents.

Multi-RTKs have also been an avenue of investigation for groups outside of the PPTC. Gobin et al. showed that imatinib, which targets PDGFRα, PDGFR-β, EGFR, IGF1R, and several other receptors, inhibits tumor growth in mouse models of osteosarcoma [181]. Sampson et al. demonstrated that crizotinib, which is FDA-approved for ROS1-positive non-small cell lung cancer and ALK-positive solid tumors, had significant antitumor effects in vitro and in vivo, at least in part, through inhibition of Met [182].

Other investigators have also studied MDM2 inhibition. Treatment with Nutlin-3 caused disruption of p53-MDM2 binding, resulting in decreased degradation of p53. Of note, efficacy of the agent is dependent on wild-type p53, which is seen in only a small subset of osteosarcoma. Osteosarcoma xenografts with wild-type p53 were treated with Nutlin-3 and experienced significant tumor growth inhibition [183].

Beyond the therapeutic targets already mentioned, several groups have identified targets of interest that have not been investigated by the PPTC. One such target is ezrin, a cytoskeletal linker protein that connects actin cytoskeleton to plasma membrane proteins [184]. High ezrin expression in an osteosarcoma mouse model resulted in pulmonary metastasis and has been associated with poor survival in dogs with naturally occurring osteosarcoma, as well as pediatric patients [185]. When an ezrin-dependent metastatic mouse model was treated with NSC305787 and NSC66839, both small molecule inhibitors of ezrin, overall survival was increased with both drugs, though it was only statistically significant for NSC305787 [186]. Additionally, targeting of the ezrin-regulated mTOR/S6K1/4E-BP1 pathway with rapamycin and CCI-779 resulted in pulmonary metastasis inhibition and prolonged survival in vivo [187]. Another group demonstrated that sorafenib treatment in vivo resulted in decreased tumor volume and metastasis, in part through downregulation of the ezrin pathway [188].

Another approach to inhibit osteosarcoma metastasis is to target chemokines and their receptors. CXCR3 expression has been described in a variety of malignancies, including osteosarcoma [189], and is thought to play a role in metastasis [190, 191]. Treatment with AMG487, a small molecule inhibitor of CXCR3, resulted in a significant reduction in metastatic burden in a metastatic model utilizing the SaO2-LM7 cell line [192].

Heat shock protein 90 (HSP90) stabilizes and activates a multitude of proteins. Many of these proteins are involved in constitutive signaling and responses to stress [193, 194]. Cancer cells use this complex to protect mutated and overexpressed oncoproteins from misfolding and degradation. When osteosarcoma xenograft models were treated with single-agent alvespimycin, an HSP90 inhibitor, or in combination with imetelstat, a telomerase inhibitor, there was significant reduction in tumor volume [195]. Similarly, Ory et al. demonstrated that treatment with the HSP90 inhibitor PF4942847, alone, or in combination with zoledronic acid, resulted in significant tumor growth inhibition and decreased metastases [196].

Although recurrent Wnt/β-catenin mutations have not been described in osteosarcoma, the role of this pathway in osteosarcoma biology is an area of investigation. Tegavivint, a small molecule inhibitor of β-catenin, had strong antitumor effects in both primary and metastatic tumors in a osteosarcoma xenografts [197]. Interestingly, treatment with BHQ880, a monoclonal antibody against dickkopf-1 (DKK-1), which is an inhibitor of Wnt signaling, resulted in decreased tumor growth and metastases [116].

A subset of osteosarcoma is characterized by MYC overexpression. Historically, MYC has been considered undruggable; however, recent studies have demonstrated various strategies for indirect inhibition. In osteosarcoma PDXs, treatment with AT7519 resulted in significant tumor growth inhibition in MYC-amplified tumors, likely through inhibition of CDK9 [22]. Another strategy is to target the transcriptional activity of MYC through bromodomain and extra-terminal domain (BET) inhibition; however, in osteosarcoma models, Baker et al. showed that treatment with the BET inhibitor JQ1 induced apoptosis independent of MYC [198].

STAT3 activation is thought to play a role in tumor cell survival and proliferation in human and canine osteosarcoma [199]. Toosendanin, a STAT3 inhibitor, suppressed osteosarcoma cell growth, invasion, and angiogenesis in vitro. Furthermore, toosendanin treatment resulted in decreased tumor growth, reduction of metastasis, and prolonged survival of osteosarcoma xenografts [200]. In another study, administration of pectolinarigenin to osteosarcoma xenografts blocked STAT3 activation and impaired tumor growth and metastasis [79].

Wee1 is a mediator of the G2/M cell-cycle checkpoint, and inhibition of Wee1 by adavosertib has been reported to enhance the effects of cytotoxic chemotherapy [201, 202]. Kreahling et al. demonstrated that adavosertib had significant antitumor activity in osteosarcoma xenografts, both as a single agent and in combination with gemcitabine [105]. Most of the studies done to date in preclinical models have not specifically matched targeted therapies to the subsets of osteosarcomas that have specific alterations. Given the heterogeneity of osteosarcoma described previously, it is possible that stratification for targeted therapy based on genomic characteristics of a tumor could increase the rate of response. Such a "genome-informed" approach was tested recently [22]. Whole-genome sequencing was used to identify recurrent copy number alterations in subsets of osteosarcoma PDX models. Based on this analysis, six candidate pathways were identified for targeting: MYC amplification, with the CDK9 inhibitor AT7519; CCNE amplification with the CDK2 inhibitor dinaciclib; VEGFA amplification with the VEGFR inhibitor sorafenib; CDK4 amplification or FOXM1 amplification with the CDK4/6 inhibitor palbociclib; AURKB amplification with the AURKB inhibitor AZD1152; and AKT gain or PTEN loss with the AKT inhibitor MK2206. PDX models treated with the matched drug based on copy number analysis demonstrated significantly higher tumor growth inhibition compared to a "nonmatched" approach. Translation of this approach to clinical trials has still not been done but may represent a promising new approach to treatment of this disease. Another potentially highly effective strategy is to leverage the availability of mouse and canine models together with human osteosarcoma. Such an approach has recently been described [203]. Lastly, it should be noted that the high mutational burden of osteosarcomas could potentially make them susceptible to immunotherapies. To date, such approaches have not been successful, but it is possible that future studies could find ways to reactivate the immune system for therapeutic benefit in osteosarcoma.

References Cited

1. Bridge JA et al (1997) Cytogenetic findings in 73 osteosarcoma specimens and a review of the literature. Cancer Genet Cytogenet 95(1):74–87
2. Bayani J et al (2003) Spectral karyotyping identifies recurrent complex rearrangements of chromosomes 8, 17, and 20 in osteosarcomas. Genes Chromosomes Cancer 36(1):7–16
3. Tarkkanen M et al (1995) Gains and losses of DNA sequences in osteosarcomas by comparative genomic hybridization. Cancer Res 55(6):1334–1338
4. Hoogerwerf WA, Hawkins AL, Perlman EJ, Griffin CA (1994) Chromosome analysis of nine osteosarcomas. Genes Chromosom Cancer 9(2):88–92
5. Wei G et al (1999) CDK4 gene amplification in osteosarcoma: reciprocal relationship with INK4A gene alterations and mapping of 12q13 amplicons. Int J Cancer 80(2):199–204
6. Lonardo F, Ueda T, Huvos AG, Healey J, Ladanyi M (1997) p53 and MDM2 alterations in osteosarcomas: correlation with clinicopathologic features and proliferative rate. Cancer 79(8):1541–1547
7. Ikeda S et al (1989) Amplification of both c-myc and c-raf-1 oncogenes in a human osteosarcoma. Jpn J Cancer Res 80(1):6–9
8. Gamberi G et al (1998) C-myc and c-fos in human osteosarcoma: prognostic value of mRNA and protein expression. Oncology 55(6):556–563
9. Miller CW, Aslo A, Won A, Tan M, Lampkin B, Koeffler HP (1996) Alterations of the p53, Rb and MDM2 genes in osteosarcoma. J Cancer Res Clin Oncol 122(9):559–565
10. Martin JW, Squire JA, Zielenska M (2012) The genetics of osteosarcoma. Sarcoma 2012:627254
11. Morrow JJ, Khanna C (2015) Osteosarcoma genetics and epigenetics: emerging biology and candidate therapies. Crit Rev Oncog 20(3–4):173–197
12. Ladanyi M, Park CK, Lewis R, Jhanwar SC, Healey JH, Huvos AG (1993) Sporadic amplification of the MYC gene in human osteosarcomas. Diagn Mol Pathol 2(3):163–167
13. Tarkkanen M et al (1999) DNA sequence copy number increase at 8q: a potential new prognostic marker in high-grade osteosarcoma. Int J Cancer 84(2):114–121
14. Maire G, Yoshimoto M, Chilton-MacNeill S, Thorner PS, Zielenska M, Squire JA (2009) Recurrent RECQL4 imbalance and increased gene expression levels are associated with structural chromosomal instability in sporadic osteosarcoma. Neoplasia 11(3):260–268., 3p following 268
15. Cheng L et al (2019) Integration of genomic copy number variations and chemotherapy-response biomarkers in pediatric sarcoma. BMC Med Genet 12(Suppl 1):23
16. Smida J et al (2010) Genomic alterations and allelic imbalances are strong prognostic predictors in osteosarcoma. Clin Cancer Res 16(16):4256–4267
17. Smida J et al (2017) Genome-wide analysis of somatic copy number alterations and chromosomal breakages in osteosarcoma. Int J Cancer 141(4):816–828
18. Malkin D et al (1990) Germ line p53 mutations in a familial syndrome of breast cancer, sarcomas, and other neoplasms. Science (80-) 250(4985):1233–1238
19. Wong FL et al (1997) Cancer incidence after retinoblastoma. Radiation dose and sarcoma risk. JAMA 278(15):1262–1267
20. Chen X et al (2014) Recurrent somatic structural variations contribute to tumorigenesis in pediatric osteosarcoma. Cell Rep 7(1):104–112
21. Perry JA et al (2014) Complementary genomic approaches highlight the PI3K/mTOR pathway as

a common vulnerability in osteosarcoma. Proc Natl Acad Sci 111(51):E5564–E5573
22. Sayles LC et al (2019) Genome-informed targeted therapy for osteosarcoma. Cancer Discov 9(1):46–63
23. Suehara Y et al (2019) Clinical genomic sequencing of pediatric and adult osteosarcoma reveals distinct molecular subsets with potentially targetable alterations. Clin Cancer Res 25(21):6346–6356
24. Kovac M et al (2015) Exome sequencing of osteosarcoma reveals mutation signatures reminiscent of BRCA deficiency. Nat Commun 6:1–9
25. Negri GL et al (2019) Integrative genomic analysis of matched primary and metastatic pediatric osteosarcoma. J Pathol 249(3):319–331
26. Brady SW et al (2019) The clonal evolution of metastatic osteosarcoma as shaped by cisplatin treatment. Mol Cancer Res 17(4):895–906
27. Oh JH, Kim H-S, Kim H-H, Kim WH, Lee SH (2006) Aberrant methylation of p14ARF gene correlates with poor survival in osteosarcoma. Clin Orthop Relat Res 442:216–222
28. Ottaviano L et al (2010) Molecular characterization of commonly used cell lines for bone tumor research: a trans-European EuroBoNet effort. Genes Chromosomes Cancer 49(1):40–51
29. Lorenz S et al (2016) Unscrambling the genomic chaos of osteosarcoma reveals extensive transcript fusion, recurrent rearrangements and frequent novel TP53 aberrations. Oncotarget 7(5):5273–5288
30. Ren L et al (2015) Characterization of the metastatic phenotype of a panel of established osteosarcoma cells. Oncotarget 6(30):29469–29481
31. Mohseny AB et al (2011) Functional characterization of osteosarcoma cell lines provides representative models to study the human disease. Lab Investig 91(8):1195–1205
32. Lauvrak SU et al (2013) Functional characterisation of osteosarcoma cell lines and identification of mRNAs and miRNAs associated with aggressive cancer phenotypes. Br J Cancer 109(8):2228–2236
33. Yuan J et al (2009) Osteoblastic and osteolytic human osteosarcomas can be studied with a new xenograft mouse model producing spontaneous metastases. Cancer Investig 27(4):435–442
34. Bauer HCF, Brosjö O, Broström LÅ, Nilsson OS, Reinholt FP, Tribukait B (1986) Growth and ploidy of human osteosarcoma xenografts in serial passage in nude mice. Eur J Cancer Clin Oncol 22(7):821–830
35. Kimura K et al (2002) Establishment of human osteosarcoma cell lines with high metastatic potential to lungs and their utilities for therapeutic studies on metastatic osteosarcoma. Clin Exp Metastasis 19(6):477–485
36. Manara MC et al (2000) Reversal of malignant phenotype in human osteosarcoma cells transduced with the alkaline phosphatase gene. Bone 26(3):215–220
37. Jia SF, Worth LL, Kleinerman ES (1999) A nude mouse model of human osteosarcoma lung metastases for evaluating new therapeutic strategies. Clin Exp Metastasis 17(6):501–506
38. Duan X, Jia S-F, Zhou Z, Langley RR, Bolontrade MF, Kleinerman ES (2004) Association of alphavbeta3 integrin expression with the metastatic potential and migratory and chemotactic ability of human osteosarcoma cells. Clin Exp Metastasis 21(8):747–753
39. Zhang Y et al (2018) Down-regulation of Skp2 expression inhibits invasion and lung metastasis in osteosarcoma. Sci Rep 8(1):14294
40. Billiau A, Edy VG, Heremans H et al (1977) Human interferon: mass production in a newly established cell line, MG-63. Antimicrob Agents Chemother 12:11–15
41. Heremans H, Billiau A, Cassiman JJ et al (1978) In vitro cultivation of human tumor tissues. II. Morphological and virological characterization of three cell lines. Oncology 35:246–252
42. Lorenz S, Barøy T, Sun J et al Unscrambling the genomic chaos of osteosarcoma reveals extensive transcript fusion, recurrent rearrangements and frequent novel TP53 aberrations. Oncotarget 7. Epub ahead of print 2016. https://doi.org/10.18632/oncotarget.6567
43. Masuda H, Miller C, Koeffler HP et al (1987) Rearrangement of the p53 gene in human osteogenic sarcomas. Proc Natl Acad Sci U S A 84:7716–7719
44. Ottaviano L, Schaefer K-L, Gajewski M et al (2010) Molecular characterization of commonly used cell lines for bone tumor research: a trans-European EuroBoNet effort. Genes Chromosomes Cancer 49:40–51
45. Lauvrak SU, Munthe E, Kresse SH et al (2013) Functional characterisation of osteosarcoma cell lines and identification of mRNAs and miRNAs associated with aggressive cancer phenotypes. Br J Cancer 109:2228–2236
46. Su Y, Luo X, He B-C et al (2009) Establishment and characterization of a new highly metastatic human osteosarcoma cell line. Clin Exp Metastasis 26:599–610
47. Ren L, Mendoza A, Zhu J et al (2015) Characterization of the metastatic phenotype of a panel of established osteosarcoma cells. Oncotarget 6:29469–29481
48. Mohseny AB, Machado I, Cai Y et al (2011) Functional characterization of osteosarcoma cell lines provides representative models to study the human disease. Lab Investig 91:1195–1205
49. Kimura K, Nakano T, Park Y-B et al (2002) Establishment of human osteosarcoma cell lines with high metastatic potential to lungs and their utilities for therapeutic studies on metastatic osteosarcoma. Clin Exp Metastasis 19:477–485
50. Yuan J, Ossendorf C, Szatkowski JP et al (2009) Osteoblastic and osteolytic human osteosarcomas can be studied with a new xenograft mouse model producing spontaneous metastases. Cancer Investig 27:435–442

51. Pontén J, Saksela E (1967) Two established in vitro cell lines from human mesenchymal tumours. Int J Cancer 2:434–447
52. Zhang Y, Xiong Y, Yarbrough WG (1998) ARF promotes MDM2 degradation and stabilizes p53: ARF-INK4a locus deletion impairs both the Rb and p53 tumor suppression pathways. Cell 92:725–734
53. Manara MC, Baldini N, Serra M et al (2000) Reversal of malignant phenotype in human osteosarcoma cells transduced with the alkaline phosphatase gene. Bone 26:215–220
54. Diller L, Kassel J, Nelson CE et al (1990) p53 functions as a cell cycle control protein in osteosarcomas. Mol Cell Biol 10:5772–5781
55. Kubbutat MH, Jones SN, Vousden KH (1997) Regulation of p53 stability by Mdm2. Nature 387:299–303
56. Moneo V, Serelde BG, Fominaya J et al (2007) Extreme sensitivity to Yondelis (Trabectedin, ET-743) in low passaged sarcoma cell lines correlates with mutated p53. J Cell Biochem 100:339–348
57. Tiram G, Segal E, Krivitsky A et al (2016) Identification of dormancy-associated MicroRNAs for the design of osteosarcoma-targeted dendritic polyglycerol nanopolyplexes. ACS Nano 10:2028–2045
58. Gvozdenovic A, Arlt MJE, Campanile C et al (2013) CD44 enhances tumor formation and lung metastasis in experimental osteosarcoma and is an additional predictor for poor patient outcome. J Bone Miner Res 28:838–847
59. Sabile AA, Arlt MJE, Muff R et al (2012) Cyr61 expression in osteosarcoma indicates poor prognosis and promotes intratibial growth and lung metastasis in mice. J Bone Miner Res 27:58–67
60. Sabile AA, Arlt MJE, Muff R et al (1832) Caprin-1, a novel Cyr61-interacting protein, promotes osteosarcoma tumor growth and lung metastasis in mice. Biochim Biophys Acta 2013:1173–1182
61. Duan X, Jia S-F, Zhou Z et al (2004) Association of alphavbeta3 integrin expression with the metastatic potential and migratory and chemotactic ability of human osteosarcoma cells. Clin Exp Metastasis 21:747–753
62. Rhim JS, Cho HY, Huebner RJ (1975) Non-producer human cells induced by murine sarcoma virus. Int J Cancer 15:23–29
63. Botter SM, Arlt MJE, Born W et al (2015) Mammalian models of bone sarcomas. In: Heymann D (ed) Bone cancer. Elsevier Inc., Amsterdam, Boston, Heidelberg, London, New York, Oxford, Paris, San Diego, San Francisco, Singapore, Sydney, Tokyo, pp 347–364
64. Rhim JS, Park DK, Arnstein P et al (1975) Transformation of human cells in culture by N-methyl-N'-nitro-N-nitrosoguanidine. Nature 256:751–753
65. Luu HH, Kang Q, Jong KP et al (2005) An orthotopic model of human osteosarcoma growth and spontaneous pulmonary metastasis. Clin Exp Metastasis 22:319–329
66. Zhang Y, Zvi YS, Batko B et al (2018) Down-regulation of Skp2 expression inhibits invasion and lung metastasis in osteosarcoma. Sci Rep 8:14294
67. Samid D, Mandler R (1989) Human osteosarcoma cells transformed by ras-oncogenes: a new model for in vivo studies of pulmonary metastasis. Clin Biotechnol 1:21–26
68. Berlin O, Samid D, Donthineni-Rao R et al (1993) Development of a novel spontaneous metastasis model of human osteosarcoma transplanted orthotopically into bone of athymic mice. Cancer Res 53:4890–4895
69. Fodstad O, Brøgger A, Bruland O et al (1986) Characteristics of a cell line established from a patient with multiple osteosarcoma, appearing 13 years after treatment for bilateral retinoblastoma. Int J Cancer 38:33–40
70. Eikenes L, Tari M, Tufto I et al (2005) Hyaluronidase induces a transcapillary pressure gradient and improves the distribution and uptake of liposomal doxorubicin (Caelyx) in human osteosarcoma xenografts. Br J Cancer 93:81–88
71. Su Y et al (2009) Establishment and characterization of a new highly metastatic human osteosarcoma cell line. Clin Exp Metastasis 26(7):599–610
72. Luu HH et al (2005) An orthotopic model of human osteosarcoma growth and spontaneous pulmonary metastasis. Clin Exp Metastasis 22(4):319–329
73. Muff R et al (2015) Genomic instability of osteosarcoma cell lines in culture: impact on the prediction of metastasis relevant genes. PLoS One 10(5):e0125611
74. Tiram G et al (2016) Identification of dormancy-associated MicroRNAs for the Design of Osteosarcoma-Targeted Dendritic Polyglycerol Nanopolyplexes. ACS Nano 10(2):2028–2045
75. Nomura M et al (2019) Tegavivint and the β-catenin/ALDH axis in chemotherapy-resistant and metastatic osteosarcoma. J Natl Cancer Inst 111:1–12
76. Majzner RG et al (2019) CAR T cells targeting B7-H3, a pan-cancer antigen, demonstrate potent preclinical activity against pediatric solid tumors and brain tumors. Clin Cancer Res 25(8):2560–2574
77. Husmann K et al (2013) Matrix Metalloproteinase 1 promotes tumor formation and lung metastasis in an intratibial injection osteosarcoma mouse model. Biochim Biophys Acta 1832(2):347–354
78. Butch ER et al (2019) Positron emission tomography detects in vivo expression of Disialoganglioside GD2 in mouse models of primary and metastatic osteosarcoma. Cancer Res 79(12)3112–3124
79. Zhang T et al (2016) Natural product pectolinarigenin inhibits osteosarcoma growth and metastasis via SHP-1-mediated STAT3 signaling inhibition. Cell Death Dis 7(10):e2421
80. Sabile AA et al (2012) Cyr61 expression in osteosarcoma indicates poor prognosis and promotes intratibial growth and lung metastasis in mice. J Bone Miner Res 27(1):58–67
81. Broadhead ML, Dass CR, Choong PFM (2011) Systemically administered PEDF against primary

and secondary tumours in a clinically relevant osteosarcoma model. Br J Cancer 105(10):1503–1511
82. Igarashi K et al (2017) Effective metabolic targeting of human osteosarcoma cells in vitro and in orthotopic nude-mouse models with recombinant Methioninase. Anticancer Res 37(9):4807–4812
83. Xu WT, Bian ZY, Fan QM, Li G, Tang TT (2009) Human mesenchymal stem cells (hMSCs) target osteosarcoma and promote its growth and pulmonary metastasis. Cancer Lett 281(1):32–41
84. Berlin O, Samid D, Donthineni-Rao R, Akeson W, Amiel D, Woods VL (1993) Development of a novel spontaneous metastasis model of human osteosarcoma transplanted orthotopically into bone of athymic mice. Cancer Res 53(20):4890–4895
85. Eikenes L, Tari M, Tufto I, Bruland OS, de Lange Davies C (2005) Hyaluronidase induces a transcapillary pressure gradient and improves the distribution and uptake of liposomal doxorubicin (Caelyx) in human osteosarcoma xenografts. Br J Cancer 93(1):81–88
86. Hausser H-J, Brenner RE (2005) Phenotypic instability of Saos-2 cells in long-term culture. Biochem Biophys Res Commun 333(1):216–222
87. Masters JRW (2000) Human cancer cell lines: fact and fantasy. Nat Rev Mol Cell Biol 1(3):233–236
88. Drexler HG, Uphoff CC, Dirks WG, MacLeod RAF (2002) Mix-ups and mycoplasma: the enemies within. Leuk Res 26(4):329–333
89. Wang H et al (2006) Comparative analysis and integrative classification of NCI60 cell lines and primary tumors using gene expression profiling data. BMC Genomics 7:166
90. Jung J, Seol HS, Chang S (2018) The generation and application of patient-derived xenograft model for cancer research. Cancer Res Treat 50(1):1–10
91. Mendoza A et al (2010) Modeling metastasis biology and therapy in real time in the mouse lung. J Clin Invest 120(8):2979–2988
92. Bayles I et al (2019) Ex vivo screen identifies CDK12 as a metastatic vulnerability in osteosarcoma. J Clin Invest 129(10):4377–4392
93. Morrow JJ et al (2018) Positively selected enhancer elements endow osteosarcoma cells with metastatic competence. Nat Med 24(2):176–185
94. Shultz LD et al (2005) Human lymphoid and myeloid cell development in NOD/LtSz-scid IL2R gamma null mice engrafted with mobilized human hemopoietic stem cells. J Immunol 174(10):6477–6489
95. Ishii S et al (1982) Analysis of osteoid-forming activity of human osteosarcoma implanted into nude mice. Int Orthop 6(4):215–223
96. Meyer WH, Houghton JA, Houghton PJ, Webber BL, Douglass EC, Look AT (1990) Development and characterization of pediatric osteosarcoma xenografts. Cancer Res 50(9):2781–2785
97. Meohas W, Granato RA, Guimarães JAM, Dias RB, Fortuna-Costa A, Duarte MEL (2018) Patient-derived xenografts as a preclinical model for bone sarcomas. Acta Ortop Bras 26(2):98–102
98. Stewart E et al (2017) Orthotopic patient-derived xenografts of paediatric solid tumours. Nature 549(7670):96–100
99. Bruheim S, Bruland OS, Breistol K, Maelandsmo GM, Fodstad Ø (2004) Human osteosarcoma xenografts and their sensitivity to chemotherapy. Pathol Oncol Res 10(3):133–141
100. Llombart-Bosch A, Carda C, Boix J, Pellin A, Peydro-Olaya A (1988) Value of nude mice xenografts in the expression of cell heterogeneity of human sarcomas of bone and soft tissue. Pathol Res Pract 183(6):683–692
101. Fujisaki T, Wada T, Takahashi M, Yamawaki S, Ishii S (1995) In vitro chemosensitivity assay for human osteosarcoma using tumor xenografts. Clin Orthop Relat Res 313:279–285
102. Blattmann C et al (2015) Establishment of a patient-derived orthotopic osteosarcoma mouse model. J Transl Med 13:136
103. Igarashi K et al (2018) Tumor-targeting Salmonella typhimurium A1-R combined with recombinant methioninase and cisplatinum eradicates an osteosarcoma cisplatinum-resistant lung metastasis in a patient-derived orthotopic xenograft (PDOX) mouse model: decoy, trap and kill chemothe. Cell Cycle 17(6):801–809
104. Goldstein SD, Hayashi M, Albert CM, Jackson KW, Loeb DM (2015) An orthotopic xenograft model with survival hindlimb amputation allows investigation of the effect of tumor microenvironment on sarcoma metastasis. Clin Exp Metastasis 32(7):703–715
105. Kreahling JM et al (2013) Wee1 inhibition by MK-1775 leads to tumor inhibition and enhances efficacy of gemcitabine in human sarcomas. PLoS One 8(3):e57523
106. Kuijjer ML et al (2011) mRNA expression profiles of primary high-grade central osteosarcoma are preserved in cell lines and xenografts. BMC Med Genet 4:66
107. Broström LA, Crnalic S, Löfvenberg R, Stenling R, Boquist L (1996) Structure, growth and cell proliferation of human osteosarcoma and malignant fibrous histiocytoma xenografts in serial transplantation in nude mice. APMIS 104(11):775–783
108. Kresse SH, Meza-Zepeda LA, Machado I, Llombart-Bosch A, Myklebost O (2012) Preclinical xenograft models of human sarcoma show nonrandom loss of aberrations. Cancer 118(2):558–570
109. Kito F et al (2018) Establishment and characterization of novel patient-derived osteosarcoma xenograft and cell line. In Vitro Cell Dev Biol Anim 54(7):528–536
110. Loh AHP et al (2019) Combinatorial screening using orthotopic patient derived xenograft-expanded early phase cultures of osteosarcoma identify novel therapeutic drug combinations. Cancer Lett 442:262–270
111. Crnalic S, Hakansson I, Boquist L, Lofvenberg R, Brostrom LA (1997) A novel spontaneous metastasis model of human osteosarcoma developed using orthotopic transplantation of intact tumor

tissue into tibia of nude mice. Clin Exp Metastasis 15(2):164–172
112. Whiteford CC et al (2007) Credentialing preclinical pediatric xenograft models using gene expression and tissue microarray analysis. Cancer Res 67(1):32–40
113. Monsma DJ et al (2012) Genomic characterization of explant tumorgraft models derived from fresh patient tumor tissue. J Transl Med 10(1):1–18
114. Delgado-Baeza E, Miralles-Flores C, Gonzalez-Medievo W (1991) Heterogeneity of xenografted osteosarcoma: a human sarcoma transplanted into nude mice. Acta Orthop 62(5):410–414
115. Igarashi K et al (2018) Temozolomide combined with irinotecan regresses a cisplatinum-resistant relapsed osteosarcoma in a patient-derived orthotopic xenograft (PDOX) precision-oncology mouse model. Oncotarget 9(8):7774–7781
116. Goldstein SD, Trucco M, Bautista Guzman W, Hayashi M, Loeb DM (2016) A monoclonal antibody against the Wnt signaling inhibitor dickkopf-1 inhibits osteosarcoma metastasis in a preclinical model. Oncotarget 7(16):21114–21123
117. Igarashi K et al (2017) A patient-derived orthotopic xenograft (PDOX) mouse model of a cisplatinum-resistant osteosarcoma lung metastasis that was sensitive to temozolomide and trabectedin: implications for precision oncology. Oncotarget 8(37):62111–62119
118. Higuchi T et al (2019) Trabectedin and irinotecan combination regresses a cisplatinum-resistant osteosarcoma in a patient-derived orthotopic xenograft nude-mouse model. Biochem Biophys Res Commun 513(2):326–331
119. Murakami T et al (2017) Tumor-targeting Salmonella typhimurium A1-R regresses an osteosarcoma in a patient-derived xenograft model resistant to a molecular-targeting drug. Oncotarget 8(5):8035–8042
120. Brennecke P et al (2014) CXCR4 antibody treatment suppresses metastatic spread to the lung of intratibial human osteosarcoma xenografts in mice. Clin Exp Metastasis 31(3):339–349
121. Rainusso N et al (2019) Generation of patient-derived tumor xenografts from percutaneous tumor biopsies in children with bone sarcomas. Pediatr Blood Cancer 66(4):e27579
122. Houghton PJ, Cheshire PJ, Myers L, Stewart CF, Synold TW, Houghton JA (1992) Evaluation of 9-dimethylaminomethyl-10-hydroxycamptothecin against xenografts derived from adult and childhood solid tumors. Cancer Chemother Pharmacol 31(3):229–239
123. Houghton PJ et al (2007) The pediatric preclinical testing program: description of models and early testing results. Pediatr Blood Cancer 49(7):928–940
124. Halim A, Bagley RG, Keler T (2016) Abstract 5032: glycoprotein NMB (gpNMB) overexpression is prevalent in human cancers: pancreatic cancer, non-small cell lung cancer, head and neck cancer, and osteosarcoma. Cancer Res 76(14 Supplement):5032–5032
125. Tse KF et al (2006) CR011, a fully human monoclonal antibody-auristatin E conjugate, for the treatment of melanoma. Clin Cancer Res 12(4):1373–1382
126. Kolb EA et al (2014) Initial testing (stage 1) of glembatumumab vedotin (CDX-011) by the pediatric preclinical testing program. Pediatr Blood Cancer 61(10):1816–1821
127. Maki RG (2010) Small is beautiful: insulin-like growth factors and their role in growth, development, and cancer. J Clin Oncol 28(33):4985–4995
128. Scotlandi K, Picci P (2008) Targeting insulin-like growth factor 1 receptor in sarcomas. Curr Opin Oncol 20(4):419–427
129. Rikhof B, de Jong S, Suurmeijer AJH, Meijer C, van der Graaf WTA (2009) The insulin-like growth factor system and sarcomas. J Pathol 217(4):469–482
130. Kolb EA et al (2008) Initial testing (stage 1) of a monoclonal antibody (SCH 717454) against the IGF-1 receptor by the pediatric preclinical testing program. Pediatr Blood Cancer 50(6):1190–1197
131. Wang Y et al (2010) A fully human insulin-like growth factor-I receptor antibody SCH 717454 (Robatumumab) has antitumor activity as a single agent and in combination with cytotoxics in pediatric tumor xenografts. Mol Cancer Ther 9(2):410–418
132. Houghton PJ et al (2010) Initial testing of a monoclonal antibody (IMC-A12) against IGF-1R by the Pediatric Preclinical Testing Program. Pediatr Blood Cancer 54(7):921–926
133. Kolb EA et al (2012) Combination testing (Stage 2) of the Anti-IGF-1 receptor antibody IMC-A12 with rapamycin by the pediatric preclinical testing program. Pediatr Blood Cancer 58(5):729–735
134. Kolb EA et al (2011) Initial testing (stage 1) of the IGF-1 receptor inhibitor BMS-754807 by the pediatric preclinical testing program. Pediatr Blood Cancer 56(4):595–603
135. Cao Y et al (2014) Insulin-like growth factor 1 receptor and response to anti-IGF1R antibody therapy in osteosarcoma. PLoS One 9(8):e106249
136. Kieran MW, Kalluri R, Cho Y-J (2012) The VEGF pathway in cancer and disease: responses, resistance, and the path forward. Cold Spring Harb Perspect Med 2(12):a006593
137. Maris JM et al (2008) Initial testing of the VEGFR inhibitor AZD2171 by the pediatric preclinical testing program. Pediatr Blood Cancer 50(3):581–587
138. Fox E et al (2010) A phase 1 trial and pharmacokinetic study of cediranib, an orally bioavailable pan-vascular endothelial growth factor receptor inhibitor, in children and adolescents with refractory solid tumors. J Clin Oncol 28(35):5174–5181
139. Maris JM et al (2008) Initial testing (stage 1) of sunitinib by the pediatric preclinical testing program. Pediatr Blood Cancer 51(1):42–48
140. Keir ST et al (2010) Initial testing (stage 1) of the multi-targeted kinase inhibitor sorafenib by the

pediatric preclinical testing program. Pediatr Blood Cancer 55(6):1126–1133
141. Dubois SG et al (2011) Phase I and pharmacokinetic study of sunitinib in pediatric patients with refractory solid tumors: a children's oncology group study. Clin Cancer Res 17(15):5113–5122
142. Kim A et al (2015) Phase 2 trial of sorafenib in children and young adults with refractory solid tumors: a report from the Children's Oncology Group. Pediatr Blood Cancer 62(9):1562–1566
143. Grignani G et al (2012) A phase II trial of sorafenib in relapsed and unresectable high-grade osteosarcoma after failure of standard multimodal therapy: an Italian Sarcoma Group study. Ann Oncol Off J Eur Soc Med Oncol 23(2):508–516
144. Harrison DJ et al (2019) Evaluation of the multi-kinase inhibitor regorafenib in the Pediatric Preclinical Testing Consortium osteosarcoma, rhabdomyosarcoma, and Ewing sarcoma in vivo models. J Clin Oncol 37(15_suppl):10038
145. Davis LE et al (2019) Randomized double-blind phase II study of Regorafenib in patients with metastatic osteosarcoma. J Clin Oncol 37(16):1424–1431
146. Chen Q, Zhou Z, Shan L, Zeng H, Hua Y, Cai Z (2015) The importance of Src signaling in sarcoma. Oncol Lett 10(1):17–22
147. Kolb EA et al (2008) Initial testing of dasatinib by the pediatric preclinical testing program. Pediatr Blood Cancer 50(6):1198–1206
148. Hingorani P, Zhang W, Gorlick R, Kolb EA (2009) Inhibition of Src phosphorylation alters metastatic potential of osteosarcoma in vitro but not in vivo. Clin Cancer Res 15(10):3416–3422
149. Aplenc R et al (2011) Pediatric phase I trial and pharmacokinetic study of dasatinib: a report from the children's oncology group phase I consortium. J Clin Oncol 29(7):839–844
150. Houghton PJ et al (2008) Initial testing (stage 1) of the mTOR inhibitor rapamycin by the pediatric preclinical testing program. Pediatr Blood Cancer 50(4):799–805
151. Houghton PJ et al (2010) Stage 2 combination testing of rapamycin with cytotoxic agents by the Pediatric Preclinical Testing Program. Mol Cancer Ther 9(1):101–112
152. Spunt SL et al (2011) Phase I study of temsirolimus in pediatric patients with recurrent/refractory solid tumors. J Clin Oncol 29(21):2933–2940
153. Schuetze SM et al (2012) Results of a phase II study of sirolimus and cyclophosphamide in patients with advanced sarcoma. Eur J Cancer 48(9):1347–1353
154. Chawla SP et al (2011) Results of the phase III, placebo-controlled trial (SUCCEED) evaluating the mTOR inhibitor ridaforolimus (R) as maintenance therapy in advanced sarcoma patients (pts) following clinical benefit from prior standard cytotoxic chemotherapy (CT). J Clin Oncol 29(15_suppl):10005
155. Chawla SP et al (2012) Phase II study of the mammalian target of rapamycin inhibitor ridaforolimus in patients with advanced bone and soft tissue sarcomas. J Clin Oncol 30(1):78–84
156. Houghton PJ et al (2012) Initial testing (stage 1) of the mTOR kinase inhibitor AZD8055 by the pediatric preclinical testing program. Pediatr Blood Cancer 58(2):191–199
157. Zhang J, Yu X-H, Yan Y-G, Wang C, Wang W-J (2015) PI3K/Akt signaling in osteosarcoma. Clin Chim Acta 444:182–192
158. Gorlick R et al (2012) Testing of the Akt/PKB inhibitor MK-2206 by the Pediatric Preclinical Testing Program. Pediatr Blood Cancer 59(3):518–524
159. Carol H et al (2010) Initial testing (stage 1) of the Akt inhibitor GSK690693 by the pediatric preclinical testing program. Pediatr Blood Cancer 55(7):1329–1337
160. Chandhanayingyong C, Kim Y, Staples JR, Hahn C, Lee FY (2012) MAPK/ERK signaling in osteosarcomas, Ewing sarcomas and chondrosarcomas: therapeutic implications and future directions. Sarcoma 2012:1–8
161. Kolb EA et al (2010) Initial testing (stage 1) of AZD6244 (ARRY-142886) by the Pediatric Preclinical Testing Program. Pediatr Blood Cancer 55(4):668–677
162. Eroglu Z et al (2015) A randomised phase II trial of selumetinib vs selumetinib plus temsirolimus for soft-tissue sarcomas. Br J Cancer 112(10):1644–1651
163. Subbiah V et al (2017) Phase Ib/II study of the safety and efficacy of combination therapy with multikinase VEGF inhibitor Pazopanib and MEK inhibitor Trametinib in advanced soft tissue sarcoma. Clin Cancer Res 23(15):4027–4034
164. Gorlick R et al (2012) Initial testing (stage 1) of the cyclin dependent kinase inhibitor SCH 727965 (dinaciclib) by the pediatric preclinical testing program. Pediatr Blood Cancer 59(7):1266–1274
165. Maris JM et al (2010) Initial testing of the aurora kinase A inhibitor MLN8237 by the Pediatric Preclinical Testing Program (PPTP). Pediatr Blood Cancer 55(1):26–34
166. Mossé YP et al (2019) A phase II study of alisertib in children with recurrent/refractory solid tumors or leukemia: Children's oncology group phase I and pilot consortium (ADVL0921). Clin Cancer Res 25(11):3229–3238
167. Lowery CD et al (2019) Broad Spectrum activity of the checkpoint kinase 1 inhibitor Prexasertib as a single agent or Chemopotentiator across a range of preclinical pediatric tumor models. Clin Cancer Res 25(7):2278–2289
168. Smith MA et al (2012) Initial testing of JNJ-26854165 (Serdemetan) by the pediatric preclinical testing program. Pediatr Blood Cancer 59(2):329–332
169. Lock R et al (2008) Initial testing (stage 1) of the BH3 mimetic ABT-263 by the pediatric preclinical testing program. Pediatr Blood Cancer 50(6):1181–1189

170. Gorlick R et al (2009) Initial testing (stage 1) of lapatinib by the pediatric preclinical testing program. Pediatr Blood Cancer 53(4):594–598
171. Carol H et al (2013) Initial testing of the MDM2 inhibitor RG7112 by the Pediatric Preclinical Testing Program. Pediatr Blood Cancer 60(4):633–641
172. Smith MA et al (2012) Initial testing of the investigational NEDD8-activating enzyme inhibitor MLN4924 by the pediatric preclinical testing program. Pediatr Blood Cancer 59(2):246–253
173. Kolb EA et al (2012) Initial testing (stage 1) by the pediatric preclinical testing program of RO4929097, a γ-secretase inhibitor targeting notch signaling. Pediatr Blood Cancer 58(5):815–818
174. Houghton P, Erickson SW, Teicher B, Smith MA, Han R, Kurmasheva R (2019) In vivo evaluation of the LSD1 inhibitor SP-2577 against Ewing sarcoma and rhabdomyosarcoma preclinincal models: A report from the Pediatric Preclinical Testing Consortium (PPTC) [abstract], in Proceedings: AACR Annual Meeting 2019
175. Pignochino Y et al (2013) The combination of Sorafenib and Everolimus abrogates mTORC1 and mTORC2 upregulation in osteosarcoma preclinical models. Clin Cancer Res 19(8):2117–2131
176. Grignani G et al (2015) Sorafenib and everolimus for patients with unresectable high-grade osteosarcoma progressing after standard treatment: a non-randomised phase 2 clinical trial. Lancet Oncol 16(1):98–107
177. Gobin B et al (2014) NVP-BEZ235, a dual PI3K/mTOR inhibitor, inhibits osteosarcoma cell proliferation and tumor development in vivo with an improved survival rate. Cancer Lett 344(2):291–298
178. Gobin B et al (2015) BYL719, a new α-specific PI3K inhibitor: single administration and in combination with conventional chemotherapy for the treatment of osteosarcoma. Int J Cancer 136(4):784–796
179. Huang J-C et al (2018) NVP-BEZ235 synergizes cisplatin sensitivity in osteosarcoma. Oncotarget 9(12):10483–10496
180. Manara MC et al (2010) NVP-BEZ235 as a new therapeutic option for sarcomas. Clin Cancer Res 16(2):530–540
181. Gobin B et al (2014) Imatinib mesylate exerts antiproliferative effects on osteosarcoma cells and inhibits the tumour growth in immunocompetent murine models. PLoS One 9(3):e90795
182. Sampson ER et al (2011) The orally bioavailable met inhibitor PF-2341066 inhibits osteosarcoma growth and osteolysis/matrix production in a xenograft model. J Bone Miner Res 26(6):1283–1294
183. Vassilev LT et al (2004) In vivo activation of the p53 pathway by small-molecule antagonists of MDM2. Science 303(5659):844–848
184. Bretscher A, Reczek D, Berryman M (1997) Ezrin: a protein requiring conformational activation to link microfilaments to the plasma membrane in the assembly of cell surface structures. J Cell Sci 110(24):3011–3018
185. Khanna C et al (2004) The membrane-cytoskeleton linker ezrin is necessary for osteosarcoma metastasis. Nat Med 10(2):182–186
186. Bulut G et al (2012) Small molecule inhibitors of ezrin inhibit the invasive phenotype of osteosarcoma cells. Oncogene 31(3):269–281
187. Wan X, Mendoza A, Khanna C, Helman LJ (2005) Rapamycin inhibits ezrin-mediated metastatic behavior in a murine model of osteosarcoma. Cancer Res 65(6):2406–2411
188. Pignochino Y et al (2009) Sorafenib blocks tumour growth, angiogenesis and metastatic potential in preclinical models of osteosarcoma through a mechanism potentially involving the inhibition of ERK1/2, MCL-1 and ezrin pathways. Mol Cancer 8:118
189. Laverdiere C et al (2005) Messenger RNA expression levels of CXCR4 correlate with metastatic behavior and outcome in patients with osteosarcoma. Clin Cancer Res 11(7):2561–2567
190. Kawada K et al (2004) Pivotal role of CXCR3 in melanoma cell metastasis to lymph nodes. Cancer Res 64(11):4010–4017
191. Walser TC et al (2006) Antagonism of CXCR3 inhibits lung metastasis in a murine model of metastatic breast cancer. Cancer Res 66(15):7701–7707
192. Pradelli E et al (2009) Antagonism of chemokine receptor CXCR3 inhibits osteosarcoma metastasis to lungs. Int J Cancer 125(11):2586–2594
193. Wandinger SK, Richter K, Buchner J (2008) The Hsp90 chaperone machinery. J Biol Chem 283(27):18473–18477
194. Zhao R et al (2005) Navigating the chaperone network: an integrative map of physical and genetic interactions mediated by the hsp90 chaperone. Cell 120(5):715–727
195. Hu Y, Bobb D, He J, Hill DA, Dome JS (2015) The HSP90 inhibitor alvespimycin enhances the potency of telomerase inhibition by imetelstat in human osteosarcoma. Cancer Biol Ther 16(6):949–957
196. Ory B et al (2016) Blocking HSP90 addiction inhibits tumor cell proliferation, metastasis development, and synergistically acts with Zoledronic acid to delay osteosarcoma progression. Clin Cancer Res 22(10):2520–2533
197. Nomura M et al (2019) Tegavivint and the β-catenin/ALDH axis in chemotherapy-resistant and metastatic osteosarcoma. J Natl Cancer Inst 111(11):1216–1227
198. Baker EK et al (2015) BET inhibitors induce apoptosis through a MYC independent mechanism and synergise with CDK inhibitors to kill osteosarcoma cells. Sci Rep 5:10120
199. Fossey SL et al (2009) Characterization of STAT3 activation and expression in canine and human osteosarcoma. BMC Cancer 9:81
200. Zhang T et al (2017) Toosendanin demonstrates promising antitumor efficacy in osteosarcoma by targeting STAT3. Oncogene 36(47):6627–6639
201. Hirai H et al (2010) MK-1775, a small molecule Wee1 inhibitor, enhances anti-tumor efficacy of

various DNA-damaging agents, including 5-fluorouracil. Cancer Biol Ther 9(7):514–522
202. Rajeshkumar NV et al (2011) MK-1775, a potent Wee1 inhibitor, synergizes with gemcitabine to achieve tumor regressions, selectively in p53-deficient pancreatic cancer xenografts. Clin Cancer Res 17(9):2799–2806
203. Davis LE et al (2017) Integration of genomic, transcriptomic and functional profiles of aggressive osteosarcomas across multiple species. Oncotarget 8(44):76241–76256

Genomics and the Immune Landscape of Osteosarcoma

2

Chia-Chin Wu and J. Andrew Livingston

Abstract

Conventional osteosarcoma (OS) is a high-grade intraosseous malignancy with production of osteoid matrix; however, a deeper dive into the underlying genetics reveals genomic complexity and instability that result in significant tumor heterogeneity. While early karyotyping studies demonstrated aneuploidy with chromosomal complexity and structural rearrangements, further investigations have identified few recurrent genetic alterations with the exception of the tumor suppressors *TP53* and *RB1*. More recent studies utilizing next-generation sequencing (NGS; whole-exome sequencing, WES; and whole-genome sequencing, WGS) reveal a genomic landscape predominantly characterized by somatic copy number alterations rather than point/indel mutations. Despite its genomic complexity, OS has shown variable immune infiltrate and limited immunogenicity. In the current chapter, we review the hallmarks of OS genomics across recent NGS studies and the immune profile of OS including a large institutional cohort of OS patients with recurrent and metastatic disease. Understanding the genomic and immune landscape of OS may provide opportunities for translation in both molecularly targeted therapies and novel immuno-oncology approaches.

Keywords

Osteosarcoma · Genomics · Next-generation sequencing (NGS) · Chromothripsis · Telomere lengthening · Immune profiling

Introduction

Osteosarcoma (OS) is the most common primary malignancy of the bone predominantly occurring in adolescents with a second peak in incidence as secondary OS among older adults [44]. For patients presenting with localized disease at diagnosis, standard multi-agent chemotherapy combined with surgical resection yields long-term survival rates of ~70% [6, 44]. Metastatic disease either at diagnosis or at the time of recurrence portends a poor prognosis with survival of 20–30% [28, 42]. Thus, there has been a long-

C.-C. Wu
Department of Genomic Medicine, The University of Texas MD Anderson Cancer Center, Houston, TX, USA
e-mail: cwu9@mdanderson.org

J. A. Livingston (✉)
Department of Sarcoma Medical Oncology, The University of Texas MD Anderson Cancer Center, Houston, TX, USA

Department of Pediatrics, The University of Texas MD Anderson Cancer Center, Houston, TX, USA
e-mail: JALivingston@mdanderson.org

E. S. Kleinerman, R. Gorlick (eds.), *Current Advances in the Science of Osteosarcoma*, Advances in Experimental Medicine and Biology 1258, https://doi.org/10.1007/978-3-030-43085-6_2

standing interest in understanding the underlying biology of OS tumorigenesis, evolution, and metastasis in order to identify novel treatment strategies and improve survival outcomes.

Recent progress made in next-generation sequencing (NGS) and molecular genetic studies of osteosarcoma has broadened our view of the genetic hallmarks of the disease and potential therapeutic approaches for patients. The point mutation burden of OS is around 1.1~1.5 per Mb [13, 49, 71], making it the highest mutation burden among pediatric solid tumors but intermediate overall and significantly lower than melanoma or non-small cell lung cancer (Fig. 2.1). The OS genome is characterized by genomic complexity and instability, enriched with rearrangements, and somatic copy number alterations. Figure 2.2 shows the MD Anderson Cancer Center osteosarcoma (MDACC OS) cohort has a higher level of rearrangements than most of other tumor types. This suggests that rearrangements and copy number alterations are major driving forces contributing to OS oncogenesis. In addition, it has become clear that genome instability has a significant impact on the interaction between the tumor cells and immune system [47]. In this chapter, we review the molecular genetics of OS, which are associated with genome instability and immune landscape, based on the findings of recent whole-genome/whole-exome sequencing (WGS/WES) studies (Table 2.1).

Genomic Landscape of Osteosarcoma

Key Altered Genes and Pathways Associated with OS Genome Instability and Oncogenesis

OS is characterized by complex genome instability and high level of genetic heterogeneity [4, 13]. The majority of the resultant genetic alterations are associated with copy number changes and genome rearrangement. Genome instability can lead to changes in both the cancer genome and the tumor microenvironment. Elucidating the mechanisms of genome instability in OS would thus aid in our understanding of tumorigenesis, evolution, progression, and metastasis in order to develop new therapeutic approaches [76]. This section reviews key altered genes and pathways associated with OS genome instability and oncogenesis identified in recent WGS/WES studies. The

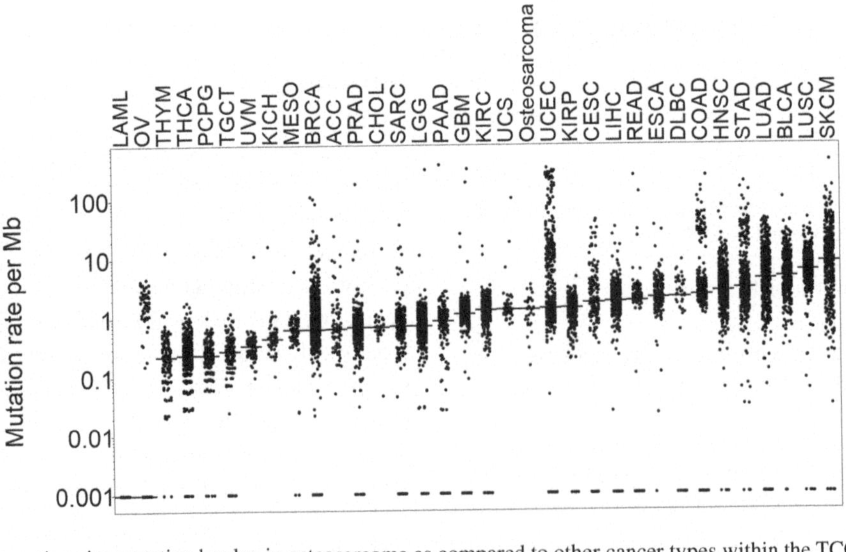

Fig. 2.1 Somatic point mutation burden in osteosarcoma as compared to other cancer types within the TCGA [61]

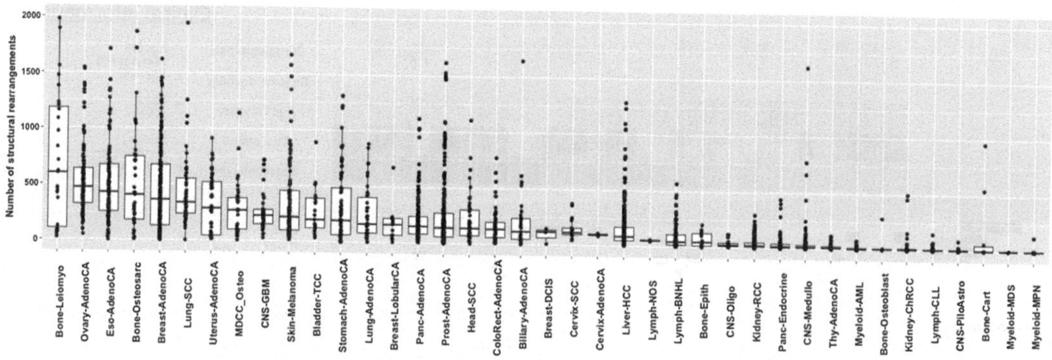

Fig. 2.2 Frequency of structural variants in OS (bone osteosarcoma) and other human cancer types within the ICGC [37]

Table 2.1 NGS studies in osteosarcoma

Citation	Patient population	Sequenced specimens	Key findings
Chen et al. [13]	Pediatric	34 WGS	1. Identified kataegis in 50% of the tumors 2. Discovered new insights into alterations (rearrangements) in TP53 to promote the OS oncogenesis
Perry et al. [49]	Pediatric and adult	13 WGS/59 WES	1. Identified recurrent mutations in the PI3K/mTOR pathway, and proposed it as an OS therapeutic target 2. Kataegis were detected in almost all cases
Kovac et al. [34]	Pediatric	31 WES/92 SNP array	1. Identified recurrent BRCA1/2 inactivation and showed that BRCA alterations may be associated with sensitivity to PARP inhibition in OS cells
Bousquet et al. [8]	Pediatric	7 WES	1. Confirmed the presence of genetic alterations of the TP53 and RB1 genes
Behjati et al. [4]	Pediatric and adult	47 WGS/7RNAseq	1. Identified recurrent mutations in IGF signaling as a potential therapeutic target in OS treatments. 2. Identified the chromothripsis pattern in 30% of the cases
Wu et al. [71]	Pediatric and adult	36 WGS/54 RNAseq	1. Genomic complexity of OS may be associated with cooperative alterations of TP53, alternative telomere lengthening (ALT), and whole-genome doubling (WGD) 2. Younger patients showed enrichment in rearrangements associated with chromothripsis 3. Several observed immunogenomic features may contribute to the limited immunotherapy response in OS including transcript suppression of neoantigens by nonsense-mediated decay (NMD), significant negative correlation between copy number loss and immune infiltration, and significant negative correlation between the gene expression/copy number of PARP2 and the immune infiltration

most frequently altered genes and their associated cancer signaling pathways of our MDACC OS cohort of recurrent and metastatic OS patients (MDACC OS) are shown in Fig. 2.3. Importantly, the majority of these pathways are associated with the underlying genome instability that is a hallmark of OS.

TP53

TP53, a tumor suppressor gene, codes for a protein that can respond to diverse cellular stresses and thereby induce cell cycle arrest, apoptosis, senescence, and DNA repair. Somatic mutations in TP53 are one of the most frequent alterations in human cancers, in which the majority of

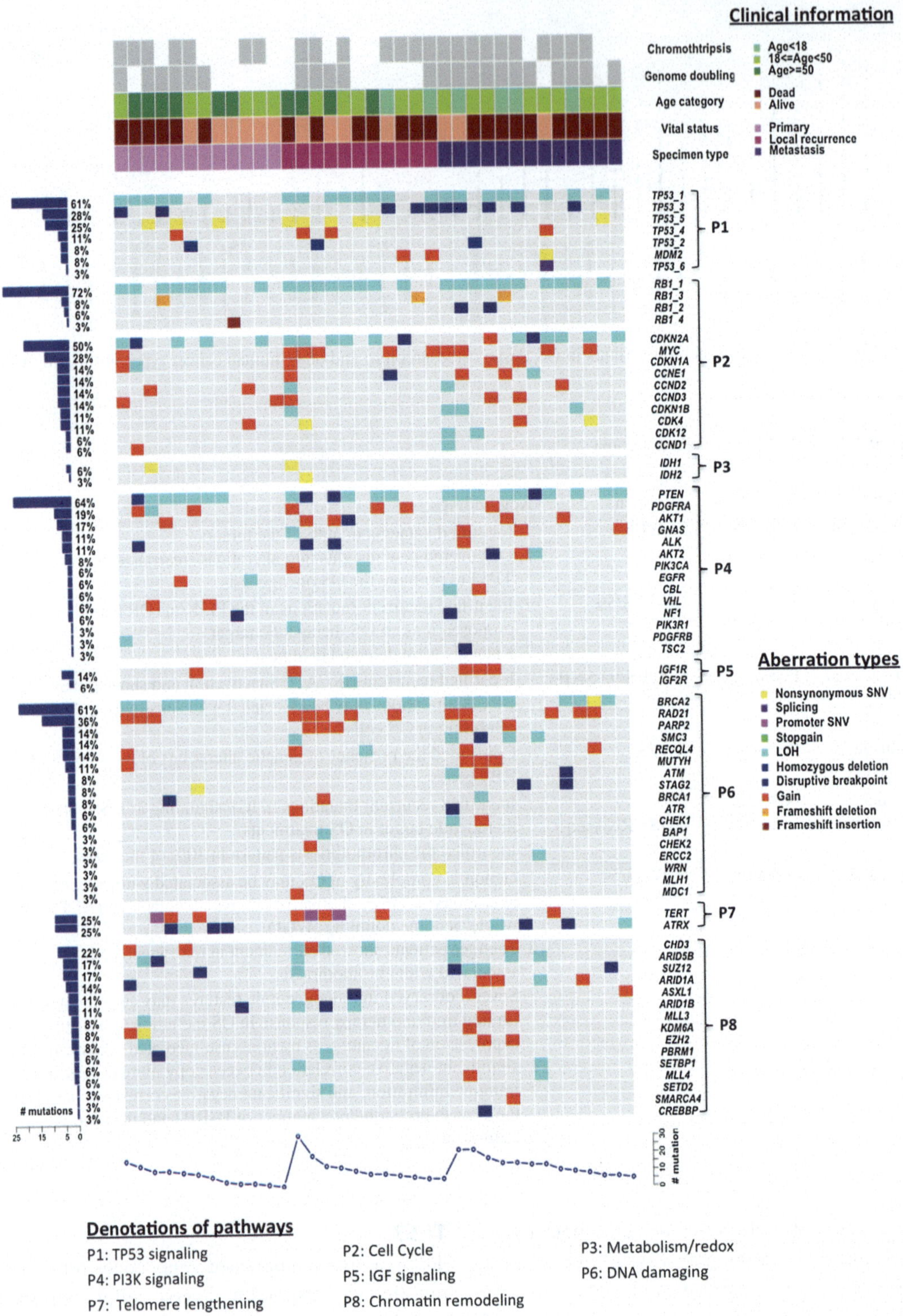

Fig. 2.3 The mutation landscape of the MDACC OS cohort. Genomic alteration identified by WGS for selected genes and key pathways

genetic alterations across cancer types are missense substitutions [50]. With WES and targeted sequencing, it had previously been estimated that only 20%~50% of osteosarcomas carry mutations in the p53 pathway, and other portion of the tumors were identified as so-called TP53 wild type [33, 79]). However, in a WGS study of 34 pediatric OS samples, Chen et al. [13] discovered new insights into alterations in TP53 to promote the OS oncogenesis. In their cohort, they identified 55% of TP53 mutations were caused by structural variations, whose breakpoints were mostly confined to the first intron of the gene [13], thereby inactivating TP53. In the 36 samples of our MDACC OS cohort, we also identified 9 samples (25%) as having TP53 structural variations. Therefore, it now is suspected that up to 75–90% of OS patients harbor various types of TP53 genetic alterations [13, 49, 71], which is the most prevalent genetic alteration in OS.

Loss of the TP53 pathways that disable the cell's ability to respond to DNA damage mediates genome instability and triggers OS oncogenesis [40]. Several TP53-deficient cell lines and genetically engineered mouse models also have been developed to model OS oncogenesis and indicated the causal relation between TP53 alterations and OS initiation/genome instability [23, 66]. Taken together, these mechanistic studies and associations observed from sequenced patient samples identify TP53 alterations as having the strongest association with genome instability and oncogenesis in OS.

RB1 and Other DNA Damage Repair Pathways

Retinoblastoma transcriptional corepressor 1 (RB1), a key regulator of cell cycle progression by controlling the G1/S phase transition, is another prevalent genetic alteration in OS. Alterations in RB pathway can prevent cell cycle arrest in response to DNA damage to induce genome instability and promote oncogenesis [65]. Alterations in RB1 have been identified in 50%–78% of OS across NGS studies [13, 33, 49, 71]. Unlike TP53, the depletion of RB alone was not sufficient to induce OS formation in mouse models, and studies speculated that RB alterations may synergize with TP53 inactivation during OS oncogenesis [52].

Breast cancer susceptibility genes (BRCA1/2) encode nuclear phosphoproteins that are involved in molecular signaling in transcription, DNA repair of double-stranded breaks, and recombination, thereby playing a role in maintaining genomic stability and acting as a tumor suppressor [73]. Alterations in these genes are known to be responsible for inherited breast and ovarian cancers. In OS, Kovac et al. [33] identified BRCA1/2 inactivation in 112 (91%) and 96 (78%) of their 123 samples, primarily caused by copy number alterations. They also showed that BRCA alterations in OS cell lines are associated with sensitivity to PARP inhibition, a strategy that was shown to induce cell cycle arrest and apoptosis in BRCA1-, BRCA2-, and PALB2-deficient breast cancers [62]. We also identified alterations (mostly copy number LOH) in BRCA in 89% of our 36 MDACC OS samples [71]. By analyzing the mutation spectrum of their sample cohort, Kovac [33] also identified COSMIC signature 3 and signature 5 in their WES data. Signature 5 is associated with an age-related mutational process, whereas signature 3 is characterized by a pattern enriching of C > G substitutions that is strongly associated with BRCA1 and BRCA2 mutations in breast, pancreatic, and ovarian cancers [1]. However, the mutation spectrum of the MDACC OS WGS sample cohort is dominated by C > T substitutions, C > A, and T > C substitutions. We identified two prevalent mutation signatures: COSMIC signature 5 and signature 8. Behjati et al. [4] also identified signature 5 and 8 are the most prevalent mutation signatures in their OS WGS cohort. We found that signature 8 is significantly associated with worse prognosis, but its etiology is still unknown. The difference of the mutation signature analysis results may be related to the lower mutation burden observed in OS WES data, compared to OS WGS data. In addition, some studies recently proposed that the genetic association of mutation signatures would be tissue-specific or cancer type-specific [7, 26]. Therefore, the association between signature 3 and BRCA in OS warrants further validation in other OS cohorts.

Telomere Lengthening Pathways

Telomeres can protect chromosomes of a cell from DNA damage but become shorter with each cell division, eventually leading to senescence or apoptosis [40]. During oncogenesis, cancer cells frequently activate either telomerase-dependent or telomerase-independent elongation mechanisms in order to protect against telomere shortening in accelerated cell division cycle and maintain unlimited growth and clonal evolution of genomically unstable cells [40]. Through the process of telomere lengthening, cancer cells can accumulate large amounts of genome alterations.

Promoter mutations of TERT, an active component of telomerase, were previously identified in 1 of 23 (4.3%) OS patient samples [31]. No TERT mutation was found in current OS NGS studies [4, 13, 33, 49] except our MDACC OS cohort. We identified TERT promoter mutations (chr5: 1295228 C > T) in two patient samples, but we found the mutations and expression of TERT are not significantly associated with longer telomeres. Therefore, an association between TERT mutations with telomere lengthening in OS has not been well-established.

A telomerase-independent mechanism termed alternative lengthening of telomeres (ALT) also has been frequently identified in several cancer types [20]. Cancers utilizing ALT often have lost function of ATRX, a chromatin remodeling protein, and/or DAXX, a death domain associated protein, through DNA mutation or deletion. Chen's study [13] identified five samples with point mutations in ATRX and five with focal deletions or structural variation affecting the coding region of the gene. They also found samples with ATRX alterations tend to have a greater number of telomeric reads estimated from the sequencing data. Our MDACC OS cohort also identified seven patients with deleterious alterations in ATRX (7/36, 20%) as well as one patient with copy number loss in DAXX who all had telomere lengths greater than the cohort median. Lower expression levels of ATRX were also significantly correlated with longer telomere lengths. We also found that patient samples with the longest telomere length carried alterations in both TP53 and ATRX, supporting the permissive context in which TP53 alterations can allow for activation of ALT in OS. In addition, our MDACC dataset also showed that the expression levels of known telomere maintenance genes, including HNRNPA2B1, WRN, and HUS1, were also significantly correlated with telomere length [14]. However, the exact mechanisms surrounding telomere maintenance in the ALT pathway are unclear, and the effects of the telomere-related mutations on ALT are still needed to be explored. Based upon these findings, there is a growing interest in investigating ATR inhibitors or other agents that target DNA damage response in OS.

IGF Signaling/PI3K-mTOR

The insulin-like growth factor (IGF) signaling includes three ligands (INS, IGF1, and IGF2), three receptors (IR, IGF1R, and IGF2R), as well as six IGF-binding proteins (IGFBPs), which provide a potent proliferative signaling system that can block apoptosis and stimulate growth and differentiation in many cell types. Numerous studies have demonstrated the role of IGF signaling in the development and progression of various cancer types as well as its role in resistance to chemotherapeutic agents [38]. Given the association between IGF signaling and bone growth [72], disorders of IGF signaling are thought to be implicated in OS pathogenesis. Recently, Behjati et al. [4] identified recurrent alterations in IGF signaling as a potential therapeutic target in OS treatments. They found alterations of IGF signaling in 8 of 112 (7%) WES and WGS samples and validated the observation with IGF1R amplifications observed in 14% of 87 OS samples using fluorescence in situ hybridization (FISH). We also identified alterations of IGF signaling in 7 of the 36 MDACC OS samples (20%) (Fig. 2.3). Interestingly, some studies have shown that alteration in TP53 and DNA repair defects in tumor cells may activate IGF1R signaling [3, 69]; however, additional studies are still needed to explore this cause-effect relation. In addition, Perry et al. [49] also found recurrent mutations in the downstream signaling pathways of IGF signaling, the PI3K/mTOR pathway, in 14 of the 59 OS samples, a similar rate to the Behjati cohort which identified downstream pathway alterations including PI3K or MAPK signaling in 27% of tumors. Perry et al. [49] also demonstrated OS cell lines are responsive to pharmacologic and genetic inhibition of the PI3K/mTOR pathway both in vitro and in vivo and proposed this pathway as a therapeutic target for the treatment of OS. Our MDACC cohort similarly identified a high frequency of alterations in the PI3K/mTOR pathway in 28 of the 36 (78%) OS samples (Fig. 2.3). Across these cohort studies, these findings support further investigation of IGF1R, PI3K, or mTOR inhibition in patients with OS which may have greater activity or relevance in a biomarker selected patient population.

Other Genomic Events Associated with OS Genome Instability

While several of the aforementioned pathways can mediate genomic complexity and instability in OS, other genetic or epigenetic alterations and events have been implicated in increasing genome instability in OS. This section will review several genomic events associated with genomic complexity and instability of OS which have been further elucidated by current WGS/WES studies.

Genome Doubling

During clonal evolution in oncogenesis, genomically unstable cells continually lose and gain whole and/or parts of chromosomes to provide potent selective pressure for clonal expansion. However, genome instability beyond a certain threshold is likely to cause cancer cells with unviable karyotypes [76]. Therefore, OS cells need to maintain viability of their TP53/RB1 mutation-induced unstable genomes through multiple mechanisms. Whole-genome doubling (WGD) is one mechanism that can increase viability of cancer cells with significant chromosomal instability [18, 80, 81]). By applying the allele-specific copy number profiles inferred from WGS data and the algorithm modified from the previously published studies [11, 18], we identified WGD in 58% (22/38) of samples in the MDACC OS cohort, a comparable frequency to what has been observed in colorectal and breast cancer [11]. In addition, we also found that OS samples with WGD tend to have a higher number of rearrangements and copy number alterations than those without WGD. Furthermore, 50% (18/36) of patients had losses of heterozygosity (LOH) in TP53 and/or RB1 along with WGD. Given the inherent lower likelihood of losing two copies after WGD, these findings support that TP53 and RB1 aberrations likely occurred prior to WGD [11].

Chromothripsis

Chromothripsis is the genomic process by which massive genomic rearrangements are acquired in a single catastrophic event [21]. Chromothripsis may generate genetic drivers in oncogenesis through DNA copy number gain and loss as well as rearrangements, such as translocations. Chromothripsis is associated with both somatic and germ line TP53 mutations in pediatric medulloblastoma and acute myeloid leukemia [51]. In addition, chromothripsis can be associated with telomere crisis induced by telomere shortening in accelerated cancer cell cycle division [39]. In this study, the authors showed that telomere loss promotes end-to-end chromosome fusions and dicentric chromosomes during mitosis and undergoes breakage-fusion-bridge cycles, eventually resulting in hundreds of DNA breaks [39]. However, these associations warrant further investigation and validation, particularly in relation to OS pathogenesis. Chromothripsis has been observed at varying frequencies (20–89%) in OS patient samples of all ages across OS studies [4, 13, 71]. To date, the etiological factors and mechanisms underlying OS chromothripsis are still unknown, and no specific genomic regions and genes were found to be significantly associated with chromothripsis in OS samples. Our MDACC OS studies recently found that there was a trend for younger patients to have rearrangements that are clustered and associated with chromothripsis as compared with older patients. This result was also observed in the TARGET OS cohort dataset. This suggests that oncogenesis may be more driven by catastrophic chromothripsis events in young OS patients as compared to older adults.

Kataegis

Kataegis is a pattern of localized hypermutation (enriched of C > T and C > G changes at TpC dinucleotides) associated with the APOBEC deaminases [55]. Chen et al. [13] and Perry et al. [49] found approximately 50–85% of their OS patient samples showed kataegis patterns. Approximately 60% of the MDACC OS samples also showed the kataegis patterns. No important cancer genes were found to be recurrently located in the kataegis regions identified in OS samples. However, most of these kataegis patterns occurred in no-coding regions, some of which may include cis- and trans-regulatory elements or may be transcribed into functional non-coding RNA molecules, such as transfer RNA, ribosomal RNA, and regulatory RNAs. More investigations on the association of these kataegis regions with OS oncogenesis are necessary in future studies.

The Osteosarcoma Immune Landscape and Immunogenomic Interplay

Interactions between the immune system and tumor play an important role in effective tumor control. Aberrations in this interaction can lead to ineffective tumor surveillance, enhance tumor growth, and enable metastatic disease progression. For this reason, there has been a long-standing interest in targeting this interaction and modulating the host's immune response as a strategy to eliminate cancer. Targeting immune checkpoints, such as cytotoxic T-lymphocyte-associated antigen 4 (CTLA-4) and programmed cell death 1 (PD-1)/ligand 1 (PD-L1), has been an overwhelmingly successful step forward for immunotherapy in the treatment of cancer. Immune cells, including CD8+ tumor-infiltrating lymphocytes, are initially attracted to tumor cells by the presence of tumor-specific antigens which are encoded by somatic alterations in cancer cells. Tumors can escape immune-surveillance by modulating antigen expression and upregulating inhibitory immune checkpoints to lead to immune cell apoptosis, anergy, and tolerance. Immune checkpoint inhibitors (ICI) block such signals in order to activate an antitumor immune response. The success of ICIs in the clinic, yielding durable responses in a subset of patients with previously incurable metastatic disease, such as melanoma and lung cancer, has revived enthusiasm for immunotherapy and established a new paradigm for cancer treatment [16]. The Food and Drug Administration (FDA) has approved a number of ICIs including anti-PD-1, anti-PD-L1, and anti-CTLA4 antibodies among others for the treatment of a wide range of malignancies. Despite their remarkable success, only a subset of cancer patients benefit from these therapies, and responses are varied across patients and cancer types. Therefore, there is a growing need to understand mechanisms of the resistance to ICI and identify predictive biomarkers for personalized immunotherapy approaches.

Osteosarcoma demonstrates significant genetic complexity and genome instability with resultant high levels of genomic rearrangements and the highest point mutation burden as compared to other pediatric cancers, suggesting that these genomics factors may yield neoantigens capable of eliciting an immune response. However, despite this rationale, recent clinical trials using immune checkpoint inhibitors in OS have been disappointing. Therefore, this section will outline the OS immune landscape and genomic features that may contribute to resistance to ICI and other immunotherapy agents in OS.

OS Immune Landscape

Transcriptome profiles derived from bulk RNAseq and other methods have been used to study features of the tumor and the microenvironment that are associated with tumor response or resistance to ICI in several cancer types including OS. Most often, these studies compare the transcriptome of responders vs. nonresponders to identify key differentially expressed genes that may account for response or resistance. While several studies have explored the transcriptional signatures linked to ICI responses across various clinical patient cohorts [25], the small sample size of most of these studies limits their generalizability [43]. This is particularly true for rare cancers such as OS. One particular challenge in OS is the overall lack of responders to ICI and the concurrent lack of high-quality patient samples to undertake such a study. Given the limited number of responses on clinical trials to date such as the SARC028 study of pembrolizumab in soft tissue and bone sarcoma [59], such a transcriptome analysis of responders vs. nonresponders has not been feasible. Therefore, we focused on a larger cohort of OS patients with poor risk – those with recurrent metastatic disease – as this cohort is thought to represent OS patients who would be considered for treatment with ICI. The MDACC OS cohort also includes four pretreatment specimens from patients treated with anti-PD-L1 in combination with anti-CTLA4 therapy, all of whom did not respond to treatment.

In the ongoing search for biomarkers, immune infiltration levels (and, in particular, tumor-infiltrating CD8+ T cells) and PD-1/PD-L1 expressions have been shown to be associated with response to ICIs. However, expression is variable, and the significance across studies and across tumor types remains controverted as a considerable number of patients with high levels of immune infiltration and high PD-1/PD-L1 expression have poor responses or fail to respond to treatment [56]. These points aside, several studies have recently been conducted in OS evaluating the prevalence and prognostic significance of immune infiltrate and PD-1/PD-L1 expression. Shen et al. [57] first measured RNA expression levels for PD-L1 in 38 OS samples using quantitative real-time RT-PCR and found that high levels of PD-L1 are expressed in a subset of OS and that PD-L1 expression is positively correlated with tumor-infiltrating lymphocytes. Koirala et al. [32] and Palmerini et al. [48] also identified the similar results in much larger OS cohorts. Further, they identified an association between CD8+ infiltrate and superior overall survival, whereas infiltration with dendritic cells and macrophages as well as PD-L1 expression was associated with a poor prognosis. Within the MDACC OS cohort, we did not observe an association between immune infiltrate, CD8+ TIL, or PD-L1 expression, and overall survival. This may be due in part to the poor risk nature of the patients we included for study. Further translational studies are needed to determine if either immune infiltrate or PD-L1 expression correlates with clinical benefit from ICIs in the treatment of OS.

To understand the immune infiltrate level of OS in a broader context, we compared the immune infiltration score [74] derived from the bulk RNAseq data of our cohort against other tumor types profiled in the Cancer Genome Atlas (TCGA), the 87 TARET OS samples, the 7 OS samples from the Behjati's study [4], as well as 4 patients with metastatic OS who were treated with ICIs but exhibited no objective responses [71]. All four OS cohorts have comparable median immune score (Fig. 2.4). We observed that these four OS cohorts have intermediate median immune scores that are lower than many of the cancer types that have shown clinical benefit and treatment responses to ICIs with high immune infiltrate levels such as melanoma (TCGA-SKCM) and lung cancer (TCGA-LUAD and TCGA-LUSC) but are higher than those that have shown minimal activity with current immunotherapy approaches such as uveal melanoma (TCGA-UVM) (Fig. 2.4). When compared to other sarcoma subtypes, the median immune scores of dedifferentiated liposarcoma (TCGA-DDLPS) and undifferentiated pleomorphic sarcoma (TCGA-UPS) – two soft tissue sarcoma subtypes where ICI are active – are higher than OS samples. We also found less than 10% of OS samples whose immune infiltration levels were among the highest quartile across tumor types. These results suggest that while most OS specimens may have insufficient immune infiltrate to elicit meaningful responses

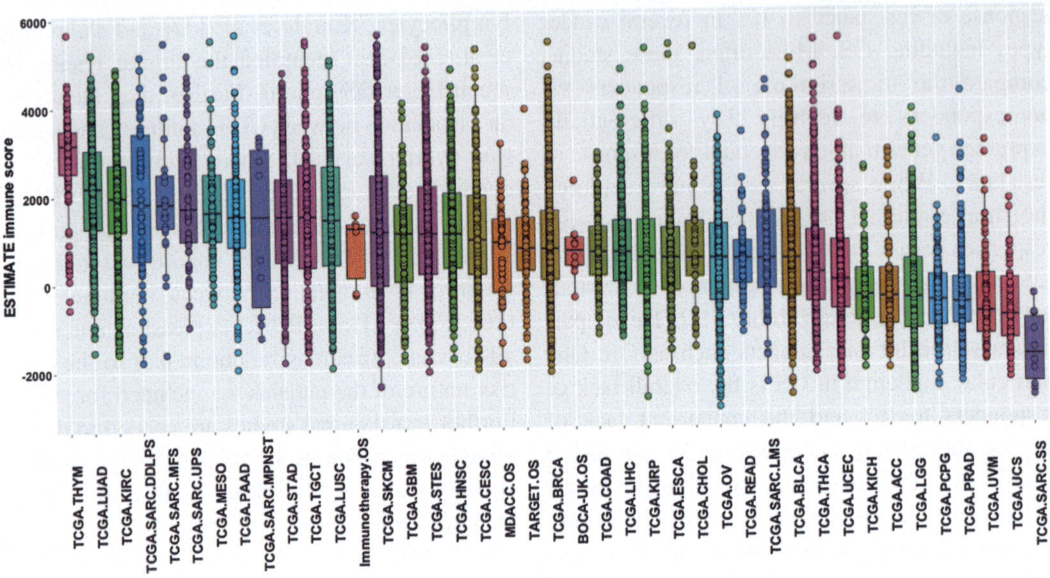

Fig. 2.4 Immune scores in TCGA tumor types and the four OS cohorts. ESTIMATE scores derived from RNAseq show intermediate immune score for OS across cohorts as compared to the TCGA

to ICI, there is a subset of patients who would be predicted to benefit. However, immune infiltrate alone is inadequate to predict treatment response.

Bulk RNAseq data were also used to infer profiles of different infiltrated immune cells [12, 36, 63]. We characterized the composition of various infiltrated immune cells across our MDACC OS samples [71] and identified three clusters of our samples: immune-high, intermediate, and low (Fig. 2.5). The immune-low and the immune-high samples, respectively, have the lowest and highest levels of all immune cell types including CD8+ lymphocytes. Several known tumor-intrinsic immunosuppressive pathways were found to be deregulated between the immune-high and low cluster samples, such as PD-1 signaling, CTLA4 pathway, and IFNG signaling, suggesting that the immune-high tumors may upregulate immune-suppressive signals that inhibit T-cell activation.

OS Genomic Features Associated with the Immune Response

Immune modulation in cancer cells is recognized as a hallmark of cancer initiation and progression, implying that tumor cell-intrinsic factors are associated with tumor response/resistance to ICIs [56]. Therefore, this section reviews several OS genomic features revealed by WES/WGS/RNA-seq studies that may be associated with OS response/resistance to ICIs. These factors may explain in part why the majority of OS patients have failed to benefit from treatment with ICIs and also present opportunities for novel therapeutic approaches.

Neoantigens

During oncogenesis, tumors accumulate thousands of genetic alterations, including point mutations, indels, and rearrangements. Some of them alter the amino acid sequence of the encoded proteins, called neoantigens, which do not present in normal cells [30]. The immune system can discriminate self from these non-self-antigens expressed by cancer cells and activate immune response to kill cancer cells. ICI can enhance and strengthen the immune response to non-self-antigens and promote antitumor activity.

Correlations between tumor-specific antigen and tumor response to checkpoint blockade

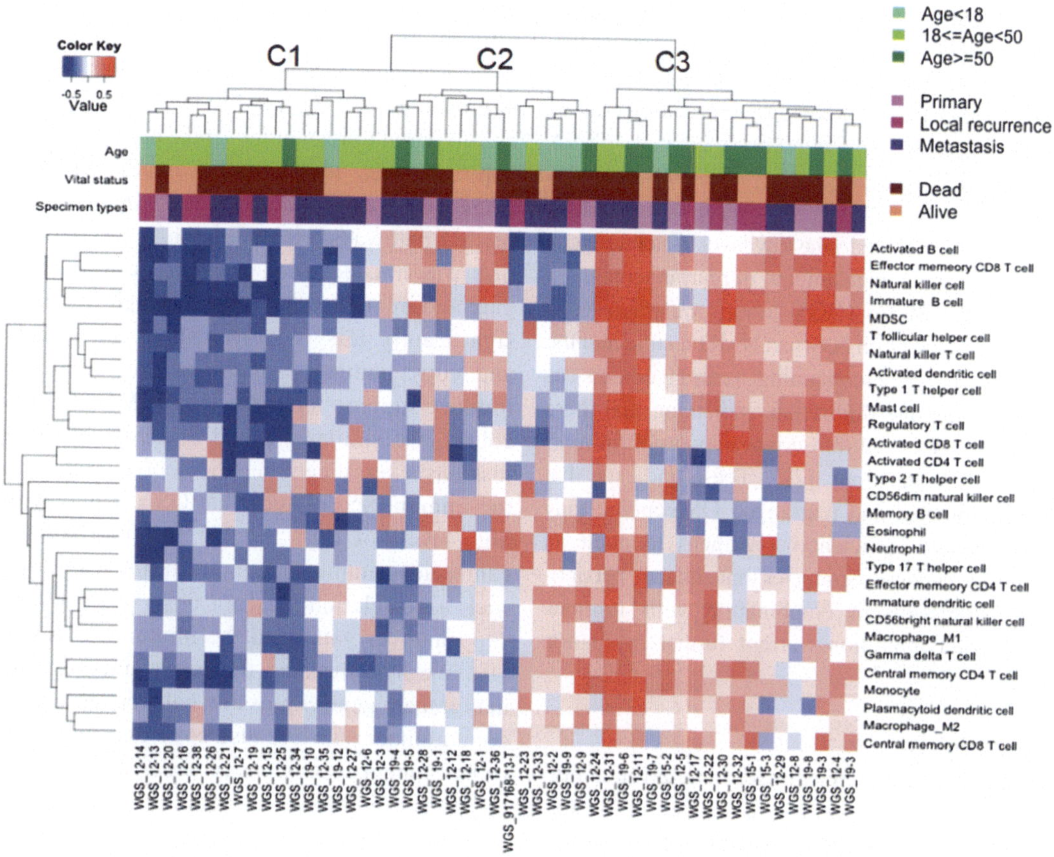

Fig. 2.5 Immune cell profiling of the MDACC OS cohort. RNAseq identifies three unique clusters with low (C1), intermediate (C2), and high (C3) immune infiltrate

therapies have largely been focused on tumor mutation burden, specifically point mutations. Several studies showed that nonsynonymous point mutation burden is one of the most reliable predictive biomarkers associated with responsiveness to ICIs in both melanoma and lung cancer [77, 78]. Recently, Turajlic et al. [64] investigated whether the frameshift nature of indel mutations can create novel open reading frames and a large quantity of neoantigens, which might contribute to the immunogenic response. They found renal cell carcinomas, one of the cancer types that have clinical benefit and response to immune checkpoint blockade related to high immune infiltrate levels, have the highest proportion and number of indel mutations, compared to other cancer types in TCGA [64]. Their analysis also showed that frameshift indel count is significantly associated with response to ICIs.

Although OS has a much lower point mutation burden compared to those of melanoma and lung cancer (Fig. 2.3), OS demonstrates significant genomic complexity with the high levels of genomic rearrangements that could potentially generate high-level neoantigens. However, we found that most of mutations detected by WGS (i.e., whole genome DNA-seq) were not detected by RNAseq in our MDACC OS cohort samples [71]. Unexpressed mutations tended to occur in genes that have low expression or whose variant allele frequencies were low. The limited overlap of point mutations identified in both WGS and RNAseq also has been observed in non-small cell lung cancer and glioblastoma [15, 45]. In addition, we observed that very few predicted rearrangements involving coding regions were expressed, suggesting that most of the rearrangements are truncated or harbor premature termination codons [71]. Therefore, the nonsense-mediated mRNA decay (NMD) pathway, which selectively degrades mRNAs harboring premature termination codons [9], may contribute to the low level of neoantigens generated from large amount of rearrangements in OS. In the MDACC OS cohort, we observed a positive association between NMD factors and the number of gene-containing rearrangements as well as immune infiltration, indicating that there may be substantial transcript suppression in rearranged OS genomes [71]. These indicate that highly mutated and rearranged OS genome may not generate sufficient neoantigens to elicit an immune response. Strategies that enhance neoantigen expression in combination with immune checkpoint inhibition warrant further evaluation in OS.

Aneuploidy

In addition to increasing the mutational burden, genome instability in cancer cells also leads to chromosome copy number alterations that are categorized in two major classes: whole or arm level copy number changes known as aneuploidy or focal copy number changes [5]. While focal copy number changes that involve tumor suppressor genes or oncogenes are often considered actionable targets in cancer therapy, the functional relationship between aneuploidy and oncogenesis is not well understood. Several recent studies have shown that aneuploidy is associated with the immune suppression across multiple cancer types [17, 47, 53, 60]. Of interest, Davoli et al. [17] found that chromosome and arm level of copy number alterations have a greater contribution to immune suppression than focal level of copy number alterations. They hypothesize that chromosome and arm level of copy number alterations can impact gene expression of a large number of genes and may thus impair or deregulate cellular signaling needs for cytotoxic immune cell infiltration. However, their studies did not specifically explore the impact of copy number gain and/or loss on immune infiltration. In melanoma, Roh et al. [53] identified a higher burden of copy number loss in nonresponders to CTLA-4 and PD-1 blockade as compared to responders and found that these copy number losses were associated with decreased expression of genes in immune-related pathways. However, the same association was not identified between copy number gain and immune response. Although frequent, significantly less is known about aneuploidy and copy number changes as they relate to immune suppression in OS. In the MDACC OS cohort, we similarly found that copy number loss has a significant negative correlation with the immune scores such that OS samples with high levels of copy number loss had significantly less immune cell infiltrate. Similarly, such a correlation was not observed between copy number gains and the immune scores in OS. We hypothesized that copy number loss may impact gene expression balance more than copy number gain because copy loss may lead to permanent loss of many genes and eventually impact immune response.

Genetic Alterations and Pathways

A range of genetic/epigenetic aberrations in tumor cells can influence various aspects of the immune landscape, such as activation of immunosuppressive immune cells, regulation of dendritic cell activation and T-cell priming, instigation of tumor resistance to T-cell attack, and deregulation of immune checkpoint molecule expression [30, 43, 54, 68]. Alterations and deregulation of multiple oncogenic pathways such as MAPK/PTEN/PI3K, WNT/beta-catenin, and JAK/STAT (termed genetic T-cell exclusion) have specifically been shown to be associated with resistance to ICIs. Aberrations in antigen processing/presentation pathway or interferon-gamma signaling have also been implicated in primary resistance to immunotherapy [2, 56, 58, 75].

In our MDACC OS cohort, we also found many of these tumor-intrinsic immunosuppressive pathways such as IFNG, MAPK/PI3K/mTOR, and JAK/STAT and antigen presenting pathways are dysregulated between the immune-high and low samples that were identified by RNAseq (sect. 3.1). Some of these pathways were also found to be deregulated between OS and normal samples through the analysis of RNAseq data [81]). These pathways are currently being explored as potential combination strategies that could extend the benefits of ICIs for OS treatments.

In our MDACC OS cohort, we also integrated genomic aberrations and transcriptomic analyses to identify genes whose aberrations and gene expression are significantly associated with immune infiltration, such as TP53 and PARP2. Among other findings, we identified a negative association between PARP2, a druggable target involved in the DNA damage response, amplification (35% samples), and gene expression with the immune infiltration score (such that OS samples with high PARP2 expression had significantly lower immune infiltrate). While prior studies have demonstrated the sensitivity of OS cell lines to PARP inhibitors [19, 33], these studies largely focused on the role of BRCA2 deletion in DNA damage response and its synergy with PARP inhibitions. However, PARP2 amplification was found in the samples of these OS studies [4, 33], and the association of PARP2 amplification with immune response in OS was not previously identified. PARP appears to play an important role in modulating the immune response. PARP inhibition can increase intratumoral CD8+ T cells and drive production of IFNg and TNFa in murine ovarian tumors [24] and can upregulate of PD-L1 expression, providing further rationale for its combination of immune checkpoint blockade [27]. Several trials are currently underway evaluating PARP inhibitors in combination with immune checkpoint inhibitors across a range of solid tumors but have not yet been undertaken in OS [10].

Translational Applications

Across NGS studies, few recurrent potentially actionable alterations have been characterized including PI3K/AKT/mTOR pathway, IGF signaling, and the potential role of PARP inhibitors among a subset of OS with BRCA signatures. However, many of these targets also contribute to immune modulation and therefore may be relevant in combination with immunotherapy for the treatment of OS. For example, mutations in PI3K or AKT can lead in constitutive PD-L1 expression [33], whereas mTOR inhibition can enhance the production of CD8+ memory T cells [35]. PARP inhibition has been shown to increase intratumoral CD8+ T cells, drive production of IFNγ and TNFα [24], and upregulate PD-L1 expression independent of its role in the DNA damage response [27] . In the MDACC OS cohort, PARP2 was functionally associated with the MHC class I antigen presentation pathway further supporting the rationale for exploring PARP + immunotherapy combinations in osteosarcoma. Data to support combining IGF signaling inhibitors with immunotherapies are limited but may warrant further exploration [67]. A practical challenge facing targeted therapy + immunotherapy combinations in the treatment of osteosarcoma is the lack of single agent efficacy data for either immune checkpoint inhibitors or many of the available targeted therapies in unselected patient populations. Immunotherapy combinations with VEGFR and/or mTOR inhibitors may be an appropriate starting point for clinical trials given their activity in metastatic osteosarcoma and early promising combination data in selected soft tissue sarcoma subtypes [70].

Conclusion

NGS studies in OS have yielded additional insight into its genomic complexity and heterogeneity. Predominant genomic features such as aneuploidy and pathway alterations as well as limited neoantigen expression influence the immune landscape of OS and result in a similarly diverse and heterogeneous immune spectrum of tumors. However, these immunosuppressive mechanisms in OS may themselves present opportunities for novel therapeutic combinations.

References

1. Alexandrov LB, Nik-Zainal S, Wedge DC et al (2013) Signatures of mutational processes in human cancer. Nature 500(7463):415–421
2. Bai J, Gao Z, Li X et al (2017) Regulation of PD-1/PD-L1 pathway and resistance to PD-1/PD-L1 blockade. Oncotarget 8(66):110693–110707
3. Bayram F, Bitgen N, Donmez-Altuntas H et al (2014) Increased genome instability and oxidative DNA damage and their association with IGF-1 levels in patients with active acromegaly. Growth Hormon IGF Res 24(1):29–34
4. Behjati S, Tarpey PS, Haase K et al (2017) Recurrent mutation of IGF signaling genes and distinct patterns of genomic rearrangement in osteosarcoma. Nat Commun 8:ncomms15936
5. Beroukhim R, Mermel CH, Porter D et al (2010) The landscape of somatic copy-number alteration across human cancers. Nature 463(7283):899–905
6. Bielack SS, Kempf-Bielack B, Delling GN et al (2002) Prognostic factors in high-grade osteosarcoma of the extremities or trunk: an analysis of 1,702 patients treated on neoadjuvant cooperative osteosarcoma study group protocols. J Clin Oncol 20:776–790
7. Blokzijl F, de Ligt J, Jager M et al (2016) Tissue-specific mutation accumulation in human adult stem cells during life. Nature 538(7624):260–264
8. Bousquet M, Noirot C, Accadbled F et al (2016) Whole-exome sequencing in osteosarcoma reveals important heterogeneity of genetic alterations. Ann Oncol 27:738–744
9. Brogna S, Wen J (2009) Nonsense-mediated mRNA decay (NMD) mechanisms. Nat Struct Mol Biol 16(2):107–113
10. Burgess MA, Bolejack V, Tine BAV et al (2017) Multicenter phase II study of pembrolizumab (P) in advanced soft tissue (STS) and bone sarcomas (BS): final results of SARC028 and biomarker analyses. J Clin Oncol 35:11008–11008
11. Carter SL, Cibulskis K, Helman E et al (2012) Absolute quantification of somatic DNA alterations in human cancer. Nat Biotechnol 30(5):413–421
12. Charoentong P, Finotello F, Angelova M et al (2017) Pan-cancer Immunogenomic analyses reveal genotype-immunophenotype relationships and predictors of response to checkpoint blockade. Cell Rep 18(1):248–262
13. Chen X, Bahrami A, Pappo A et al (2014) Recurrent somatic structural variations contribute to tumorigenesis in pediatric osteosarcoma. Cell Rep 7:104–112
14. Chen Y-J, Hakin-Smith V, Teo M et al (2006) Association of mutant TP53 with alternative lengthening of telomeres and favorable prognosis in glioma. Cancer Res 66:6473–6476
15. Coudray A, Battenhouse AM, Bucher P et al (2018) Detection and benchmarking of somatic mutations in cancer genomes using RNA-seq data. bioRxiv 2018:249219
16. Couzin-Frankel J (2013) Breakthrough of the year 2013. Cancer immunotherapy. Science 342(6165):1432–1433
17. Davoli T, Uno H, Wooten EC et al (2017) Tumor aneuploidy correlates with markers of immune evasion and with reduced response to immunotherapy. Science 355(6322):pii:eaaf8399
18. Dewhurst SM, McGranahan N, Burrell RA et al (2014) Tolerance of whole-genome doubling propagates chromosomal instability and accelerates cancer genome evolution. Cancer Discov 4(2):175–185
19. Engert F, Kovac M, Baumhoer D et al (2017) Osteosarcoma cells with genetic signatures of BRCAness are susceptible to the PARP inhibitor talazoparib alone or in combination with chemotherapeutics. Oncotarget 8(30):48794–48806
20. Flynn RL, Cox KE, Jeitany M et al (2015) Alternative lengthening of telomeres renders cancer cells hypersensitive to ATR inhibitors. Science 347:273–277
21. Forment JV, Kaidi A, Jackson SP (2012) Chromothripsis and cancer: causes and consequences of chromosome shattering. Nat Rev Cancer 12(10):663–670
22. Gianferante DM, Mirabello L, Savage SA (2017) Germline and somatic genetics of osteosarcoma – connecting aetiology, biology and therapy. Nat Rev Endocrinol 13(8):480–491
23. He Y, de Castro LF, Shin MH et al (2015) p53 loss increases the osteogenic differentiation of bone marrow stromal cells. Stem Cells 33(4):1304–1319
24. Huang J, Wang L, Cong Z et al (2015) The PARP1 inhibitor BMN 673 exhibits immunoregulatory effects in a Brca1(−/−) murine model of ovarian cancer. Biochem Biophys Res Commun 463(4):551–556
25. Hugo W, Zaretsky JM, Sun L et al (2016) Genomic and transcriptomic features of response to anti-PD-1 therapy in metastatic melanoma. Cell 165(1):35–44
26. Jager M, Blokzijl F, Kuijk E et al (2019) Deficiency of global genome nucleotide excision repair explains mutational signature observed in cancer. Genome Res 29(7):1067–1077
27. Jiao S, Xia W, Yamaguchi H et al (2017) PARP inhibitor upregulates PD-L1 expression and enhances cancer-associated immunosuppression. Clin Cancer Res 23:3711–3720
28. Kager L, Zoubek A, Pötschger U et al (2003) Primary metastatic osteosarcoma: presentation and outcome of patients treated on neoadjuvant Cooperative Osteosarcoma Study Group protocols. J Clin Oncol 21:2011–2018
29. Kansara M, Teng MW, Smyth MJ et al (2014) Translational biology of osteosarcoma. Nat Rev Cancer 14(11):722–735
30. Keenan TE, Burke KP, Van Allen EM (2019) Genomic correlates of response to immune checkpoint blockade. Nat Med 25(3):389–402
31. Killela PJ, Reitman ZJ, Jiao Y et al (2013) TERT promoter mutations occur frequently in gliomas and a subset of tumors derived from cells with low

rates of self-renewal. Proc Natl Acad Sci U S A 110(15):6021–6026
32. Koirala P, Roth ME, Gill J et al (2016) Immune infiltration and PD-L1 expression in the tumor microenvironment are prognostic in osteosarcoma. Sci Rep 6:30093
33. Kovac M, Blattmann C, Ribi S et al (2015) Exome sequencing of osteosarcoma reveals mutation signatures reminiscent of BRCA deficiency. Nat Commun 6:8940
34. Lastwika KJ, Wilson W 3rd, Li QK et al (2015) Control of PD-L1 expression by oncogenic activation of the AKT/mTOR pathway in non-small cell lung cancer. Cancer Res 76(2):227–238
35. Li Q, Rao RR, Araki K et al (2011) A central role for mTOR kinase in homeostatic proliferation induced CD8+ T cell memory and tumor immunity. Immunity 34(4):541–553
36. Li B, Severson E, Pignon JC et al (2016) Comprehensive analyses of tumor immunity: implications for cancer immunotherapy. Genome Biol 17:174
37. Li Y, Roberts ND, Weischenfeldt J et al (2017) Patterns of structural variation in human cancer. bioRxiv. https://doi.org/10.1101/181339v1
38. Li YS, Liu Q, He HB et al (2019) The possible role of insulin-like growth factor-1 in osteosarcoma. Curr Probl Cancer 43(3):228–235
39. Maciejowski J, Li Y, Bosco N et al (2015) Chromothripsis and kataegis induced by telomere crisis. Cell 163:1641–1654
40. Maciejowski J, de Lange T (2017) Telomeres in cancer: tumour suppression and genome instability. Nat Rev Mol Cell Biol 18(3):175–186
41. McEachron TA, Triche TJ, Sorenson L et al (2018) Profiling targetable immune checkpoints in osteosarcoma. Onco Targets Ther 7(12):e1475873
42. Meyers PA, Heller G, Healey JH et al (1993) Osteogenic sarcoma with clinically detectable metastasis at initial presentation. J Clin Oncol 11:449–453
43. Miao D, Margolis CA, Vokes NI et al (2018) Genomic correlates of response to immune checkpoint blockade in microsatellite-stable solid tumors. Nat Genet 50(9):1271–1281
44. Mirabello L, Troisi RJ, Savage SA (2009) Osteosarcoma incidence and survival rates from 1973 to 2004: data from the Surveillance, Epidemiology, and End Results Program. Cancer 115:1531–1543
45. Mouw KW, Goldberg MS, Konstantinopoulos PA et al (2017) DNA damage and repair biomarkers of immunotherapy response. Cancer Discov 7(7):675–693
46. O'Brien TD, Jia P, Xia J et al (2015) Inconsistency and features of single nucleotide variants detected in whole exome sequencing versus transcriptome sequencing: a case study in lung cancer. Methods 83:118–127
47. Ock CY, Hwang JE, Keam B et al (2017) Genomic landscape associated with potential response to anti-CTLA-4 treatment in cancers. Nat Commun 8(1):1050
48. Palmerini E, Agostinelli C, Picci P et al (2017) Tumoral immune-infiltrate (IF), PD-L1 expression and role of CD8/TIA-1 lymphocytes in localized osteosarcoma patients treated within protocol ISG-OS1. Oncotarget 8(67):111836–111846
49. Perry JA, Kiezun A, Tonzi P et al (2014) Complementary genomic approaches highlight the PI3K/mTOR pathway as a common vulnerability in osteosarcoma. Proc Natl Acad Sci 111:E5564–E5573
50. Petitjean A, Achatz MI, Borresen-Dale AL et al (2007) TP53 mutations in human cancers: functional selection and impact on cancer prognosis and outcomes. Oncogene 26(15):2157–2165
51. Rausch T, Jones David TW, Zapatka M et al (2012) Genome sequencing of pediatric medulloblastoma links catastrophic DNA rearrangements with TP53 mutations. Cell 148:59–71
52. Rickel K, Fang F, Tao J (2017) Molecular genetics of osteosarcoma. Bone 102:69–79
53. Roh W, Chen PL, Reuben A et al (2017) Integrated molecular analysis of tumor biopsies on sequential CTLA-4 and PD-1 blockade reveals markers of response and resistance. Sci Transl Med 9(379):pii: eaah3560
54. Sade-Feldman M, Jiao YJ, Chen JH et al (2017) Resistance to checkpoint blockade therapy through inactivation of antigen presentation. Nat Commun 8(1):1136
55. Seplyarskiy VB, Soldatov RA, Popadin KY et al (2016) APOBEC-induced mutations in human cancers are strongly enriched on the lagging DNA strand during replication. Genome Res 26(2):174–182
56. Sharma P, Hu-Lieskovan S, Wargo JA et al (2017) Primary, adaptive, and acquired resistance to cancer immunotherapy. Cell 168(4):707–723
57. Shen JK, Cote GM, Choy E et al (2014) Programmed cell death ligand 1 expression in osteosarcoma. Cancer Immunol Res 2(7):690–698
58. Sun C, Mezzadra R, Schumacher TN (2018) Regulation and function of the PD-L1 checkpoint. Immunity 48(3):434–452
59. Tawbi HA, Burgess M, Bolejack V et al (2017) Pembrolizumab in advanced soft-tissue sarcoma and bone sarcoma (SARC028): a multicentre, two-cohort, single-arm, open-label, phase 2 trial. Lancet Oncol 18(11):1493–1501
60. Taylor AM, Shih J, Ha G et al (2018) Genomic and functional approaches to understanding cancer aneuploidy. Cancer Cell 33(4):676–689
61. The Cancer Genome Atlas (2013) The cancer genome atlas pan-cancer analysis project. Nat Genet 45(10):1113–1120
62. Thompson LH (2012) Recognition, signaling, and repair of DNA double-strand breaks produced by ionizing radiation in mammalian cells: the molecular choreography. Mutat Res 751:158–246
63. Thorsson V, Gibbs DL, Brown SD et al (2018) The immune landscape of cancer. Immunity 48(4):812–830

64. Turajlic S, Litchfield K, Xu H et al (2017) Insertion-and-deletion-derived tumour-specific neoantigens and the immunogenic phenotype: a pan-cancer analysis. Lancet Oncol 18(8):1009–1021
65. van Harn T, Foijer F, van Vugt M et al (2010) Loss of Rb proteins causes genomic instability in the absence of mitogenic signaling. Genes Dev 24(13):1377–1388
66. Walkley CR, Qudsi R, Sankaran VG et al (2008) Conditional mouse osteosarcoma, dependent on p53 loss and potentiated by loss of Rb, mimics the human disease. Genes Dev 22(12):1662–1676
67. Wargo JA, Cooper ZA, Flaherty KT (2014) Universes collide: combining immunotherapy with targeted therapy for cancer. Cancer Discov 4(12):1377–1386
68. Wellenstein MD, de Visser KE (2018) Cancer-cell-intrinsic mechanisms shaping the tumor immune landscape. Immunity 48(3):399–416
69. Werner H, Sarfstein R, LeRoith D et al (2016) Insulin-like growth factor 1 signaling axis meets p53 genome protection pathways. Front Oncol 6:159
70. Wilky BA, Trucco MM, Subhawong TK et al (2019) Axitinib plus pembrolizumab in patients with advanced sarcomas including alveolar soft-part sarcoma: a single-centre, single-arm, phase 2 trial. Lancet Oncol 20(6):837–848
71. Wu CC, Beird HC, Livingston JA et al (2019) Immuno-genomic landscape of osteosarcoma. Nat Commun 11(1):1008
72. Yakar S, Rosen CJ, Beamer WG et al (2002) Circulating levels of IGF-1 directly regulate bone growth and density. J Clin Invest 110(6):771–781
73. Yoshida K, Miki Y (2004) Role of BRCA1 and BRCA2 as regulators of DNA repair, transcription, and cell cycle in response to DNA damage. Cancer Sci 95(11):866–871
74. Yoshihara K, Shahmoradgoli M, Martinez E et al (2013) Inferring tumour purity and stromal and immune cell admixture from expression data. Nat Commun 4:2612
75. Zhao X, Subramanian S (2017) Oncogenic pathways that affect antitumor immune response and immune checkpoint blockade therapy. Pharmacol Ther 181:76–84
76. Sansregret L, Vanhaesebroeck B, Swanton C (2018) Determinants and clinical implications of chromosomal instability in cancer. Nature Reviews Clinical Oncology 15 (3):139–150
77. Brown SD, Warren RL, Gibb EA et al (2014) Neo-antigens predicted by tumor genome meta-analysis correlate with increased patient survival. Genome Research 24 (5):743–750
78. Schumacher TN, Schreiber RD (2015) Neoantigens in cancer immunotherapy. Science 348 (6230):69–74
79. Wunder JS, Gokgoz N, Parkes R et al(2005) TP53 mutations and outcome in osteosarcoma: a prospective, multicenter study. Journal of Clinical Oncology 23 (7):1483–1490
80. Holland AJ, Cleveland DW (2009) Boveri revisited: chromosomal instability, aneuploidy and tumorigenesis. Nature Reviews Molecular Cell Biology 10 (7):478–487
81. Yang Y, Zhang Y, Qu X et al (2016) Identification of differentially expressed genes in the development of osteosarcoma using RNA-seq. Oncotarget 7 (52)

RECQ DNA Helicases and Osteosarcoma

3

Linchao Lu, Weidong Jin, and Lisa L. Wang

Abstract

The RECQ family of DNA helicases is a conserved group of enzymes that plays an important role in maintaining genomic stability. Humans possess five RECQ helicase genes, and mutations in three of them – *BLM*, *WRN*, and *RECQL4* – are associated with the genetic disorders Bloom syndrome, Werner syndrome, and Rothmund-Thomson syndrome (RTS), respectively. These syndromes share overlapping clinical features, and importantly they are all associated with an increased risk of cancer. Patients with RTS have the highest specific risk of developing osteosarcoma compared to all other cancer predisposition syndromes; therefore, RTS serves as a relevant model to study the pathogenesis and molecular genetics of osteosarcoma. The "tumor suppressor" function of the RECQ helicases continues to be an area of active investigation. This chapter will focus primarily on the known cellular functions of RECQL4 and how these may relate to tumorigenesis, as well as ongoing efforts to understand RECQL4's functions in vivo using animal models. Understanding the RECQ pathways will provide insight into avenues for novel cancer therapies in the future.

Keywords

RECQ · RECQL4 · DNA helicase · Rothmund-Thomson syndrome · RTS · Bloom syndrome · Werner syndrome · Osteosarcoma · Genomic instability

Introduction

The roles of the RECQ helicases in cancer and specifically the role of RECQL4 in osteosarcoma (OS) are areas of active investigation. While it is known that constitutional mutations in the *RECQ* genes predispose patients to developing cancer, the exact mechanisms of tumorigenesis remain to be fully explored. As basic science research continues to reveal the normal cellular functions of the RECQ helicases, application of this knowledge to OS pathogenesis will provide avenues for future investigation into targeted therapies for this disease. This chapter will primarily focus on what is currently known about the *RECQL4* DNA helicase gene, which is mutated in the OS predisposition disorder Rothmund-Thomson syndrome (RTS).

L. Lu (✉) · W. Jin · L. L. Wang (✉)
Department of Pediatrics, Section of Hematology/Oncology, Texas Children's Cancer Center, Baylor College of Medicine, Houston, TX, USA
e-mail: linchaol@bcm.edu; weidong.jin@bcm.edu; llwang@bcm.edu

E. S. Kleinerman, R. Gorlick (eds.), *Current Advances in the Science of Osteosarcoma*, Advances in Experimental Medicine and Biology 1258, https://doi.org/10.1007/978-3-030-43085-6_3

RECQ Family of DNA Helicases and Cancer Predisposition

The RECQ DNA helicases are a family of proteins that are important in maintaining genomic integrity. DNA helicases are ubiquitous molecular motor proteins that harness the chemical free energy of ATP hydrolysis to catalyze the unwinding of duplex DNA and as such play important roles in nearly all aspects of nucleic acid metabolism, including replication, repair, recombination, and transcription [115]. The RECQ helicases belong to the SF2 superfamily of DNA helicases that unwind DNA in a 3′ ↑ 5′ direction in an ATP- and Mg^{2+}-dependent fashion [5, 8]. As such, they contain a conserved region that includes the seven characteristic helicase motifs (I, Ia, II, III, IV, V, and VI) that define this family of helicases and that are important for coupling ATP hydrolysis to the separation of DNA strands. The first RECQ helicase was discovered in *Escherichia coli* (*E. coli*) in a screen for resistance to thymineless death [81]. Subsequently, RECQ proteins have been identified in multiple species. These evolutionarily conserved proteins are defined by their common central helicase motif, a highly conserved region of approximately 400 amino acids (Fig. 3.1) [8, 55]. The number of RECQ helicases increases from lower to higher organisms. Bacteria such as *E. coli* have one (RecQ), as do yeast (*Sgs1* in *Saccharomyces cerevisiae* and *Rqh1* in *Schizosaccharomyces pombe*), while *Caenorhabditis elegans* has two and *Arabidopsis thalianas* has seven RECQ helicases [58].

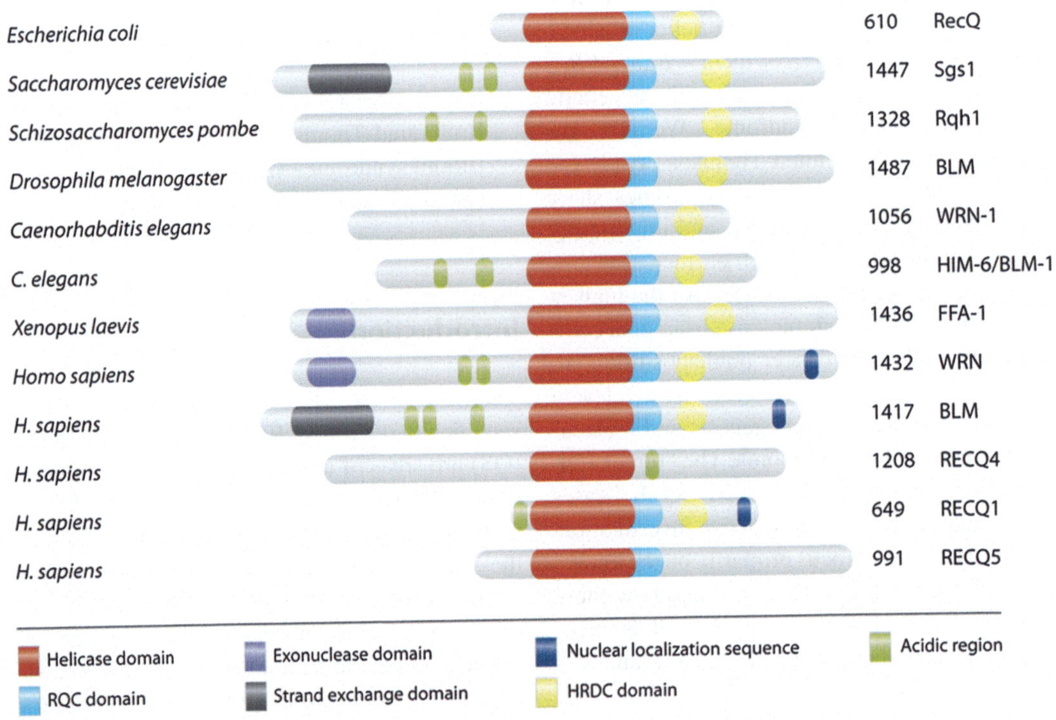

Fig. 3.1 Structural features of RecQ helicases. The RecQ proteins have several structural domains that are conserved from bacteria through humans. All RecQ proteins have a core helicase domain. Most RecQ proteins also contain conserved helicase and RNAse D C-terminal (HRDC) and RecQ C-terminal (RQC) domains that are thought to mediate interactions with nucleic acid and other proteins, respectively. Many RecQ proteins have acidic regions that enable protein-protein interactions, and some of the RecQ proteins have nuclear localization sequences. WRN and FFA-1 protein are unique in that they also contain an exonuclease domain. Sgs1 and Blm are the first characterized members of this family of proteins containing a functional strand exchange domain in their N-terminus. The number of amino acids in each protein is indicated on the right. (Reprinted with permission from Bernstein et al. [8])

In humans, there are five RECQ helicases (Fig. 3.1). Three of these, WRN, BLM, and RECQL4, are associated with human diseases [79]. Mutations in the *WRN* gene [137] cause Werner syndrome [73], and mutations in the *BLM* gene [30] are responsible for Bloom syndrome [36]. Mutations in *RECQL4* are associated with three overlapping disorders: RTS, RAPADILINO syndrome, and Baller-Gerold syndrome (BGS) [56, 101, 117]. Although *RECQL* and *RECQL5* have not thus far been associated with any human genetic disorders, both have been linked to human tumorigenesis [23, 28, 127]. In one study, rare germ line truncating mutations in the *RECQL* gene were shown to be associated with an increased risk of breast cancer in two populations of high-risk patients [23]. A few small studies have demonstrated that specific single-nucleotide polymorphisms in RECQL5 are more common in OS patients [28, 139], and decreased expression of *RECQL5* in OS tumors may be associated with disease progression [127].

All of the human RECQ disorders are cancer predisposition syndromes, but they have varying cancer profiles (Table 3.1). Patients with Werner syndrome display features of premature aging, such as diabetes, coronary artery disease, cataracts, and osteoporosis. They are susceptible primarily to thyroid cancer, melanoma, meningioma, soft tissue sarcomas, and OS. In a study of the spectrum of cancers in Werner syndrome patients, OS was found to comprise 8% of all neoplasms [62]. In contrast, patients with Bloom syndrome are susceptible to all types of cancers seen in the general population, but at a much higher frequency and at an earlier age. These include leukemia and lymphomas and epithelial cancers of the colon, breast, head and neck, and cervix, as well as OS, which accounted for 2% of the first 100 cases of cancers reported in the Bloom Registry [22, 37]. Among the *RECQL4*-associated disorders, patients with RTS have a very high and *specific* risk for OS, in addition to nonmelanoma skin cancers (squamous and basal cell carcinomas). In one clinical cohort study of 41 RTS patients, 30% had a diagnosis of OS [122]. Patients with RAPADILINO syndrome and *RECQL4* mutations are also at risk for cancer, most commonly lymphomas as well as OS [102]. These patients share many of the same phenotypes as RTS patients, including small stature, limb deformities, radial ray defects, and absent patellae. Interestingly, these patients do not display poikiloderma, which is a defining feature of RTS. BGS is the least well-characterized of the *RECQL4* disorders. These patients are characterized by craniosynostosis and radial ray defects, as well as poikiloderma in some patients. So far only a few cases have been described to have *RECQL4* mutations, and cancer has only been

Table 3.1 Human RECQ helicase syndromes

Disease	Main clinical features	Cancer predisposition	Gene location
Bloom syndrome	Small stature, photosensitive rash, immunodeficiency	Multiple tumor types, including leukemia, lymphoma, solid tumors	*BLM* 15q26.1
Werner syndrome	Premature aging, cataracts, diabetes, atherosclerosis	Soft tissue sarcomas, skin (melanoma), thyroid cancer, osteosarcoma	*WRN* 8p11
Rothmund-Thomson syndrome	Poikiloderma, radial ray and other skeletal defects, alopecia	*Osteosarcoma*, skin cancer (squamous and basal cell carcinomas)	*RECQL4* 8q24.3
RAPADILINO syndrome	Small stature, radial ray and limb deformities, palatal defects, absent patella	Lymphoma and osteosarcoma	*RECQL4* 8q24.3
Baller-Gerold Ssyndrome	Craniosynostosis, radial ray defects, poikiloderma	Possibly lymphoma	*RECQL4* 8q24.3

described in one patient who developed a midline NK cell lymphoma [26]. Overall there have been over 60 *RECQL4* mutations identified among these three disorders [116]. Exact genotype-phenotype correlations with respect to specific mutations and resultant phenotypes, including cancer, remain to be elucidated.

As a group, the RECQ helicases are considered "caretakers" of the genome and as such do not necessarily directly drive tumorigenesis but prevent genomic instability that results in accumulation of structural changes in oncogenes or tumor suppressors that could then lead to cancer [17]. This protection of genome stability is achieved through their various roles in DNA replication, repair, and telomere maintenance. It is also possible that the RECQ helicases could play a more direct role in affecting tumorigenesis. While the exact molecular mechanisms of tumor suppression have yet to be worked out fully, deficiency of the WRN, BLM, and RECQL4 proteins in humans clearly predisposes to the development of cancer.

Structure and Functions of the RECQL4 DNA Helicase

The role of RECQL4 in DNA replication has been extensively studied, and it appears that while RECQL4 may participate in many cellular functions, its primary role is in the initiation of DNA replication [46, 74, 94, 113, 126, 132, 133]. This is achieved primarily through its N-terminal domain (amino acids 1–370) which shares homology to the yeast replication factor Sld2 in *S. cerevisiae* and Drc1 in *S. pombe* [72, 74, 94], both of which are important for establishing replication forks during the initiation of DNA replication. After phosphorylation by cyclin-dependent kinases, Sld2 binds Dpb11, a key mediator of the formation of the active replicative helicase complex on replication origins and a crucial factor in the initiation of DNA replication [51, 108, 119]. In *Xenopus*, it has been shown that xRECQL4 belongs to the replication initiation complex and helps to promote loading of replication factors at the origins, after pre-replication complex formation [94]. The N-terminal amino acid region 1–596 of RECQL4 interacts directly with xCut5 (frog orthologue of Dpb11), which is responsible for recruiting DNA polymerases to the sites of replication [74]. RECQL4 has also been shown to interact with multiple DNA replication factors, such as MCM10, MCM2-7, CTF4, CDC45, GINS, and SLD5 which are essential for initiation of DNA replication [46, 47, 57, 132], as well as TopBP1, the vertebrate orthologue of Dpb11 [87]. The C-terminus of RECQL4 including the helicase domain also appears to play a role in replication under stressed conditions. Human pre-B lymphocyte cells with mutant RECQL4 lacking the C-terminus were shown to have replication defects only after ionizing radiation, perhaps by allowing replication forks to negotiate the radiation-damaged DNA templates [59]. Because RECQL4 is overexpressed in many types of sporadic cancers (see below), the effect of overexpression of RECQL4 on replication has also been studied. Although overexpression of RECQL4 alone did not affect replication, when RECQL4 was fused to a subunit of the origin recognition complex-ORC4 protein, overexpression of this fusion protein induced increased binding of RECQL4 to late replication origins in early S phase and recruitment of replication initiation factors [99]. As a result, early activation of replication was observed in genes with late replication origins, leading to elevated replication stress caused by replication-transcription conflicts [99]. Therefore, the binding of RECQL4 to replication origins needs to be tightly regulated to ensure a normal replication process. In addition to initiation of DNA replication, RECQL4 may also play a role in replication fork restart given its high affinity to Holliday junction substrates demonstrated by in vitro binding assays via N-terminal amino acid residues 320–400 [96].

RECQL4 has been shown to bind additional nucleic acid substrates in vitro, including guanine quadruplex (G4) structures [54]. G4 is a type of secondary structure formed in guanine-rich sequences and is found in replication origins, gene promoter regions, and telomeric DNA sequences [41]. BLM, WRN, and RECQL4 have all been shown to be important for telomere maintenance [20]. Both BLM and WRN helicases bind and unwind G4 DNA substrates [77], while RECQL4 only binds but has no detectable unwinding activity [54]. Gene expression analyses using fibroblasts from both Bloom and Werner syndrome patients showed that BLM and WRN regulate transcription through G4 DNA sequences [85, 109]. The biological function of RECQL4 at G4 sites needs further investigation given the importance of G4 sequences in normal physiological processes as well as in tumorigenesis. In addition to the abovementioned functional domains, the N-terminus of RECQL4 also contains several localization regions, including two nuclear localization domains [9], a region of acetylation by p300 which regulates nuclear to cytoplasmic localization [27], and a predicted mitochondrial localization signal in amino acids 1–84 [25].

Initially, researchers were unable to demonstrate actual DNA unwinding activity by RECQL4 using a variety of DNA substrates [69, 134]. Helicase activity was finally demonstrated for RECQL4 by several groups [11, 91, 107, 131], which was likely masked in previous assays by the strong annealing activity of the enzyme. In vitro biochemical data suggested that RECQL4 possesses another N-terminal region contributing to DNA unwinding besides the well-known conserved helicase domain [131], although known helicase motifs and nucleotide binding sites were not found to be present in this region. The in vivo function of this extra helicase domain requires further investigation. In addition to the helicase domain, other protein domains in RecQ helicases, including the helicase-and-RNase D C-terminal (HRDC) and RecQ-C-terminal (RQC) domains, are also important for helicase unwinding activity. However, RECQL4 lacks the structurally conserved HRDC domain which is felt to be important for interactions with nucleic acids (Fig. 3.1) [8, 80]. Human RECQL4 also appears to lack the structurally conserved RQC domain that is important for zinc and DNA binding and for helicase activity. However, through bioinformatic and biochemical analyses, Mojumdar et al. identified a functional RQC domain in human RECQL4 that is essential for these activities [72, 78]. In addition, the crystal structure of a human RECQL4 fragment (residues 449–1111), including the helicase domain and the majority of the C-terminus, revealed that a RECQL4 zinc binding domain (R4ZBD, residues 836–1045) resides downstream of the helicase domain and is important for DNA unwinding activity in a biochemical DNA helicase activity assay [50]. Interestingly, the last 92 residues of human RECQL4 have also been shown to play an important role in helicase activity by increasing DNA binding [50].

In addition to its role in DNA replication, RECQL4 has also been implicated to function in various aspects of DNA repair, including double-strand break (DSB) repair [61, 66, 67, 90, 97, 103], nucleotide excision repair (NER) [19, 31], and base excision repair (BER) [95]. RECQL4 plays important roles in both homologous recombination (HR)-dependent and nonhomologous end-joining (NHEJ)-mediated repair of DSBs. RECQL4 has been shown to interact physically with the Ku70/Ku80 heterodimer [97], which forms a complex with DNA-PKcs to play a central role in NHEJ-mediated DSB repair. During HR-dependent DSB repair, RECQL4 has been shown to interact physically by co-immunoprecipitation with RAD51, a key protein involved in the HR pathway of DSB repair, and to associate with RAD51 by immunofluorescence in DNA damage foci [61, 90, 103]. Lu et al. reported that RECQL4 participates in 5′ end resection of DSBs, the first step in HR-mediated DSB repair [67]. RECQL4 interacts with the MRE11-RAD50-NBS1 (MRN) complex and increases the recruitment of CtIP which stimulates end resection by the MRN complex [67]. Interestingly, the participation of RECQL4 in both pathways was shown to be cell cycle dependent and was regulated by the phosphorylation of RECQL4 by cyclin-dependent kinases CDK1/CDK2. RECQL4 stimulates NHEJ in G1 phase and promotes HR-mediated DSB repair in S and G2 phases when CDK1/CDK2 activity is high [66]. RECQL4 has also been shown to interact with BLM helicase, which like RECQL4 probably has many functions in the cell, the most important of which is its role in HR. This interaction was strengthened in S-phase and after ionizing radiation treatment in human cells, indicating that RECQL4 coordinates with BLM to function in DNA replication and DNA damage repair [104]. Ribosomal protein S3 (RPS3), a component of 40S small subunit of the ribosome contributing to protein translation, has also been shown to interact with the N-terminus of RECQL4 and modulates its activity during DNA damage repair [89]. RECQL4 helicase activity appears to be essential for the end resection and HR-dependent repair of DSBs [67]. However, a knock-in mouse model (*Recql4*^{K525A}), mimicking human *RECQL4*^{K508M}, displayed normal development and normal life span compared to wild-type littermates [12]. Cells derived from these mice had no significant difference in growth rate after treatment with genotoxic agents [12]. This discrepancy between human cells and mouse models requires more detailed investigation. Nevertheless, taken together, the data suggest that lack of RECQL4 functional activity in DNA repair can lead to increased DSBs, DNA replication stress, genomic instability, and cancer development.

The NER pathway is a major mediator of repair of UV damage, and RECQL4 has been shown to colocalize with XPA, a key protein involved in NER, and to interact with XPA directly by GST pull-down assay [31]. The BER pathway is the main mechanism for repair of oxidative DNA lesions, and RECQL4 was also found to colocalize and functionally interact with key proteins involved in BER, including APE1, FEN1, and DNA polymerase β, after treatment with H_2O_2 [95]. Werner et al. showed that after H_2O_2 treatment, RECQL4 translocates from the cytoplasm to the nucleus and forms nuclear foci in normal human fibroblasts. After recovery from oxidant damage, viable RTS patient fibroblasts underwent irreversible growth arrest and had significantly decreased DNA synthesis [124]. Woo et al. also showed that in response to oxidative stress, RECQL4 had altered cellular localization to the nucleolus and using a T7 phage display screen showed that RECQL4 C-terminus interacts with the single-strand break repair protein, poly(ADP-ribose) polymerase-1 (PARP-1) [125]. PARP-1 is activated in response to a wide variety of DNA-damaging agents and modulates the cellular sensitivity to γ-irradiation [68].

The response of *RECQL4* mutant cells to different genotoxic agents has been investigated by several groups; these have included UV and ionizing radiation (IR), hydrogen peroxide, topoisomerases inhibitors, and chemotherapy agents such as doxorubicin and cisplatin [10, 19, 31, 49, 59, 103, 124]. However, the results have been somewhat inconsistent between studies, likely reflecting the use of different primary cells or cell lines (transformed cells vs. untransformed cells, RTS patient cells vs. *RECQL4* knockdown cells), different assays to determine sensitivity, and different *RECQL4* mutations present in the cells. For example, some studies have demonstrated significant increased sensitivity to UV radiation [88, 100, 105], while others have shown moderate or no increase in sensitivity [49, 59]. Using CRISPR-Cas9, Kohzaki et al. deleted the C-terminus of RECQL4 after the NLS domain, including the conserved helicase domain, in several human cancer cell lines [60]. These cells displayed hypersensitivity to IR and cisplatin, which primarily introduce DNA DSB and interstrand cross-links, respectively. In vitro cell-based DSB repair reporter assays showed that these cells displayed increased single-strand annealing activity and reduced alternative end-joining mediated pathway. They showed that RAD52 inhibition suppressed the growth of cancer cell lines in vitro and in xenograft mouse models. In addition, cisplatin treatment had an additive inhibitory effect with RAD52 inhibition on tumor cell growth, providing a potential treatment avenue for cancer patients with *RECQL4* mutations and increased RAD52 expression [60].

As mentioned earlier, RECQ proteins bind to G4 structures such as those found in telomeric DNA, and RECQL4 has been shown to play a role in telomere maintenance [38]. RTS patient cells and human cells with *RECQL4* knockdown exhibit increased fragile telomeric ends. In addition, human RECQL4 localizes to telomeres and interacts with shelterin protein telomeric repeat-binding factor 2 (TRF2) which maintains telomere integrity [38]. RECQL4 also interacts with the WRN protein and stimulates WRN's activity on telomeric D-loops. Similar to WRN and BLM, RECQL4 also appears to be able to resolve these D-loops, which is necessary for replication to take place at the telomeres, and this resolving activity is stimulated by TRF1 and TRF2 as well as the shelterin protein POT1 [38]. Also similar to WRN and BLM, RECQL4 seems to be more active on telomeric D-loops that contain 8-oxoguanine base lesions, indicative of oxidative damage. Unlike WRN, however, RECQL4 also has a clear preference for unwinding D-loops that contain thymine glycol (Tg) lesions, which are the most common oxidation product of the thymine base, and this activity is stimulated by TRF2 [34]. Thus, mutations in *RECQL4* could result in dysfunctional telomeres, which are well known to play a role in both tumor suppression and tumor progression, depending on the cellular milieu, particularly with respect to the checkpoint status of the cells [130].

In addition to these nuclear functions, RECQL4 has also been shown to localize in the cytosol [27, 134] as well as in the mitochondria [16, 21, 25, 120]. Yin et al. showed that RECQL4 interacts with cytosolic ubiquitin ligases UBR1 and UBR2 which function in the N-end rule pathway by ubiquitination and degradation of proteins [134]. Dietschy et al. demonstrated that RECQL4 can be acetylated by histone acetyltransferase p300 resulting in the cytosolic translocation of RECQL4 from the nucleus [27], providing a mechanism to modulate RECQL4 nuclear activities. In the mitochondria, loss of RECQL4 led to abnormalities in mitochondrial DNA (mtDNA) as well as mitochondrial function caused by reduced replication of mtDNA [16, 25] or caused by reduced proofreading and polymerization functions of mitochondrial DNA polymerase-γ (PolγA) [40]. Interestingly, a *RECQL4* mutation frequently reported in RAPADILINO patients who are predisposed to lymphoma and osteosarcoma disrupts the interaction between RECQL4 and mitochondrial p32 protein [120] while also enhancing the interaction between RECQL4 and mitochondrial helicase PEO1, leading to increased replication of mtDNA. Both increased or decreased mtDNA content could cause abnormal mitochondrial function demonstrated by abnormal mitochondrial metabolism and glycolysis [40, 120].

In addition to the abovementioned cellular functions, RECQL4 was also recently demonstrated to play a role in mitosis. RECQL4 was shown to be a microtubule-associated protein and to participate in the maintenance of chromosome alignment during mitosis [135]. It was identified among the proteins with a nuclear localization sequence (NLS) that can be pulled down by Taxol-stabilized microtubules in mitotic *Xenopus* egg extracts. RECQL4-depleted HeLa cells as well as RTS fibroblasts exhibited spindle abnormalities, including misaligned chromosomes and increased micronuclei. Interestingly, using immunoprecipitation with tagged proteins and GST pull-down assays in human cells, RECQL4 was shown to interact with aurora kinase B (AURKB) and to modulate its protein stability by reducing ubiquitination of AURKB [33], an essential protein that modulates mitosis by regulating chromosome alignment and segregation.

Rothmund-Thomson Syndrome (RTS): Nature's Model of Osteosarcoma

RTS was first described in 1868 by Dr. Auguste Rothmund, who was a German ophthalmologist. He described poikiloderma, the classic skin finding in RTS, along with rapidly developing bilateral juvenile cataracts in several families in an isolated region in the Bavarian Alps [92]. In 1921, Dr. Sydney Thomson, a British dermatologist, described a similar rash in two sisters, but instead of juvenile cataracts, they had bone abnormalities (radial ray defects) [114]. Later, Dr. William Taylor in the United States proposed that the two disorders described by Rothmund and Thomson were the same, and he proposed the eponym Rothmund-Thomson syndrome [112]. Mutations in the *RECQL4* gene in RTS were not discovered until 1999 [55, 56], 131 years after the original description by Rothmund. It is now known that approximately two-thirds of patients with RTS have mutations in the *RECQL4* gene

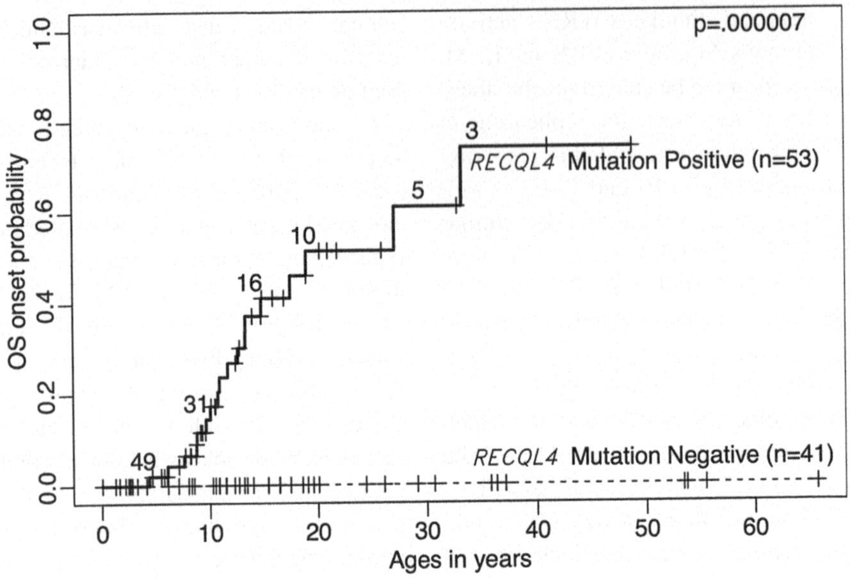

Fig. 3.2 Estimated probability of osteosarcoma onset in Rothmund-Thomson syndrome, classified by *RECQL4* mutation status. The time to OS onset was defined from the date of birth to the first diagnosis of OS. Event-time data were analyzed by Kaplan-Meier method, and the difference between the *RECQL4* mutation-positive and *RECQL4* mutation-negative patients was compared by the log-rank method

(designated Type 2 RTS). The other one-third of patients who lack *RECQL4* mutations are designated as Type 1 RTS. Mutations in the *ANAPC1* gene, which encodes the APC1 protein, a component of the anaphase-promoting complex/cyclosome (APC/C), have recently been identified as causative in a subset of Type 1 RTS patients [3]. Previous studies have shown that the presence of pathogenic mutations in *RECQL4* correlates significantly with risk of developing OS (Fig. 3.2) [121]. None of the patients with Type 1 RTS developed OS, while every RTS patient with OS had *RECQL4* mutations. These pathogenic mutations included nonsense, frameshift, splice site, and intronic deletions. Unlike other hereditary cancer syndromes known to predispose patients to OS, such as Li-Fraumeni syndrome and hereditary retinoblastoma, where the causative genes, p53 and RB, respectively, are commonly mutated in sporadic OS [14], mutations in *RECQL4* have not been detected in sporadic OS tumors [86]. Thus, *RECQL4* does not appear to be a direct target for somatic mutations in sporadic OS. However, the extremely high and specific risk for OS in Type 2 RTS patients suggests that the RECQL4 helicase plays a clear role in OS tumor suppression, making RTS a relevant model for the study of human OS pathogenesis.

In addition to cancer of the bone, patients with RTS also have prominent developmental defects of the bone. In a study of 28 RTS patients who underwent skeletal surveys, 75% were found to have major skeletal abnormalities, including radial, ulnar, or thumb agenesis/hypoplasia, radioulnar and radiohumeral synostoses, abnormal metaphyseal trabeculation, brachymesophalangy, and osteopenia [75]. This risk correlated with presence of *RECQL4* mutations. Understanding the role that RECQL4 plays in normal skeletal development will provide additional insight into the specific risk for OS, since many developmental pathways, such as the Wnt, Hedgehog, and Notch signaling pathways, not only are critical for normal skeletal development [39, 44, 110] but also play important roles in tumorigenesis [7, 18, 52, 111, 118].

Early case reports suggested that OS arising in RTS patients may be different from sporadic OS, i.e., arising in unusual or multiple (multifocal) sites [29]. In addition, because of the implicated role of *RECQL4* in DNA damage repair, clinicians may consider decreasing chemotherapy doses up-front for RTS patients diagnosed with OS. However, a study of 12 RTS patients with OS showed that their tumors had features that mirrored OS in the general population with regard to location of primary tumor (distal long bones), histology (conventional OS), histologic response to neoadjuvant chemotherapy, and overall outcomes [42]. The major difference was that the age of onset was younger in the RTS cohort compared to sporadic OS, which is not surprising given the genetic predisposition of RTS patients to OS. Some patients developed mucositis requiring dose modifications, particularly to doxorubicin (no more than 25% decrease), but there is no current method to determine a priori who will experience increased toxicities. Therefore, current recommendations are to treat with standard doses of chemotherapy and to adjust according to the patient's individual course. The similarities between OS in RTS and sporadic OS support the further study of the contribution of the RECQL4 pathways in the pathogenesis of OS.

Understanding the Role of RECQL4 in Osteosarcoma Development Using Mouse Models

Recql4 Global Knockout Mouse Models

In order to understand the function of RECQL4 in OS tumorigenesis in vivo, three mouse models of global *Recql4* disruption have been generated. In the first model, exons 5–8 of *Recql4* upstream of the conserved helicase domain (exons 9–15) were replaced with *PGKneo* and *LacZ* cassettes [45]. Homozygous mutants died during early embryonic stage E3.5–6.5. Although there was no information about transcripts and protein levels of *Recql4* in the paper, presumably this targeting strategy generated a null mutation as a result of nonsense mediated decay. The second mouse model by Hoki et al. targeted exon 13 of the helicase domain of *Recql4* with a neomycin cassette [43]. These homozygous mutants were viable at birth, but 95% of them died within 2 weeks. The remaining 5% exhibited growth retardation, skin atrophy, hair abnormalities, and tissue hypoplasia, such as severely reduced bone trabeculae and fewer and smaller villi of the small intestine. The MEFs from these mutants showed reduced proliferation. However, there was no malignancy reported in these mice. The third global mouse model was generated by replacing exons 9–13 in the conserved helicase domain of *Recql4* with a *PGK-HPRT* cassette [71]. Homozygous mutants were born alive with normal Mendelian ratio, but 16% of them died within 24 hours of birth. The remaining mutants exhibited tail pigmentation defects by 12 months, and palatal patterning defects were seen in all examined animals. Furthermore, 6% of these mutants developed limb defects at birth, ranging from preaxial polydactyly of hindlimbs to forelimb aplasia. Interestingly, 5% of these mutants developed OS or lymphoma by 20 months, while heterozygous and wild-type mice had no tumor formation, although this difference was not found to be statistically significant.

Recql4 Conditional (Bone-Specific) Mouse Models

Because the previous global *Recql4* knockout mouse models failed to recapitulate the high risk of OS seen in RTS patients with *RECQL4* mutations, skeletal-specific conditional knockout mouse models have been developed to assess the effect of Recql4 deficiency in the bone. Lu et al. developed a conditional knockout model of *Recql4* in early skeletal progenitor cell system by crossing these *Recql4* mice with *Prx1*-Cre transgenic mice. Resultant mutants developed foreshortened limbs, digit defects, abnormal growth plates and joints, and craniosynostosis, recapitulating the major skeletal defects seen in RTS patients. Mouse tissues lacking Recql4 displayed increased DNA damage and elevated p53 activation, leading to increased cell death, reduced cell proliferation, and increased senescence. These defects were partially rescued by concurrent inactivation of *p53*, indicating that p53 activation may contribute to the skeletal phenotypes seen in RTS patients. RTS human fibroblasts were also shown to have increased p53 phosphorylation and expression of downstream target genes of p53 [24, 25]. Similarly, depletion of RECQL4 in primary human fibroblasts causes increased DNA damage and cellular senescence as well as p53 activation and increased expression of target genes [65].

Ng et al. developed another conditional knockout model using *Osx*-Cre to inactivate *Recql4* in osteoblast progenitor cells at a later stage of skeletal development, and they observed reduced body weight and decrease in trabecular and cortical bone [83]. Mice lacking *Recql4* in the osteocytes and a subset of osteoblasts showed no striking developmental skeletal abnormalities [83], indicating that RECQL4 plays a more important developmental role in the early stages of osteoblast differentiation. Unlike human RTS patients, however, these homozygous *Recql4* conditional knockout mice did not develop OS. Interestingly, mice with homozygous loss of both *Recql4* and *p53* in the osteoblast progenitor cells showed delayed osteosarcoma development and significantly longer survival compared to p53 homozygous loss alone, indicating that Recql4 may actually be necessary for OS development in mice [83]. The mouse models developed to date have not been able to recapitulate the high incidence of OS seen in RTS patients, and further work is in progress to understand these differences and to dissect the molecular mechanisms underlying OS development in RTS patients.

In order to more closely mirror the human disease, induced pluripotent stem cell (iPSC) techniques have been used to model RECQ syndromes using patient-derived somatic cells including peripheral blood mononuclear cells and dermal fibroblasts. Werner syndrome iPSCs have been generated by several groups [15, 35, 98, 123], and they exhibit normal karyotypes and stable chromosomes after long-term culture [98]. In addition, human embryonic stem cells were also used to generate *WRN*-deficient cells which were further differentiated into human mesenchymal stem cells (MSCs), demonstrating that WRN is essential for maintaining heterochromatin stability and that loss of WRN in human MSCs leads to disorganization of heterochromatin and increased senescence [138]. Thus far, an iPSC line has been generated from dermal fibroblasts derived from a *RECQL4* heterozygous carrier [48], and work is ongoing to establish iPSC lines differentiated into osteoblasts from RTS patient fibroblasts with biallelic *RECQL4* mutations in order to identify the molecular mechanisms underlying the high risk of OS in RTS patients.

Clinical Implications for Understanding RECQ Gene Defects and Potentially Targeting RECQ-Related Pathways for Cancer Therapy

Based on the roles of the RECQ proteins in normal cellular proliferation, DNA damage response, DNA repair, and telomere maintenance, there is growing interest in exploring inhibition of these functions in susceptible cancer cell types. Small molecule inhibitors of the WRN [1] and BLM [84] proteins have been identified as potential antiproliferative cancer therapies. Both of these molecules were identified through in vitro helicase activity screens. The WRN inhibitor, a small molecule inhibitor identified from the National Cancer Institute Diversity Set, designated NSC 19630 [2], was shown to inhibit cell proliferation and to induce apoptosis in a WRN-dependent manner. It also caused increase in DSBs and accumulation of blocked replication forks in human tumor cells grown in culture. NSC 19630 also had a synergistic effect on inhibiting cell proliferation when cells were co-treated along with telomestatin, a small molecule that binds G4 structures and causes disruption of telomere-associated proteins, as well as a PARP inhibitor KU0058948. It also acted synergistically with the topoisomerase inhibitor topotecan in inducing DSBs. Investigators later characterized a structurally related compound, NSC 617145, which they demonstrated was able to sensitize cancer cells to mitomycin C, resulting in decreased cell proliferation, increased DNA damage, and chromosomal abnormalities [1]. More recently, through high-throughput CRISPR-Cas9-mediated knockout and/or RNA interference screening, the WRN helicase has been shown by several groups to be a promising synthetically lethal target in cancers with high levels of microsatellite instability (MSI), including colorectal, endometrial, ovarian, and gastric cancers [6, 13, 53, 63]. MSI is caused by an impaired DNA mismatch repair pathway leading to small insertions and/or deletions in genomic nucleotide repeats. The helicase function of WRN is essential for this synthetic lethality [13, 63], which was not observed with other RecQ helicases [13, 63]. Similarly, inactivation of WRN leads to increased DNA damage and cell death in MSI high cancer cells, but not in microsatellite stable cancer cells [6, 13, 53, 63]. Therefore, these small molecule inhibitors to the WRN protein may be useful to target cancers with high levels of MSI.

The small molecule inhibitor of BLM, ML216 [84], was found to exert its action by preventing BLM from binding to DNA. Cells treated with ML216 showed decreased proliferation as well as an increase in sister chromatid exchanges, a hallmark of Bloom syndrome. One of the proposed future uses of this BLM-specific inhibitor would be to test its efficacy in treating tumor cells that depend on the ALT (alternative lengthening of telomeres) mechanism for maintenance of telomeres, since previous work showed that the BLM orthologue Sgs1 is required for telomere maintenance in the absence of telomerase[140]. Approximately 5–10% of tumors depend on the ALT pathway for continued proliferation, including OS; therefore, further exploration of this BLM-specific inhibitor could reveal a new therapeutic strategy for targeting susceptible tumors.

Expression of *RECQL4* has been found to be upregulated in a variety of cancer types in addition to sporadic OS [70, 93], including soft tissue sarcomas [64], prostate cancer [106], cervical cancer [82], breast cancer [4, 32], gastric cancer [76], and oral cancer [136], suggesting that inactivation of RECQL4, and thus inhibition of its functions in cellular replication/viability, genome stability, DNA repair, and telomere maintenance, may be attractive as a potential adjunct to cancer therapy in susceptible tumor cells. RECQL4 may also work in coordination with other Holliday junction processing proteins, including BLM, to prevent replication fork stalling and reversal in order to maintain cancer cell fitness by resolving increased Holliday junctions in cancer cells with overexpression of RAD51 [128, 129]. Additionally, in gastric cancer cells, overexpression of *RECQL4* has been linked to increased resistance to cisplatin by physically interacting with YB1 and AKT, as well as by increasing AKT-dependent YB1 phosphorylation and expression of the downstream drug resistance gene *MDR1* [76]. These data suggest that RECQL4 may be required for rapid tumor cell proliferation and chemoresistance, providing a potential therapeutic target for cancer cells with overexpression of RECQL4.

Although sporadic OS tumors have not been found to have somatic *RECQL4* mutations, a recent study which examined germ line sequence data from over 5000 sporadic pediatric cancer patients revealed an increase in heterozygous *RECQL4* loss-of-function variants in OS patients compared to non-cancer database controls[141]. While presence of a *RECQL4* heterozygous mutation does not cause RTS with its associated high risk of OS, it may still confer an elevated OS risk in carriers compared to the general population. This has implications for genetic counseling of these patients and may also offer potential avenues for novel targeted therapies for their specific tumors. Ongoing basic science and clinical research is needed to fully understand the cellular context and molecular mechanisms by which RECQL4 exerts its actions on osteosarcomagenesis, and this will provide useful information on the basic biology of OS and open avenues for potential new therapies for OS.

References

1. Aggarwal M, Banerjee T, Sommers JA, Brosh RM (2013) Targeting an Achilles' heel of cancer with a WRN helicase inhibitor. Cell Cycle 12(20):3329–3335
2. Aggarwal M, Sommers JA, Shoemaker RH, Brosh RM Jr (2011) Inhibition of helicase activity by a

small molecule impairs Werner syndrome helicase (WRN) function in the cellular response to DNA damage or replication stress. Proc Natl Acad Sci U S A 108(4):1525–1530
3. Ajeawung NF, Nguyen TTM, Lu L, Kucharski TJ, Rousseau J, Molidperee S, Atienza J, Gamache I, Jin W, Plon SE, Lee BH, Teodoro JG, Wang LL, Campeau PM (2019) Mutations in ANAPC1, encoding a scaffold subunit of the anaphase-promoting complex, cause Rothmund-Thomson syndrome type 1. Am J Hum Genet 105(3):625–630. https://doi.org/10.1016/j.ajhg.2019.06.011
4. Arora A, Agarwal D, Abdel-Fatah TM, Lu H, Croteau DL, Moseley P, Aleskandarany MA, Green AR, Ball G, Rakha EA, Chan SY, Ellis IO, Wang LL, Zhao Y, Balajee AS, Bohr VA, Madhusudan S (2016) RECQL4 helicase has oncogenic potential in sporadic breast cancers. J Pathol 238(4):495–501. https://doi.org/10.1002/path.4681
5. Bachrati CZ, Hickson ID (2003) RecQ helicases: suppressors of tumorigenesis and premature aging. Biochem J 374(Pt 3):577–606. https://doi.org/10.1042/BJ20030491
6. Behan FM, Iorio F, Picco G, Goncalves E, Beaver CM, Migliardi G, Santos R, Rao Y, Sassi F, Pinnelli M, Ansari R, Harper S, Jackson DA, McRae R, Pooley R, Wilkinson P, van der Meer D, Dow D, Buser-Doepner C, Bertotti A, Trusolino L, Stronach EA, Saez-Rodriguez J, Yusa K, Garnett MJ (2019) Prioritization of cancer therapeutic targets using CRISPR-Cas9 screens. Nature 568(7753):511–516. https://doi.org/10.1038/s41586-019-1103-9
7. Belyea B, Kephart JG, Blum J, Kirsch DG, Linardic CM (2012) Embryonic signaling pathways and rhabdomyosarcoma: contributions to cancer development and opportunities for therapeutic targeting. Sarcoma 2012:406239
8. Bernstein KA, Gangloff S, Rothstein R (2010) The RecQ DNA helicases in DNA repair. Annu Rev Genet 44:393–417. https://doi.org/10.1146/annurev-genet-102209-163602
9. Burks LM, Yin J, Plon SE (2007) Nuclear import and retention domains in the amino terminus of RECQL4. Gene 391(1–2):26–38. https://doi.org/10.1016/j.gene.2006.11.019
10. Cabral RE, Queille S, Bodemer C, de Prost Y, Neto JB, Sarasin A, ya-Grosjean L (2008) Identification of new RECQL4 mutations in Caucasian Rothmund-Thomson patients and analysis of sensitivity to a wide range of genotoxic agents. Mutat Res 643(1–2):41–47
11. Capp C, Wu J, Hsieh TS (2009) Drosophila RecQ4 has a 3′-5' DNA helicase activity that is essential for viability. J Biol Chem 284(45):30845–30852. https://doi.org/10.1074/jbc.M109.008052
12. Castillo-Tandazo W, Smeets MF, Murphy V, Liu R, Hodson C, Heierhorst J, Deans AJ, Walkley CR (2019) ATP-dependent helicase activity is dispensable for the physiological functions of Recql4. PLoS Genet 15(7):e1008266. https://doi.org/10.1371/journal.pgen.1008266
13. Chan EM, Shibue T, McFarland JM, Gaeta B, Ghandi M, Dumont N, Gonzalez A, McPartlan JS, Li T, Zhang Y, Bin Liu J, Lazaro JB, Gu P, Piett CG, Apffel A, Ali SO, Deasy R, Keskula P, Ng RWS, Roberts EA, Reznichenko E, Leung L, Alimova M, Schenone M, Islam M, Maruvka YE, Liu Y, Roper J, Raghavan S, Giannakis M, Tseng YY, Nagel ZD, D'Andrea A, Root DE, Boehm JS, Getz G, Chang S, Golub TR, Tsherniak A, Vazquez F, Bass AJ (2019) WRN helicase is a synthetic lethal target in microsatellite unstable cancers. Nature 568(7753):551–556. https://doi.org/10.1038/s41586-019-1102-x
14. Chen X, Bahrami A, Pappo A, Easton J, Dalton J, Hedlund E, Ellison D, Shurtleff S, Wu G, Wei L, Parker M, Rusch M, Nagahawatte P, Wu J, Mao S, Boggs K, Mulder H, Yergeau D, Lu C, Ding L, Edmonson M, Qu C, Wang J, Li Y, Navid F, Daw NC, Mardis ER, Wilson RK, Downing JR, Zhang J, Dyer MA, St. Jude Children's Research Hospital-Washington University Pediatric Cancer Genome P (2014) Recurrent somatic structural variations contribute to tumorigenesis in pediatric osteosarcoma. Cell Rep 7(1):104–112. https://doi.org/10.1016/j.celrep.2014.03.003
15. Cheung HH, Liu X, Canterel-Thouennon L, Li L, Edmonson C, Rennert OM (2014) Telomerase protects werner syndrome lineage-specific stem cells from premature aging. Stem Cell Rep 2(4):534–546. https://doi.org/10.1016/j.stemcr.2014.02.006
16. Chi Z, Nie L, Peng Z, Yang Q, Yang K, Tao J, Mi Y, Fang X, Balajee AS, Zhao Y (2012) RecQL4 cytoplasmic localization: implications in mitochondrial DNA oxidative damage repair. Int J Biochem Cell Biol 44(11):1942–1951. https://doi.org/10.1016/j.biocel.2012.07.016
17. Chu WK, Hickson ID (2009) RecQ helicases: multifunctional genome caretakers. Nat Rev Cancer 9(9):644–654. https://doi.org/10.1038/nrc2682
18. Cleton-Jansen AM, Anninga JK, Briaire-de BI, Romeo S, Oosting J, Egeler RM, Gelderblom H, Taminiau AH, Hogendoorn PC (2009) Profiling of high-grade central osteosarcoma and its putative progenitor cells identifies tumourigenic pathways. Br J Cancer 101(11):1909–1918
19. Crevel G, Vo N, Crevel I, Hamid S, Hoa L, Miyata S, Cotterill S (2012) Drosophila RecQ4 is directly involved in both DNA replication and the response to UV damage in S2 cells. PLoS One 7(11):e49505. https://doi.org/10.1371/journal.pone.0049505
20. Croteau DL, Popuri V, Opresko PL, Bohr VA (2014) Human RecQ helicases in DNA repair, recombination, and replication. Annu Rev Biochem 83:519–552. https://doi.org/10.1146/annurev-biochem-060713-035428
21. Croteau DL, Rossi ML, Canugovi C, Tian J, Sykora P, Ramamoorthy M, Wang ZM, Singh DK, Akbari M, Kasiviswanathan R, Copeland

WC, Bohr VA (2012) RECQL4 localizes to mitochondria and preserves mitochondrial DNA integrity. Aging Cell 11(3):456–466. https://doi.org/10.1111/j.1474-9726.2012.00803.x
22. Cunniff C, Bassetti JA, Ellis NA (2017) Bloom's syndrome: clinical spectrum, molecular pathogenesis, and cancer predisposition. Mol Syndromol 8(1):4–23. https://doi.org/10.1159/000452082
23. Cybulski C, Carrot-Zhang J, Kluzniak W, Rivera B, Kashyap A, Wokolorczyk D, Giroux S, Nadaf J, Hamel N, Zhang S, Huzarski T, Gronwald J, Byrski T, Szwiec M, Jakubowska A, Rudnicka H, Lener M, Masojc B, Tonin PN, Rousseau F, Gorski B, Debniak T, Majewski J, Lubinski J, Foulkes WD, Narod SA, Akbari MR (2015) Germline RECQL mutations are associated with breast cancer susceptibility. Nat Genet 47(6):643–646. https://doi.org/10.1038/ng.3284
24. Davis T, Tivey HS, Brook AJ, Grimstead JW, Rokicki MJ, Kipling D (2013) Activation of p38 MAP kinase and stress signalling in fibroblasts from the progeroid Rothmund-Thomson syndrome. Age (Dordr) 35(5):1767–1783. https://doi.org/10.1007/s11357-012-9476-9
25. De S, Kumari J, Mudgal R, Modi P, Gupta S, Futami K, Goto H, Lindor NM, Furuichi Y, Mohanty D, Sengupta S (2012) RECQL4 is essential for the transport of p53 to mitochondria in normal human cells in the absence of exogenous stress. J Cell Sci 125(Pt 10):2509–2522. https://doi.org/10.1242/jcs.101501
26. Debeljak M, Zver A, Jazbec J (2009) A patient with Baller-Gerold syndrome and midline NK/T lymphoma. Am J Med Genet A 149A(4):755–759. https://doi.org/10.1002/ajmg.a.32736
27. Dietschy T, Shevelev I, Pena-Diaz J, Huhn D, Kuenzle S, Mak R, Miah MF, Hess D, Fey M, Hottiger MO, Janscak P, Stagljar I (2009) p300-mediated acetylation of the Rothmund-Thomson-syndrome gene product RECQL4 regulates its subcellular localization. J Cell Sci 122(Pt 8):1258–1267. https://doi.org/10.1242/jcs.037747
28. Dong YZ, Huang YX, Lu T (2015) Single nucleotide polymorphism in the RECQL5 gene increased osteosarcoma susceptibility in a Chinese Han population. Genet Mol Res 14(1):1899–1902. https://doi.org/10.4238/2015.March.13.18
29. el-Khoury JM, Haddad SN, Atallah NG (1997) Osteosarcomatosis with Rothmund-Thomson syndrome. Br J Radiol 70:215–218. https://doi.org/10.1259/bjr.70.830.9135453
30. Ellis NA, Groden J, Ye TZ, Straughen J, Lennon DJ, Ciocci S, Proytcheva M, German J (1995) The Bloom's syndrome gene product is homologous to RecQ helicases. Cell 83(4):655–666
31. Fan W, Luo J (2008) RecQ4 facilitates UV light-induced DNA damage repair through interaction with nucleotide excision repair factor xeroderma pigmentosum group A (XPA). J Biol Chem 283(43):29037–29044. https://doi.org/10.1074/jbc.M801928200
32. Fang H, Nie L, Chi Z, Liu J, Guo D, Lu X, Hei TK, Balajee AS, Zhao Y (2013) RecQL4 helicase amplification is involved in human breast tumorigenesis. PLoS One 8(7):e69600. https://doi.org/10.1371/journal.pone.0069600
33. Fang H, Niu K, Mo D, Zhu Y, Tan Q, Wei D, Li Y, Chen Z, Yang S, Balajee AS, Zhao Y (2018) RecQL4-Aurora B kinase axis is essential for cellular proliferation, cell cycle progression, and mitotic integrity. Oncogenesis 7(9):68. https://doi.org/10.1038/s41389-018-0080-4
34. Ferrarelli LK, Popuri V, Ghosh AK, Tadokoro T, Canugovi C, Hsu JK, Croteau DL, Bohr VA (2013) The RECQL4 protein, deficient in Rothmund-Thomson syndrome is active on telomeric D-loops containing DNA metabolism blocking lesions. DNA Repair 12(7):518–528. https://doi.org/10.1016/j.dnarep.2013.04.005
35. Gatinois V, Desprat R, Becker F, Pichard L, Bernex F, Corsini C, Pellestor F, Lemaitre JM (2019) Reprogramming of Human Peripheral Blood Mononuclear Cell (PBMC) from a patient suffering of a Werner syndrome resulting in iPSC line (REGUi003-A) maintaining a short telomere length. Stem Cell Res 39:101515. https://doi.org/10.1016/j.scr.2019.101515
36. German J (1995) Bloom's syndrome. Dermatol Clin 13(1):7–18
37. German J (1997) Bloom's syndrome. XX. The first 100 cancers. Cancer Genet Cytogenet 93(1):100–106
38. Ghosh AK, Rossi ML, Singh DK, Dunn C, Ramamoorthy M, Croteau DL, Liu Y, Bohr VA (2012) RECQL4, the protein mutated in Rothmund-Thomson syndrome, functions in telomere maintenance. J Biol Chem 287(1):196–209. https://doi.org/10.1074/jbc.M111.295063
39. Glass DA, Karsenty G (2007) In vivo analysis of Wnt signaling in bone. Endocrinology 148(6):2630–2634
40. Gupta S, De S, Srivastava V, Hussain M, Kumari J, Muniyappa K, Sengupta S (2014) RECQL4 and p53 potentiate the activity of polymerase gamma and maintain the integrity of the human mitochondrial genome. Carcinogenesis 35(1):34–45. https://doi.org/10.1093/carcin/bgt315
41. Hansel-Hertsch R, Di Antonio M, Balasubramanian S (2017) DNA G-quadruplexes in the human genome: detection, functions and therapeutic potential. Nat Rev Mol Cell Biol 18(5):279–284. https://doi.org/10.1038/nrm.2017.3
42. Hicks MJ, Roth JR, Kozinetz CA, Wang LL (2007) Clinicopathologic features of osteosarcoma in patients with Rothmund-Thomson syndrome. J Clin Oncol 25(4):370–375. https://doi.org/10.1200/JCO.2006.08.4558
43. Hoki Y, Araki R, Fujimori A, Ohhata T, Koseki H, Fukumura R, Nakamura M, Takahashi H, Noda

Y, Kito S, Abe M (2003) Growth retardation and skin abnormalities of the Recql4-deficient mouse. Hum Mol Genet 12(18):2293–2299. https://doi.org/10.1093/hmg/ddg254

44. Hu H, Hilton MJ, Tu X, Yu K, Ornitz DM, Long F (2005) Sequential roles of Hedgehog and Wnt signaling in osteoblast development. Development 132(1):49–60
45. Ichikawa K, Noda T, Furuichi Y (2002) Preparation of the gene targeted knockout mice for human premature aging diseases, Werner syndrome, and Rothmund-Thomson syndrome caused by the mutation of DNA helicases. Nihon yakurigaku zasshi Folia pharmacologica Japonica 119(4):219–226
46. Im JS, Ki SH, Farina A, Jung DS, Hurwitz J, Lee JK (2009) Assembly of the Cdc45-Mcm2-7-GINS complex in human cells requires the Ctf4/And-1, RecQL4, and Mcm10 proteins. Proc Natl Acad Sci U S A 106(37):15628–15632. https://doi.org/10.1073/pnas.0908039106
47. Im JS, Park SY, Cho WH, Bae SH, Hurwitz J, Lee JK (2015) RecQL4 is required for the association of Mcm10 and Ctf4 with replication origins in human cells. Cell Cycle 14(7):1001–1009. https://doi.org/10.1080/15384101.2015.1007001
48. Jewell BE, Liu M, Lu L, Zhou R, Tu J, Zhu D, Huo Z, Xu A, Wang D, Mata H, Jin W, Xia W, Rao PH, Zhao R, Hung MC, Wang LL, Lee DF (2018) Generation of an induced pluripotent stem cell line from an individual with a heterozygous RECQL4 mutation. Stem Cell Res 33:36–40. https://doi.org/10.1016/j.scr.2018.10.003
49. Jin W, Liu H, Zhang Y, Otta SK, Plon SE, Wang LL (2008) Sensitivity of RECQL4-deficient fibroblasts from Rothmund-Thomson syndrome patients to genotoxic agents. Hum Genet 123(6):643–653. https://doi.org/10.1007/s00439-008-0518-4
50. Kaiser S, Sauer F, Kisker C (2017) The structural and functional characterization of human RecQ4 reveals insights into its helicase mechanism. Nat Commun 8:15907. https://doi.org/10.1038/ncomms15907
51. Kamimura Y, Masumoto H, Sugino A, Araki H (1998) Sld2, which interacts with Dpb11 in Saccharomyces cerevisiae, is required for chromosomal DNA replication. Mol Cell Biol 18(10):6102–6109
52. Kansara M, Tsang M, Kodjabachian L, Sims NA, Trivett MK, Ehrich M, Dobrovic A, Slavin J, Choong PF, Simmons PJ, Dawid IB, Thomas DM (2009) Wnt inhibitory factor 1 is epigenetically silenced in human osteosarcoma, and targeted disruption accelerates osteosarcomagenesis in mice. J Clin Invest 119(4):837–851
53. Kategaya L, Perumal SK, Hager JH, Belmont LD (2019) Werner syndrome helicase is required for the survival of cancer cells with microsatellite instability. iScience 13:488–497. https://doi.org/10.1016/j.isci.2019.02.006
54. Keller H, Kiosze K, Sachsenweger J, Haumann S, Ohlenschlager O, Nuutinen T, Syvaoja JE, Gorlach M, Grosse F, Pospiech H (2014) The intrinsically disordered amino-terminal region of human RecQL4: multiple DNA-binding domains confer annealing, strand exchange and G4 DNA binding. Nucleic Acids Res 42(20):12614–12627. https://doi.org/10.1093/nar/gku993
55. Kitao S, Ohsugi I, Ichikawa K, Goto M, Furuichi Y, Shimamoto A (1998) Cloning of two new human helicase genes of the RecQ family: biological significance of multiple species in higher eukaryotes. Genomics 54(3):443–452. https://doi.org/10.1006/geno.1998.5595
56. Kitao S, Shimamoto A, Goto M, Miller RW, Smithson WA, Lindor NM, Furuichi Y (1999) Mutations in RECQL4 cause a subset of cases of Rothmund-Thomson syndrome. Nat Genet 22(1):82–84. https://doi.org/10.1038/8788
57. Kliszczak M, Sedlackova H, Pitchai GP, Streicher WW, Krejci L, Hickson ID (2015) Interaction of RECQ4 and MCM10 is important for efficient DNA replication origin firing in human cells. Oncotarget 6(38):40464–40479. https://doi.org/10.18632/oncotarget.6342
58. Kobbe D, Focke M, Puchta H (2010) Purification and characterization of RecQ helicases of plants. Methods Mol Biol 587:195–209
59. Kohzaki M, Chiourea M, Versini G, Adachi N, Takeda S, Gagos S, Halazonetis TD (2012) The helicase domain and C-terminus of human RecQL4 facilitate replication elongation on DNA templates damaged by ionizing radiation. Carcinogenesis 33(6):1203–1210. https://doi.org/10.1093/carcin/bgs149
60. Kohzaki M, Ootsuyama A, Sun L, Moritake T, Okazaki R (2019) Human RECQL4 represses the RAD52-mediated single-strand annealing pathway after ionizing radiation or cisplatin treatment. Int J Cancer. https://doi.org/10.1002/ijc.32670
61. Kumata Y, Tada S, Yamanada Y, Tsuyama T, Kobayashi T, Dong YP, Ikegami K, Murofushi H, Seki M, Enomoto T (2007) Possible involvement of RecQL4 in the repair of double-strand DNA breaks in Xenopus egg extracts. Biochim Biophys Acta 1773(4):556–564. https://doi.org/10.1016/j.bbamcr.2007.01.005
62. Lauper JM, Krause A, Vaughan TL, Monnat RJ Jr (2013) Spectrum and risk of neoplasia in Werner syndrome: a systematic review. PLoS One 8(4):e59709
63. Lieb S, Blaha-Ostermann S, Kamper E, Rippka J, Schwarz C, Ehrenhofer-Wolfer K, Schlattl A, Wernitznig A, Lipp JJ, Nagasaka K, van der Lelij P, Bader G, Koi M, Goel A, Neumuller RA, Peters JM, Kraut N, Pearson MA, Petronczki M, Wohrle S (2019) Werner syndrome helicase is a selective vulnerability of microsatellite instability-high tumor cells. Elife 8:e43333. https://doi.org/10.7554/eLife.43333
64. Linton KM, Hey Y, Saunders E, Jeziorska M, Denton J, Wilson CL, Swindell R, Dibben S, Miller CJ, Pepper SD, Radford JA, Freemont AJ (2008) Acquisition of biologically relevant gene expression

data by Affymetrix microarray analysis of archival formalin-fixed paraffin-embedded tumours. Br J Cancer 98(8):1403–1414
65. Lu H, Fang EF, Sykora P, Kulikowicz T, Zhang Y, Becker KG, Croteau DL, Bohr VA (2014) Senescence induced by RECQL4 dysfunction contributes to Rothmund-Thomson syndrome features in mice. Cell Death Dis 5:e1226. https://doi.org/10.1038/cddis.2014.168
66. Lu H, Shamanna RA, de Freitas JK, Okur M, Khadka P, Kulikowicz T, Holland PP, Tian J, Croteau DL, Davis AJ, Bohr VA (2017) Cell cycle-dependent phosphorylation regulates RECQL4 pathway choice and ubiquitination in DNA double-strand break repair. Nat Commun 8(1):2039. https://doi.org/10.1038/s41467-017-02146-3
67. Lu H, Shamanna RA, Keijzers G, Anand R, Rasmussen LJ, Cejka P, Croteau DL, Bohr VA (2016) RECQL4 promotes DNA end resection in repair of DNA double-strand breaks. Cell Rep 16(1):161–173. https://doi.org/10.1016/j.celrep.2016.05.079
68. Luo X, Kraus WL (2012) On PAR with PARP: cellular stress signaling through poly(ADP-ribose) and PARP-1. Genes Dev 26(5):417–432
69. Macris MA, Krejci L, Bussen W, Shimamoto A, Sung P (2006) Biochemical characterization of the RECQ4 protein, mutated in Rothmund-Thomson syndrome. DNA Repair 5(2):172–180. https://doi.org/10.1016/j.dnarep.2005.09.005
70. Maire G, Yoshimoto M, Chilton-MacNeill S, Thorner PS, Zielenska M, Squire JA (2009) Recurrent RECQL4 imbalance and increased gene expression levels are associated with structural chromosomal instability in sporadic osteosarcoma. Neoplasia 11(3):260–268, 263p following 268
71. Mann MB, Hodges CA, Barnes E, Vogel H, Hassold TJ, Luo G (2005) Defective sister-chromatid cohesion, aneuploidy and cancer predisposition in a mouse model of type II Rothmund-Thomson syndrome. Hum Mol Genet 14(6):813–825. https://doi.org/10.1093/hmg/ddi075
72. Marino F, Vindigni A, Onesti S (2013) Bioinformatic analysis of RecQ4 helicases reveals the presence of a RQC domain and a Zn knuckle. Biophys Chem 177-178:34–39. https://doi.org/10.1016/j.bpc.2013.02.009
73. Martin GM, Oshima J (2000) Lessons from human progeroid syndromes. Nature 408(6809):263–266
74. Matsuno K, Kumano M, Kubota Y, Hashimoto Y, Takisawa H (2006) The N-terminal noncatalytic region of Xenopus RecQ4 is required for chromatin binding of DNA polymerase alpha in the initiation of DNA replication. Mol Cell Biol 26(13):4843–4852. https://doi.org/10.1128/MCB.02267-05
75. Mehollin-Ray AR, Kozinetz CA, Schlesinger AE, Guillerman RP, Wang LL (2008) Radiographic abnormalities in Rothmund-Thomson syndrome and genotype-phenotype correlation with RECQL4 mutation status. AJR Am J Roentgenol 191(2):W62–W66. https://doi.org/10.2214/AJR.07.3619
76. Mo D, Fang H, Niu K, Liu J, Wu M, Li S, Zhu T, Aleskandarany MA, Arora A, Lobo DN, Madhusudan S, Balajee AS, Chi Z, Zhao Y (2016) Human helicase RECQL4 drives cisplatin resistance in gastric cancer by activating an AKT-YB1-MDR1 signaling pathway. Cancer Res 76(10):3057–3066. https://doi.org/10.1158/0008-5472.CAN-15-2361
77. Mohaghegh P, Karow JK, Brosh RM Jr, Bohr VA, Hickson ID (2001) The Bloom's and Werner's syndrome proteins are DNA structure-specific helicases. Nucleic Acids Res 29(13):2843–2849. https://doi.org/10.1093/nar/29.13.2843
78. Mojumdar A, De March M, Marino F, Onesti S (2017) The human RecQ4 helicase contains a functional RecQ C-terminal Region (RQC) that is essential for activity. J Biol Chem 292(10):4176–4184. https://doi.org/10.1074/jbc.M116.767954
79. Monnat RJ Jr (2010) Human RECQ helicases: roles in DNA metabolism, mutagenesis and cancer biology. Semin Cancer Biol 20(5):329–339. https://doi.org/10.1016/j.semcancer.2010.10.002
80. Morozov V, Mushegian AR, Koonin EV, Bork P (1997) A putative nucleic acid-binding domain in Bloom's and Werner's syndrome helicases. Trends Biochem Sci 22(11):417–418
81. Nakayama K, Irino N, Nakayama H (1985) The recQ gene of Escherichia coli K12: molecular cloning and isolation of insertion mutants. Mol Gen Genet MGG 200(2):266–271. https://doi.org/10.1007/bf00425434
82. Narayan G, Bourdon V, Chaganti S, rias-Pulido H, Nandula SV, Rao PH, Gissmann L, Durst M, Schneider A, Pothuri B, Mansukhani M, Basso K, Chaganti RS, Murty VV (2007) Gene dosage alterations revealed by cDNA microarray analysis in cervical cancer: identification of candidate amplified and overexpressed genes. Genes Chromosomes Cancer 46(4):373–384
83. Ng AJ, Walia MK, Smeets MF, Mutsaers AJ, Sims NA, Purton LE, Walsh NC, Martin TJ, Walkley CR (2015) The DNA helicase recql4 is required for normal osteoblast expansion and osteosarcoma formation. PLoS Genet 11(4):e1005160. https://doi.org/10.1371/journal.pgen.1005160
84. Nguyen GH, Dexheimer TS, Rosenthal AS, Chu WK, Singh DK, Mosedale G, Bachrati CZ, Schultz L, Sakurai M, Savitsky P, Abu M, McHugh PJ, Bohr VA, Harris CC, Jadhav A, Gileadi O, Maloney DJ, Simeonov A, Hickson ID (2013) A small molecule inhibitor of the BLM helicase modulates chromosome stability in human cells. Chem Biol 20(1):55–62
85. Nguyen GH, Tang W, Robles AI, Beyer RP, Gray LT, Welsh JA, Schetter AJ, Kumamoto K, Wang XW, Hickson ID, Maizels N, Monnat RJ Jr, Harris CC (2014) Regulation of gene expression by the BLM helicase correlates with the presence of

G-quadruplex DNA motifs. Proc Natl Acad Sci U S A 111(27):9905–9910. https://doi.org/10.1073/pnas.1404807111
86. Nishijo K, Nakayama T, Aoyama T, Okamoto T, Ishibe T, Yasura K, Shima Y, Shibata KR, Tsuboyama T, Nakamura T, Toguchida J (2004) Mutation analysis of the RECQL4 gene in sporadic osteosarcomas. Int J Cancer 111(3):367–372. https://doi.org/10.1002/ijc.20269
87. Ohlenschlager O, Kuhnert A, Schneider A, Haumann S, Bellstedt P, Keller H, Saluz HP, Hortschansky P, Hanel F, Grosse F, Gorlach M, Pospiech H (2012) The N-terminus of the human RecQL4 helicase is a homeodomain-like DNA interaction motif. Nucleic Acids Res 40(17):8309–8324. https://doi.org/10.1093/nar/gks591
88. Park SJ, Lee YJ, Beck BD, Lee SH (2006) A positive involvement of RecQL4 in UV-induced S-phase arrest. DNA Cell Biol 25(12):696–703
89. Patil AV, Hsieh TS (2017) Ribosomal protein S3 negatively regulates unwinding activity of RecQ like helicase 4 through their physical interaction. J Biol Chem. https://doi.org/10.1074/jbc.M116.764324
90. Petkovic M, Dietschy T, Freire R, Jiao R, Stagljar I (2005) The human Rothmund-Thomson syndrome gene product, RECQL4, localizes to distinct nuclear foci that coincide with proteins involved in the maintenance of genome stability. J Cell Sci 118(Pt 18):4261–4269. https://doi.org/10.1242/jcs.02556
91. Rossi ML, Ghosh AK, Kulikowicz T, Croteau DL, Bohr VA (2010) Conserved helicase domain of human RecQ4 is required for strand annealing-independent DNA unwinding. DNA Repair 9(7):796–804
92. Rothmund A (1868) Ueber cataracten in verbindung mit einer eigenthumlichen hautdegeneration. Albrecht von Graefes Arch Fur Ophth 14:159–182
93. Sadikovic B, Thorner P, Chilton-Macneill S, Martin JW, Cervigne NK, Squire J, Zielenska M (2010) Expression analysis of genes associated with human osteosarcoma tumors shows correlation of RUNX2 overexpression with poor response to chemotherapy. BMC Cancer 10:202. https://doi.org/10.1186/1471-2407-10-202
94. Sangrithi MN, Bernal JA, Madine M, Philpott A, Lee J, Dunphy WG, Venkitaraman AR (2005) Initiation of DNA replication requires the RECQL4 protein mutated in Rothmund-Thomson syndrome. Cell 121(6):887–898. https://doi.org/10.1016/j.cell.2005.05.015
95. Schurman SH, Hedayati M, Wang Z, Singh DK, Speina E, Zhang Y, Becker K, Macris M, Sung P, Wilson DM 3rd, Croteau DL, Bohr VA (2009) Direct and indirect roles of RECQL4 in modulating base excision repair capacity. Hum Mol Genet 18(18):3470–3483. https://doi.org/10.1093/hmg/ddp291
96. Sedlackova H, Cechova B, Mlcouskova J, Krejci L (2015) RECQ4 selectively recognizes Holliday junctions. DNA Repair 30:80–89. https://doi.org/10.1016/j.dnarep.2015.02.020
97. Shamanna RA, Singh DK, Lu H, Mirey G, Keijzers G, Salles B, Croteau DL, Bohr VA (2014) RECQ helicase RECQL4 participates in non-homologous end joining and interacts with the Ku complex. Carcinogenesis 35(11):2415–2424. https://doi.org/10.1093/carcin/bgu137
98. Shimamoto A, Kagawa H, Zensho K, Sera Y, Kazuki Y, Osaki M, Oshimura M, Ishigaki Y, Hamasaki K, Kodama Y, Yuasa S, Fukuda K, Hirashima K, Seimiya H, Koyama H, Shimizu T, Takemoto M, Yokote K, Goto M, Tahara H (2014) Reprogramming suppresses premature senescence phenotypes of Werner syndrome cells and maintains chromosomal stability over long-term culture. PLoS One 9(11):e112900. https://doi.org/10.1371/journal.pone.0112900
99. Shin G, Jeong D, Kim H, Im JS, Lee JK (2019) RecQL4 tethering on the pre-replicative complex induces unscheduled origin activation and replication stress in human cells. J Biol Chem. https://doi.org/10.1074/jbc.RA119.009996
100. Shinya A, Nishigori C, Moriwaki S, Takebe H, Kubota M, Ogino A, Imamura S (1993) A case of Rothmund-Thomson syndrome with reduced DNA repair capacity. Arch Dermatol 129(3):332–336
101. Siitonen HA, Kopra O, Kaariainen H, Haravuori H, Winter RM, Saamanen AM, Peltonen L, Kestila M (2003) Molecular defect of RAPADILINO syndrome expands the phenotype spectrum of RECQL diseases. Hum Mol Genet 12(21):2837–2844. https://doi.org/10.1093/hmg/ddg306
102. Siitonen HA, Sotkasiira J, Biervliet M, Benmansour A, Capri Y, Cormier-Daire V, Crandall B, Hannula-Jouppi K, Hennekam R, Herzog D, Keymolen K, Lipsanen-Nyman M, Miny P, Plon SE, Riedl S, Sarkar A, Vargas FR, Verloes A, Wang LL, Kaariainen H, Kestila M (2008) The mutation spectrum in RECQL4 diseases. Eur J Hum Genet 17(2):151–158
103. Singh DK, Karmakar P, Aamann M, Schurman SH, May A, Croteau DL, Burks L, Plon SE, Bohr VA (2010) The involvement of human RECQL4 in DNA double-strand break repair. Aging Cell 9(3):358–371. https://doi.org/10.1111/j.1474-9726.2010.00562.x
104. Singh DK, Popuri V, Kulikowicz T, Shevelev I, Ghosh AK, Ramamoorthy M, Rossi ML, Janscak P, Croteau DL, Bohr VA (2012) The human RecQ helicases BLM and RECQL4 cooperate to preserve genome stability. Nucleic Acids Res 40(14):6632–6648
105. Smith PJ, Paterson MC (1982) Enhanced radiosensitivity and defective DNA repair in cultured fibroblasts derived from Rothmund Thomson syndrome patients. Mutat Res 94(1):213–228
106. Su Y, Meador JA, Calaf GM, Proietti De-Santis L, Zhao Y, Bohr VA, Balajee AS (2010) Human RecQL4 helicase plays critical roles in prostate carcinogenesis. Cancer Res 70(22):9207–9217. https://doi.org/10.1158/0008-5472.CAN-10-1743

107. Suzuki T, Kohno T, Ishimi Y (2009) DNA helicase activity in purified human RECQL4 protein. J Biochem 146(3):327–335
108. Tanaka S, Komeda Y, Umemori T, Kubota Y, Takisawa H, Araki H (2013) Efficient initiation of DNA replication in eukaryotes requires Dpb11/TopBP1-GINS interaction. Mol Cell Biol 33(13):2614–2622
109. Tang W, Robles AI, Beyer RP, Gray LT, Nguyen GH, Oshima J, Maizels N, Harris CC, Monnat RJ Jr (2016) The Werner syndrome RECQ helicase targets G4 DNA in human cells to modulate transcription. Hum Mol Genet 25(10):2060–2069. https://doi.org/10.1093/hmg/ddw079
110. Tao J, Chen S, Lee B (2010) Alteration of Notch signaling in skeletal development and disease. Ann N Y Acad Sci 1192:257–268. https://doi.org/10.1111/j.1749-6632.2009.05307.x
111. Tao J, Jiang MM, Jiang L, Salvo JS, Zeng HC, Dawson B, Bertin TK, Rao PH, Chen R, Donehower LA, Gannon F, Lee BH (2014) Notch activation as a driver of osteogenic sarcoma. Cancer Cell 26(3):390–401. https://doi.org/10.1016/j.ccr.2014.07.023
112. Taylor WB (1957) Rothmund's syndrome; Thomson's syndrome; congenital poikiloderma with or without juvenile cataracts. AMA Arch Derm 75(2):236–244
113. Thangavel S, Mendoza-Maldonado R, Tissino E, Sidorova JM, Yin J, Wang W, Monnat RJ Jr, Falaschi A, Vindigni A (2010) Human RECQ1 and RECQ4 helicases play distinct roles in DNA replication initiation. Mol Cell Biol 30(6):1382–1396. https://doi.org/10.1128/MCB.01290-09
114. Thomson MS (1923) An hitherto undescribed familial disease. Br J Dermatol Syphilis 35:455–462
115. Tuteja N, Tuteja R (2004) Prokaryotic and eukaryotic DNA helicases. Essential molecular motor proteins for cellular machinery. Eur J Biochem 271(10):1835–1848. https://doi.org/10.1111/j.1432-1033.2004.04093.x
116. Van Maldergem L, Piard J, Larizza L, Wang LL (2016) RECQL4-related recessive conditions. In: Erickson RP and Winshaw-Boris A (eds) Epstein's Inborn Errors of Development: The Molecular Basis of Clinical Disorders of Morphogenesis (3 ed.) Oxford University Press, USA.
117. Van Maldergem L, Siitonen HA, Jalkh N, Chouery E, De Roy M, Delague V, Muenke M, Jabs EW, Cai J, Wang LL, Plon SE, Fourneau C, Kestila M, Gillerot Y, Megarbane A, Verloes A (2006) Revisiting the craniosynostosis-radial ray hypoplasia association: Baller-Gerold syndrome caused by mutations in the RECQL4 gene. J Med Genet 43(2):148–152. https://doi.org/10.1136/jmg.2005.031781
118. Vijayakumar S, Liu G, Rus IA, Yao S, Chen Y, Akiri G, Grumolato L, Aaronson SA (2011) High-frequency canonical Wnt activation in multiple sarcoma subtypes drives proliferation through a TCF/beta-catenin target gene, CDC25A. Cancer Cell 19(5):601–612
119. Wang H, Elledge SJ (1999) DRC1, DNA replication and checkpoint protein 1, functions with DPB11 to control DNA replication and the S-phase checkpoint in Saccharomyces cerevisiae. Proc Natl Acad Sci U S A 96(7):3824–3829
120. Wang JT, Xu X, Alontaga AY, Chen Y, Liu Y (2014) Impaired p32 regulation caused by the lymphoma-prone RECQ4 mutation drives mitochondrial dysfunction. Cell Rep 7(3):848–858. https://doi.org/10.1016/j.celrep.2014.03.037
121. Wang LL, Gannavarapu A, Kozinetz CA, Levy ML, Lewis RA, Chintagumpala MM, Ruiz-Maldanado R, Contreras-Ruiz J, Cunniff C, Erickson RP, Lev D, Rogers M, Zackai EH, Plon SE (2003) Association between osteosarcoma and deleterious mutations in the RECQL4 gene in Rothmund-Thomson syndrome. J Natl Cancer Inst 95(9):669–674
122. Wang LL, Levy ML, Lewis RA, Chintagumpala MM, Lev D, Rogers M, Plon SE (2001) Clinical manifestations in a cohort of 41 Rothmund-Thomson syndrome patients. Am J Med Genet 102(1):11–17
123. Wang S, Liu Z, Ye Y, Li B, Liu T, Zhang W, Liu GH, Zhang YA, Qu J, Xu D, Chen Z (2018) Ectopic hTERT expression facilitates reprograming of fibroblasts derived from patients with Werner syndrome as a WS cellular model. Cell Death Dis 9(9):923. https://doi.org/10.1038/s41419-018-0948-4
124. Werner SR, Prahalad AK, Yang J, Hock JM (2006) RECQL4-deficient cells are hypersensitive to oxidative stress/damage: insights for osteosarcoma prevalence and heterogeneity in Rothmund-Thomson syndrome. Biochem Biophys Res Commun 345(1):403–409. https://doi.org/10.1016/j.bbrc.2006.04.093
125. Woo LL, Futami K, Shimamoto A, Furuichi Y, Frank KM (2006) The Rothmund-Thomson gene product RECQL4 localizes to the nucleolus in response to oxidative stress. Exp Cell Res 312(17):3443–3457. https://doi.org/10.1016/j.yexcr.2006.07.023
126. Wu J, Capp C, Feng L, Hsieh TS (2008) Drosophila homologue of the Rothmund-Thomson syndrome gene: essential function in DNA replication during development. Dev Biol 323(1):130–142. https://doi.org/10.1016/j.ydbio.2008.08.006
127. Wu J, Zhi L, Dai X, Cai Q, Ma W (2015) Decreased RECQL5 correlated with disease progression of osteosarcoma. Biochem Biophys Res Commun 467(4):617–622. https://doi.org/10.1016/j.bbrc.2015.10.114
128. Xia J, Chen LT, Mei Q, Ma CH, Halliday JA, Lin HY, Magnan D, Pribis JP, Fitzgerald DM, Hamilton HM, Richters M, Nehring RB, Shen X, Li L, Bates D, Hastings PJ, Herman C, Jayaram M, Rosenberg SM (2016) Holliday junction trap shows how cells use recombination and a junction-guardian role of RecQ helicase. Sci Adv 2(11):e1601605. https://doi.org/10.1126/sciadv.1601605
129. Xia J, Mei Q, Rosenberg SM (2019) Tools to live by: bacterial DNA structures illuminate Cancer. Trends

Genet 35(5):383–395. https://doi.org/10.1016/j.tig.2019.03.001

130. Xu L, Li S, Stohr BA (2013) The role of telomere biology in cancer. Annu Rev Pathol 8:49–78

131. Xu X, Liu Y (2009) Dual DNA unwinding activities of the Rothmund-Thomson syndrome protein, RECQ4. EMBO J 28(5):568–577. https://doi.org/10.1038/emboj.2009.13

132. Xu X, Rochette PJ, Feyissa EA, Su TV, Liu Y (2009a) MCM10 mediates RECQ4 association with MCM2-7 helicase complex during DNA replication. EMBO J 28(19):3005–3014. https://doi.org/10.1038/emboj.2009.235

133. Xu Y, Lei Z, Huang H, Dui W, Liang X, Ma J, Jiao R (2009b) dRecQ4 is required for DNA synthesis and essential for cell proliferation in Drosophila. PLoS One 4(7):e6107

134. Yin J, Kwon YT, Varshavsky A, Wang W (2004) RECQL4, mutated in the Rothmund-Thomson and RAPADILINO syndromes, interacts with ubiquitin ligases UBR1 and UBR2 of the N-end rule pathway. Hum Mol Genet 13(20):2421–2430. https://doi.org/10.1093/hmg/ddh269

135. Yokoyama H, Moreno-Andres D, Astrinidis SA, Hao Y, Weberruss M, Schellhaus AK, Lue H, Haramoto Y, Gruss OJ, Antonin W (2019) Chromosome alignment maintenance requires the MAP RECQL4, mutated in the Rothmund-Thomson syndrome. Life Sci Alliance 2(1). https://doi.org/10.26508/lsa.201800120

136. Yong ZW, Zaini ZM, Kallarakkal TG, Karen-Ng LP, Rahman ZA, Ismail SM, Sharifah NA, Mustafa WM, Abraham MT, Tay KK, Zain RB (2014) Genetic alterations of chromosome 8 genes in oral cancer. Sci Rep 4:6073. https://doi.org/10.1038/srep06073

137. Yu CE, Oshima J, Fu YH, Wijsman EM, Hisama F, Alisch R, Matthews S, Nakura J, Miki T, Ouais S, Martin GM, Mulligan J, Schellenberg GD (1996) Positional cloning of the Werner's syndrome gene. Science 272(5259):258–262. https://doi.org/10.1126/science.272.5259.258

138. Zhang W, Li J, Suzuki K, Qu J, Wang P, Zhou J, Liu X, Ren R, Xu X, Ocampo A, Yuan T, Yang J, Li Y, Shi L, Guan D, Pan H, Duan S, Ding Z, Li M, Yi F, Bai R, Wang Y, Chen C, Yang F, Li X, Wang Z, Aizawa E, Goebl A, Soligalla RD, Reddy P, Esteban CR, Tang F, Liu GH, Belmonte JC (2015) Aging stem cells. A Werner syndrome stem cell model unveils heterochromatin alterations as a driver of human aging. Science 348(6239):1160–1163. https://doi.org/10.1126/science.aaa1356

139. Zhi LQ, Ma W, Zhang H, Zeng SX, Chen B (2014) Association of RECQL5 gene polymorphisms and osteosarcoma in a Chinese Han population. Tumour Biol 35(4):3255–3259. https://doi.org/10.1007/s13277-013-1425-4

140. Huang P, Pryde FE, Lester D, Maddison RL, Borts RH, Hickson ID, Louis EJ (2001) SGS1 is required for telomere elongation in the absence of telomerase. Curr Biol 11 (2):125-129. https://doi.org/10.1016/S0960-9822(01)00021-5

141. Maciaszek JL, Oak N, Chen W, Hamilton KV, McGee RB, Nuccio R, Mostafavi R, Hines-Dowell S, Harrison L, Taylor L, Gerhardt EL, Ouma A, Edmonson MN, Patel A, Nakitandwe J, Pappo AS, Azzato EM, Shurtleff SA, Ellison DW, Downing JR, Hudson MM, Robison LL, Santana V, Newman S, Zhang J, Wang Z, Wu G, Nichols KE, Kesserwan CA (2019) Enrichment of heterozygous germline RECQL4 loss-of-function variants in pediatric osteosarcoma. Cold Spring Harb Mol Case Stud 5(5):a004218. https://doi.org/10.1101/mcs.a004218

Targeting the Cancer Epigenome with Histone Deacetylase Inhibitors in Osteosarcoma

4

Christopher D. Collier, Patrick J. Getty, and Edward M. Greenfield

Abstract

Epigenetic deregulation is an emerging hallmark of cancer that enables tumor cells to escape surveillance by tumor suppressors and ultimately progress. The structure of the epigenome consists of covalent modifications of chromatin components, including acetylation by histone acetyltransferases (HATs) and deacetylation by histone deacetylases (HDACs). Targeting these enzymes with inhibitors to restore epigenetic homeostasis has been explored for many cancers. Osteosarcoma, an aggressive bone malignancy that primarily affects children and young adults, is notable for widespread genetic and epigenetic instability. This may explain why therapy directed at unique molecular pathways has failed to substantially improve outcomes in osteosarcoma over the past four decades. In this review, we discuss the potential of targeting the cancer epigenome, with a focus on histone deacetylase inhibitors (HDACi) for osteosarcoma. We additionally highlight the safety and tolerance of HDACi, combination chemotherapy with HDACi, and the ongoing challenges in the development of these agents.

Keywords

Osteosarcoma · Epigenetics · Epigenome · Histone deacetylase inhibitors · HDAC · HDACi · Lysine deacetylase inhibitors · KDAC · KDACi · Acetylation · Combination therapy · Safety

C. D. Collier (✉)
Department of Orthopaedics, University of Chicago Pritzker School of Medicine, Chicago, IL, USA

P. J. Getty
Department of Orthopaedics, University Hospitals Cleveland Medical Center, Case Western Reserve University, Cleveland, OH, USA
e-mail: patrick.getty@uhhospitals.org

E. M. Greenfield
Department of Orthopaedic Surgery, Indiana University School of Medicine and Indiana University Melvin and Bren Simon Cancer Center, Indianapolis, IN, USA
e-mail: egreenf@iu.edu

Introduction

Classic definitions of cancer describe a disease arising from genetic mutation, yet genetic mutation alone does not account for the behavioral variation observed in the vast majority of human cancers. Epigenetics, an emerging molecular explanation to bridge this gap, is the study of heritable changes in gene expression without alteration of the underlying DNA sequence. The

E. S. Kleinerman, R. Gorlick (eds.), *Current Advances in the Science of Osteosarcoma*, Advances in Experimental Medicine and Biology 1258, https://doi.org/10.1007/978-3-030-43085-6_4

epigenome consists of specific modifications of chromatin components that include DNA methylation, histone modification, nucleosome remodeling, and RNA-mediated events [25]. These modifications affect the regulation of all DNA-based processes, such as transcription, DNA repair, and replication. Consequently, abnormal epigenetic patterns or genomic alterations in chromatin regulators can make profound contributions to the pathogenesis of cancer.

The past decade has seen a growing appreciation for epigenetic complexity following the development of next-generation sequencing, international genome mapping projects [110], and improved techniques for chromatin immunoprecipitation [101]. For example, recent whole-genome sequencing suggests that childhood tumors are driven by a relatively small number of mutations compared to adult tumors – frequently in genes encoding epigenetic regulators – while copy number alterations and structural variants predominate [76]. This is particularly true for osteosarcoma, an aggressive bone malignancy that primarily affects children and young adults, where almost half of patient samples have somatic alterations in epigenetic pathways [76]. Moreover, epigenetics may be critical in osteosarcoma metastases, which are accompanied by a shift in the cancer epigenome despite minor changes in the mutational landscape [84].

To improve patient outcomes in osteosarcoma, new treatment paradigms need to be established as chemotherapy protocols have not significantly changed for over 40 years [39]. While the precise role of epigenetics in osteosarcoma remains to be defined, existing evidence suggests that epigenetic pathways are frequently disrupted and can be targeted. This was discussed in a recent report on osteosarcoma from the Children's Oncology Group, which encouraged continued research into epigenetic dysregulation [111]. In this review, we will therefore discuss the potential of targeting the cancer epigenome, with a focus on histone deacetylase inhibitors (HDACi) for osteosarcoma. We additionally highlight the safety and tolerance of HDACi, combination chemotherapy with HDACi, and the ongoing challenges in the development of these agents.

Targeting the Cancer Epigenome

Transcription is often described as the first step of gene expression and requires RNA polymerase – assisted by transcription factors and enhancers – to bind to a DNA promoter sequence and initiate RNA synthesis. A prerequisite to this process is access to the genetic information stored in chromatin, which is under epigenetic control. Chromatin is composed of nucleosomes consisting of negatively charged DNA wrapped around positively charged histone proteins. It can exist as loosely packed and transcriptionally active euchromatin, or as tightly packed and transcriptionally inactive heterochromatin [25]. The structure and function of chromatin defines the epigenome and is controlled by covalent modifications to the nucleosome complex, including methylation (DNA and histones) and acetylation (histones) [47]. DNA methyltransferases (DNMTs) and histone acetyltransferases (HATs) are examples of "writers" in this system and establish covalent marks, which are then interpreted by "readers" to remodel particular genomic regions and alter gene expression. These marks can be removed by "erasers," such as histone deacetylases (HDACs), and by the ten-eleven translocation (TET) family of 5-methylcytosine oxidases (Fig. 4.1).

Cancer cells use epigenetic pathways to their advantage by direct mutation in genes encoding chromatin-modifying enzymes or through modulated expression of epigenetic regulators [47]. For example, childhood ependymomas are defined by profoundly altered DNA methylation patterns without corresponding genetic mutation [77]. Similarly, many driver mutations in adult tumors can be traced to epigenetic genes, including mutations in histone H3 lysine 36 in sarcomas [74]. The overall effect is to shift the regulatory balance toward oncogenesis and minimize tumor suppressor activity through epigenetic suppression. To target the cancer epigenome and thereby restore cellular homeostasis, two classes of drugs are described: broad reprogrammers (so-called genomic medicines) and more classic targeted therapies designed to affect cancer-specific mutations in epigenetic genes [47].

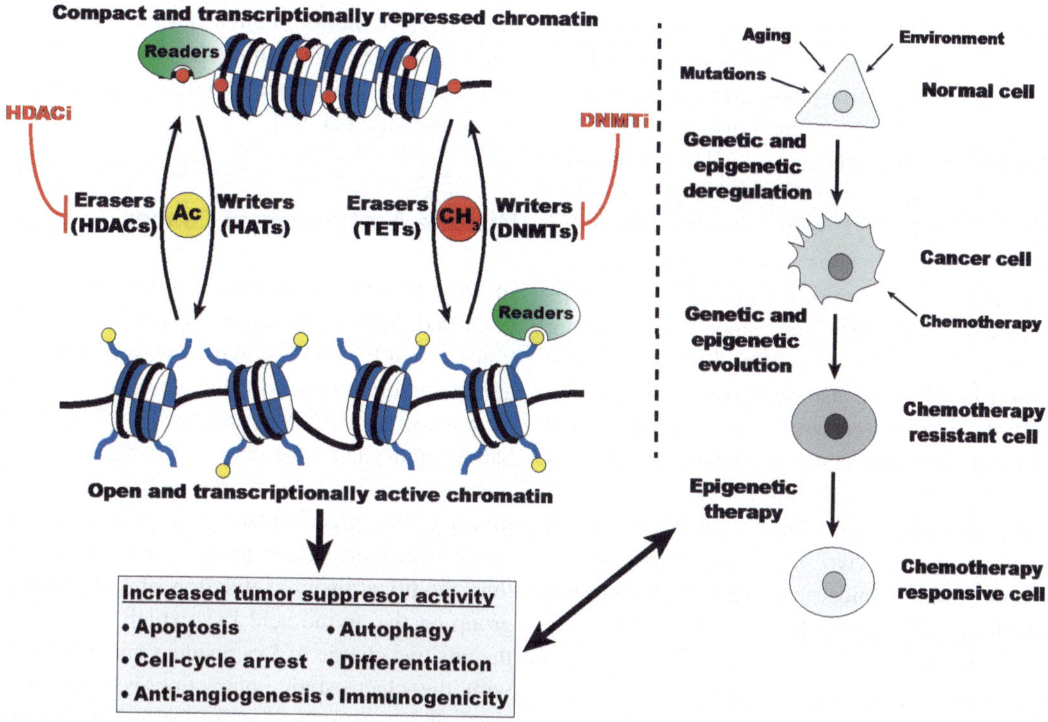

Fig. 4.1 Targeting the cancer epigenome. Negatively charged DNA is wrapped around a core of eight positively charged histone proteins to condense the genome into chromatin. The epigenome consists of specific covalent modifications of chromatin components, including DNA and histone proteins, and modulates transcriptional activity. Histone acetyltransferases (HATs) and DNA methyltransferases (DNMTs) are examples of "writers" in this system and establish covalent marks, which are then interpreted by "readers" to remodel particular genomic regions and modulate gene expression. These marks can be removed by "erasers," such as histone deacetylases (HDACs), and by the ten-eleven translocation (TET) family of 5-methylcytosine oxidases. In the appropriate setting, normal cells can be transformed into cancer cells through genetic and epigenetic deregulation. Further genetic and epigenetic changes can result in chemoresistance. The goal of epigenetic therapies like histone deacetylase inhibitors (HDACi) and DNA methyltransferase inhibitors (DNMTi) is to restore the normal epigenetic balance between oncogenes and tumor suppressor genes

Broad Reprogrammers

Broad reprogrammers include HDAC inhibitors (HDACi), DNMT inhibitors (DNMTi), and inhibitors of the bromodomain and extra-terminal motif proteins (BETi). These drugs induce genome-wide changes in gene expression and generally reverse cancer-specific patterns of gene expression [6]. HDACi and supporting evidence in osteosarcoma will be reviewed in later sections.

DNMTi act in cancer to remove hypermethylation of tumor suppressor genes, which represent the most well-described mechanism of epigenetic dysregulation in cancer. The DNMTi azacitidine and its derivative decitabine are currently approved by the US Food and Drug Administration (FDA) for treatment of myelodysplastic syndrome and acute myeloid leukemia. In osteosarcoma, Rb and p53 pathway genes are hypermethylated, and increased expression occurs in response to DNMTi treatment [85]. Accordingly, proliferation of osteosarcoma cell lines is inhibited by several DNMTi, which may also induce osteoblastic differentiation [78]. Though DNMTi demonstrate limited clinical activity in solid tumors to date, both azacitidine and decitabine are in early-phase clinical trials for osteosarcoma [93, 96].

BETi, which reversibly bind to the bromodomains of the "reader" BET proteins, comprise a third class of broad reprogrammers [47]. This group has generated considerable interest and has recently entered early-phase clinical trials for a number of hematologic malignancies and solid tumors [33]. The efficacy of BETi in osteosarcoma preclinical models is promising. For example, the BETi JQ1 reduces cell viability of osteosarcoma cell lines and xenografts through both direct inhibition of tumor cells and disruption of the osteoblastic and osteoclastic signaling required for local osteolysis and tumor progression [61]. BETi are also effective in combination with the standard-of-care agent doxorubicin and can target the metastatic phenotype in osteosarcoma [2, 84]. Similarly, the BETi JQ1 can overcome resistance to CDK12 inhibition in an osteosarcoma ex vivo model [3]. These studies support future clinical investigation, as no clinical trials are currently evaluating BETi in osteosarcoma.

Targeted Reprogrammers

The emergence of inhibitors to target specific activating mutations, such as imatinib for the Bcr-Abl tyrosine kinase in the treatment of chronic myelogenous leukemia, changed the way cancer therapeutics are developed [13]. A considerable emphasis is now placed on identifying activating mutations in cancer, which are more amendable to targeted therapy than deactivating mutations. This has led to the development of drugs inhibiting specific genetic defects in epigenetic writers, erasers, and readers [47]. The H3K27 histone N-methyltransferase (EZH2) is activated by mutations in lymphomas, and upregulation in osteosarcoma is associated with a poor prognosis [83, 120]. EZH2 inhibitors show promise in preclinical studies and are in early-phase clinical trials for lymphomas and sarcomas with positive genetic testing [94]. Further examples of targeted epigenetic therapy exist and include inhibitors of the H3K79 N-methyltransferase (DOT1L) and lysine-specific histone demethylase 1 (LSD1), which are both in early-phase clinical trials for hematologic malignancies [47]. LSD1 is overexpressed in primary osteosarcoma samples and treatment with the LSD1 inhibitor tranylcypromine reduced osteosarcoma cell growth in vitro [4]. The LDS1 inhibitor seclidemstat is currently in clinical trials for Ewing sarcoma [95].

Histone Acetylation in Cancer

In 1964, shortly after histone acetylation was first described, Allfrey and Mirsky proposed that post-translational histone acetylation was a "dynamic and reversible mechanism for activation as well as repression of RNA synthesis" [1, 129]. Over half a century later, the field of histone modifications has expanded, and there is growing recognition for the great diversity in structure and function of these added groups. The classic histone modification is acetylation of the ε-amine group on the amino acid Lys, which neutralizes the positive charge of Lys residues and interferes with the electrostatic interaction between histones and negatively charged DNA. This generally permits a chromatin state that is open and available for transcription. Similarly, histones can be modified by short-chain Lys acylations that are distinct from Lys acetylation in hydrocarbon chain length, hydrophobicity, or charge [113]. Other described histone modifications include methylation, phosphorylation, ubiquitylation, and hydroxylation, as reviewed previously [25].

Histone Acetyltransferases

Histone acetylation is mediated by three major families of HATs: GNAT (Gcn5-related *N*-acetyltransferase), MYST (Moz, Ybf2, Sas2, and Tip60), and p300/CREB-binding protein (p300/CBP) [113]. The role of HATs in cancer is controversial and is likely both HAT and cancer specific [26]. Germ line, somatic, and translocation-driven mutations in HAT genes are present in a number of malignancies, and considerable preclinical data suggests a tumor suppressive role for HATs overall [45]. However, several studies demonstrate an oncogenic role for specific HATs in some cancers, including the recent description of a novel HAT inhibitor with antitumor activity in both hematologic malignancies and prostate cancer [63]. Research considering the direct effects of HATs in osteosarcoma is limited.

Histone Deacetylases

The human genome encodes 18 HDACs, which can be grouped into two major families: the classical Zn^{2+}-dependent HDACs and NAD^{+}-dependent sirtuin HDACs (Table 4.1). Classical HDACs are further divided into four classes: class I, class IIa, class IIb, and class IV. Class I, class IIa, and class IV HDACs are primarily localized to the nucleus, while class IIb HDACs are predominately cytoplasmic [38, 89, 147]. Sirtuin deacetylases are class III HDACs and localize to different cellular compartments, including mitochondria [42]. Most, if not all, HDACs participate in shuttling between the nucleus and cytosol. The cytoplasmic class IIb HDACs, HDAC6 and HDAC10, shuttle to the nucleus to modulate histone acetylation and gene expression [72, 148]. Similarly, the nuclear class IIa and many sirtuin HDACs shuttle to the cytoplasm to deacetylate nonhistone targets [42, 147]. HDACs can therefore regulate many critical cellular processes, including survival, aging, stress response, metabolism, cell cycle, apoptosis, DNA-damage response, metastasis, autophagy, and angiogenesis, as reviewed previously [69].

Considering the importance of these pathways to oncogenesis, the role of HDACs in cancer is not surprising. This is supported by high HDAC expression in both solid and hematologic malignancies and the association of high HDAC expression with poor survival [69]. Knockdown of specific HDACs can result in decreased tumor activity in vitro and in vivo [31, 48, 108]. For example, in colorectal tumorigenesis, mice expressing inactive HDAC2 have reduced tumor rates, compared to wild-type HDAC2 mice, when crossed with mice expressing mutant adenomatous polyposis coli (Apc) [157]. These findings are probably secondary to more than epigenetic changes alone. For example, both HATs and

Table 4.1 Classification of histone deacetylases (HDACs) and their cellular function in cancer

Class	Members	Localization	Cellular function in cancer	References
I	HDAC1	Nucleus	Cell cycle, apoptosis, DNA-damage response, metastasis, angiogenesis, autophagy	[81, 82, 109, 130, 142, 150]
I	HDAC2	Nucleus	Cell cycle, apoptosis, DNA-damage response, metastasis, angiogenesis, autophagy	[81, 82, 107, 109, 130, 142]
I	HDAC3	Nucleus	Cell cycle, apoptosis, DNA-damage response, angiogenesis	[67, 81, 107, 109]
I	HDAC8	Nucleus	Apoptosis	[50]
IIA	HDAC4	Nucleus	DNA-damage response, angiogenesis	[32, 51, 115]
IIA	HDAC5	Nucleus	Cell cycle, angiogenesis	[31, 115]
IIA	HDAC7	Nucleus	Angiogenesis	[52, 128]
IIA	HDAC9	Nucleus	DNA-damage response, angiogenesis	[49, 51]
IIB	HDAC6	Cytoplasm	Cell cycle, DNA-damage response, angiogenesis, autophagy	[16, 51, 64, 75, 153]
IIB	HDAC10	Cytoplasm	Cell cycle, DNA-damage response, autophagy	[51, 68, 97, 105]
III	SIRT1	Nucleus	DNA-damage response, metastasis, angiogenesis, autophagy	[11, 43, 71, 103]
III	SIRT2	Cytoplasm	–	
III	SIRT3	Mitochondria	Autophagy	[127]
III	SIRT4	Mitochondria	–	
III	SIRT5	Mitochondria	Autophagy	[100]
III	SIRT6	Nucleus	DNA-damage response, autophagy	[122, 125]
III	SIRT7	Nucleolus	–	
IV	HDAC11	Nucleus	–	

Table 4.2 Histone deacetylase inhibitors (HDACis) currently FDA (US Food and Drug Administration)-approved for treatment of cancer (as of August 2019; https://www.accessdata.fda.gov/scripts/cder/daf)

FDA-approved HDACis	FDA approval	Approved indications
Vorinostat (Zolinza, SAHA)	2006	Cutaneous T-cell lymphoma
Romidepsin (Istodax, Depsipeptide, FK228)	2009	Cutaneous T-cell lymphoma Peripheral T-cell lymphoma
Belinostat (Beleodaq, PDX101)	2014	Peripheral T-cell lymphoma
Panobinostat (Farydak, LBH-589)	2015	Multiple myeloma

HDACs are responsible for nonhistone protein acetylation in many well-described cancer pathways, as discussed later [89]. Additionally, HDACs can be recruited by mutant multiprotein complexes in leukemia to suppress specific genes and drive cancer progression [70].

Much like HATs, evidence does exist to suggest a tumor suppressive role for HDACs in the appropriate setting. For example, low expression of HDAC10 is associated with poor survival in lung and gastric cancer patients [46, 99]. Similarly, liver-specific knockout of HDAC3 resulted in hepatocellular carcinoma, and knockout of HDAC1 and HDAC2 resulted in hematologic malignancies [7, 28]. This suggests a dual role for HATs and HDACs in the development of cancer, one that is dependent on the distinct HAT or HDAC class member, its interactions with multiprotein complexes, acetylation of nonhistone proteins, and the unique genomics of the cell type involved.

Nonhistone Protein Acetylation in Cancer

Advances in mass spectroscopy over the past two decades led to recognition of the vast number of acetylated nonhistone proteins. As a result, historically designated HATs and HDACs are increasingly referred to as lysine acetyltransferases (KATs) and lysine deacetylases (KDACs) in appreciation of the extent of nonhistone acetylation by these enzymes [30]. In this review, the terms HAT and HDACs are continued to maintain consistency while discussing the more persistently named HDACi. However, the importance of nonhistone protein acetylation – which occurs on tens of thousands of nonhistone proteins – cannot be understated [22].

A recent review on nonhistone protein acetylation detailed functional acetylation networks generated from an exhaustive literature search and made several key observations: (1) over 40% of identified acetylated nonhistone proteins are involved in transcriptional regulation; (2) only five HATs (CBP, p300, GCN5, PCAF, and Tip60) are responsible for 90% of acetylated nonhistone proteins; and (3) more than two-thirds of nonhistone acetylation sites are targets of sirtuin deacetylases [89]. Still, most cellular processes, including those critical in cancer, are influenced by acetylation through its effect on regulation of protein enzymatic activity, degradation, DNA binding, subcellular location, and interaction with other proteins. The acetylation of p53 – the first recognized nonhistone target of HATs – for example, improves its DNA binding activity and results in increased expression of p53 target genes [34]. p53 acetylation also directly competes with MDM2-mediated ubiquitination and prevents p53 degradation in proteasomes [66]. Another example is acetylation of Rb, which hinders cyclin-dependent phosphorylation and cell cycle progression [18]. Considering the importance of p53 and Rb pathways, the promotion of nonhistone acetylation by HDACi offers another pathway to target cancer.

Histone Deacetylase Inhibitors in Cancer

Many HDACi have been identified and characterized that target one or more of the classic Zn^{2+}-dependent class I, II, and IV enzymes. Inhibitors of the class III NAD^{+}-dependent sirtuins are less developed. The structure of classic HDACi requires a "surface recognition domain" linked to a "zinc-binding group" to block the active enzymatic site of HDACs [30]. The first HDACi identified, vorinostat, was derived from hydroxamate and is a nonspecific inhibitor of HDAC activity. Efforts since have focused on increasing specificity and selectivity of HDACi by incorporating other zinc-binding groups, including derivatives of benzamides, thiols, sulphamides, ketones, and trithiocarbonates [37]. As a result, class- and isoform-selective HDACi are emerging with great interest in their potential [69, 80]. Still, the most commonly used HDACi target multiple HDACs, which make it challenging to identify isoform-specific effects and tailor these agents to maximize antitumor efficacy while minimizing systemic toxicity.

Considering the broad reprogramming that occurs during HDACi therapy, it is interesting that HDACi have tumor-selective effects at all. Dawson and Kouzarides suggested that this was due to "epigenetic vulnerability" in transformed cells, resembling the well-described paradigm of "oncogene addiction" [25, 133]. In this way, tumor cells are reliant on specific epigenetic pathways for survival, while normal cells maintain redundant epigenetic regulation that allows for adaptation in the setting of epigenetic insult by HDACi [30]. Alternatively, certain cancers may co-opt HDACs to drive the expression of critical oncogenic proteins through multiprotein complexes and/or nonhistone deacetylation. HDACi therapy can therefore lead to hyperacetylation and increased activity of key proteins such as p53 and Rb, as discussed previously. Similarly, hyperacetylation of HSP90 through HDAC6 inhibition results in the degradation of oncogenic proteins, including ERBB2, BRAF, CRAF, AKT, BCR-ABL, and KIT [30]. These examples highlight the possible ways in which HDACi therapy works, but the detailed mechanisms underlying these effects remain elusive, are likely tumor-specific, and require continued investigation to develop a more complete understanding.

As of August 2019, four HDACi were FDA-approved for use in the United States (Table 4.2). Vorinostat was approved first in 2006 for cutaneous T-cell lymphoma. Romidepsin, belinostat, and panobinostat were approved since for lymphoma and multiple myeloma. Many additional agents are under development and currently in clinical trials for a variety of cancers, including sarcoma (Table 4.3). Despite promising results in hematologic malignancies, the efficacy of HDACi in clinical trials for solid tumors has been limited to date. This is probably secondary to poor tissue penetration of these agents, which is further compounded in the setting of early-phase clinical trials composed of heavily pretreated patients with refractory metastatic disease [30, 80]. Ongoing efforts have therefore focused on identifying HDACi with greater tissue penetration and selectivity, in combination with other chemotherapies, to translate promising preclinical results to the clinical setting.

Table 4.3 Histone deacetylase inhibitors (HDACis) evaluated in clinical trials for treatment of sarcoma (as of August 2019; https://www.clinicaltrials.gov)

HDACis studied for sarcoma	Clinical trial	Sarcomas included in design
Vorinostat (Zolinza, SAHA)	Phase I/II[a]	Leiomyosarcoma, soft tissue sarcoma, osteosarcoma
Romidepsin (Istodax, Depsipeptide, FK228, FR901228)	Phase I/II[a]	Soft tissue sarcoma
Belinostat (Beleodaq, PDX101)	Phase I	Soft tissue sarcoma, osteosarcoma, chondrosarcoma
Panobinostat (Farydak, LBH-589)	Phase II[a]	Soft tissue sarcoma
Mocetinostat (MGCD0103)	Phase II	Leiomyosarcoma
Chidamide	Phase II[a]	Soft tissue sarcoma
Abexinostat (PCI-24781)	Phase I/II[a]	Soft tissue sarcoma
Valproic acid (VPA)	Phase I	Soft tissue sarcoma
AR-42 (OSU-HDAC42)	Phase I	Soft tissue sarcoma
Etinostat (MS-275, SNDX-275)	Phase I	Soft tissue sarcoma, osteosarcoma, Ewing sarcoma

[a]Active clinical trial

Safety and Tolerance of Histone Deacetylase Inhibitors

Of the four FDA-approved HDACi, romidepsin and belinostat are administered intravenously, while vorinostat and panobinostat are administered orally. The half maximal inhibitory concentration (IC_{50}) of these agents is in the nanomolar range, and half-life varies from 1 to 37 hours. HDACi are generally metabolized by the liver and are used with caution in patients with hepatic impairment or in combination with hepatically cleared drugs. Renal excretion does not play a significant role in HDACi elimination. Further details concerning the pharmacodynamics and pharmacokinetics of HDACi were recently reviewed [116].

The most serious or life-threatening adverse events associated with HDACi therapy are cardiac related, including QT interval prolongation and arrhythmias. The FDA label for romidepsin and panobinostat advises baseline and periodic electrocardiogram (ECG) monitoring in patients at risk. Vorinostat had a similar warning until 2009, after additional clinical trials reported lower rates of cardiotoxicity. A review of 62 studies, with a total patient population of 3268, reported the global incidence of cardiac events for all HDACi was 28.6%, with significant variations between drugs [114]. Cardiac events were most common for romidepsin and panobinostat, but the vast majority were asymptomatic or caused mild symptoms. Overall, 90.2% of cardiac events were asymptomatic or mild, 5.4% resulted in reduction of the drug dosing, and 4.4% required interrupting treatment due to a cardiac side effect. For the 3268 patients treated in this study, there were four cardiovascular deaths (0.1%) equally distributed among the four FDA-approved HDACi. Other significant adverse events reported include myelosuppression, nausea, vomiting, diarrhea, and hepatic dysfunction [116]. With respect to osteosarcoma, it is also notable that children and adolescents should be treated with caution when using epigenetic modifiers since their epigenome may be important for development.

Newer HDACi will hopefully reduce toxicity despite early evidence suggesting that these agents share cardiac risks [116]. This may be secondary to a high degree of pharmacophore homology between the human ether-a-go-go (hERG) channel, responsible for maintaining a normal QT interval, and HDAC inhibition [117]. However, there is also evidence that HDACi more specific for class I and HDAC6 may be less cardiotoxic because of fewer transcriptional changes in cardiac-specific genes, compared to other HDACi [57]. Newer agents are therefore needed, alongside biomarkers that predict treatment response, to improve the balance between efficacy and the toxicity associated with HDACi therapy.

Preclinical Evidence for Histone Deacetylase Inhibitors in Osteosarcoma

Emergence of HDACi in Drug Screening Studies for Osteosarcoma

As a relatively rare cancer, osteosarcoma presents a unique challenge to disease-specific drug discovery because of prohibitive development cost and the scarcity of clinical trials. Attention has therefore focused on robust preclinical screening of existing drug libraries to identify promising leads ahead of clinical trials. In one such example, 54 agents were screened at clinically achievable concentrations in five osteosarcoma cell lines. The ten most effective drugs, which included HDACi romidepsin and panobinostat, were tested in two-drug combinations. For both in vitro cell line and primary cell culture experiments, combinations of HDACi and proteasome inhibitors had the greatest effect [151]. We reported similar findings in a screen of 114 FDA-approved drugs using a three-dimensional sarcosphere model of micrometastatic disease. The HDACi romidepsin and the proteasome inhibitors bortezomib and carfilzomib were among the most effective agents studied [23].

More recently, orthotopic patient-derived osteosarcoma xenografts were developed and screened against 373 drugs using expanded short-term cultures alongside commercial osteosarcoma cell lines. Osteosarcoma cells were most sensitive to HDACi, which also demonstrated additive effects when combined with gemcitabine and doxorubicin [73]. Similar efficacy for HDACi were reported in screening studies for Ewing sarcoma and synovial sarcoma [62, 102].

Using an alternative approach to drug screening, the pediatric preclinical testing consortium (PPTC) screens individual agents across multiple tumor cell lines in vitro followed by in vivo screening in subcutaneous xenograft murine models of pediatric cancers. This program routinely includes osteosarcoma xenografts and has screened more than 80 agents over the past decade, including the HDACi vorinostat and quisinostat. Vorinostat demonstrated modest inhibitory activity in vitro with no objective responses for any of the solid tumor or leukemia xenografts, including four osteosarcoma xenografts [54]. Quisinostat demonstrated potent cytotoxic activity in vitro but similarly had no objective responses in six osteosarcoma xenografts and poor activity in solid tumor xenografts overall [15]. To date, the PPTC has not evaluated the most effective HDACi identified in the beforementioned multi-agent screening studies, romidepsin and panobinostat, or considered combination therapy with HDACi. This may reconcile the conflicting results observed in PPTC and other studies. Another possibility is that the osteosarcoma samples included in the PPTC are resistant to HDACi or that tissue penetration into their in vivo model of osteosarcoma is inadequate. Still, efficacy for HDACi in osteosarcoma has been described in a number of in vitro and in vivo studies, with mechanistic insights, as detailed in the following sections [8–10, 12, 14, 19–21, 27, 29, 44, 53, 56, 58, 59, 65, 86, 87, 104, 106, 112, 119, 124, 126, 131, 132, 134, 136–141, 143, 144, 146, 149, 152, 155, 156].

Mechanisms of HDACi Efficacy in Osteosarcoma

Consistent with other cancers, many cellular processes critical to oncogenesis are impaired by HDACi therapy in osteosarcoma, including cell cycle regulation, apoptosis, invasion and migration, angiogenesis, and immunomodulation. Treatment of osteosarcoma cell lines with the HDACi trichostatin A upregulates p53 expression and promotes apoptosis in a dose-dependent manner [27]. Similarly, the HDACi sodium butyrate enhances p53 expression and decreases MDM2 expression, resulting in apoptosis in osteosarcoma cell lines and murine xenografts [141]. Direct pathways to apoptosis after HDACi therapy are also described for osteosarcoma. The extrinsic apoptotic pathway, for example, requires activation through the transmembrane Fas receptor in response to extracellular Fas ligand (FasL). Treatment of osteosarcoma cell lines with the HDACi entinostat increases signaling through the Fas receptor and inhibits growth in both cell culture and after intranasal administration of etinostat in mice with lung metastases [58]. These effects are not related to increased expression of the transmembrane Fas receptor but are the result of HDACi-induced downregulation of cellular FLIP, an inhibitor of Fas-mediated activation of caspase-8. The HDACi romidepsin and etinostat inhibit generation of FLIP mRNA and thereby sensitize osteosarcoma cells to FasL-induced cell death [106, 132]. Another HDACi, valproic acid, may further suppress growth in osteosarcoma cell lines by decreasing the expression of soluble extracellular Fas and increasing the FasL available for activation of the transmembrane Fas receptor and apoptosis [144]. Overall, these findings demonstrate the ability of HDACi to sensitize osteosarcoma tumors to apoptotic cell death.

Epigenetic modifications are implicated in the interaction of osteosarcoma cells with the surrounding microenvironment, including the metastatic phenotype [84]. Specific to HDACi, treatment of osteosarcoma cell lines with the HDACi vorinostat diminishes their ability to produce invadopodia and migrate in invasion assays, which may be mediated by downregulation of mTOR and ALDH1 genes [86]. Similarly, direct inhibition of HDAC2 by small-interfering RNAs decreased migration in osteosarcoma cell lines through an IL-6-mediated pathway [65]. Expression of IL-8, a potent promoter of cell migration and angiogenesis, is decreased by HDAC6 inhibition in osteosarcoma cells [21]. Angiogenesis is also inhibited by the HDACi valproic acid and trichostatin A by increasing the level of vascular endothelial growth factor inhibitor (VEGI) [143]. Valproic acid, when combined with the DNA methylation inhibitor hydralazine hydrochloride, further enhanced the inhibitory effect on vascular tube formation by VEGI autocrine and paracrine pathways [59].

Immunomodulatory effects of HDACi are also increasingly recognized in osteosarcoma. Administration of valproic acid to osteosarcoma cell lines induces acetylation of histones bound to MHC class I promoters and increases cell-surface expression of MHC class I-related proteins, resulting in greater NK cell-mediated cytotoxicity [146]. The HDACi etinostat acts similarly, through histone acetylation, to enhance MHC class I expression on tumor cells and also increases expression of NKG2D on NK cells, a major recognition receptor for the detection and elimination of transformed cells. The result is enhanced NK cell killing of osteosarcoma cell lines and xenografts after etinostat therapy [156]. T cell-mediated tumor cytotoxicity may also rely on HDAC expression. Selective inhibition of HDAC6 in osteosarcoma cell lines treated with HDACi tubastatin A and nexturastat A downregulates the expression of program death receptor ligand-1 (PD-L1), which activates the PD-1 pathway to prevent T cell-mediated cytotoxicity [53].

Indirect evidence highlighting the importance of HDAC function and the potential of HDACi therapy also exists, considering the targets of tumor suppressive microRNAs (miR) in osteosarcoma. Lower expression of miR-140 is observed in osteosarcoma tissues, and restoring miR-140 expression induces apoptosis in vitro and suppresses tumor growth in vivo. Bioinformatic analysis predicted the miR-140 target HDAC4, which has expression closely linked to miR-140 levels [140]. Similar effects on HDAC4 were seen with miR-145-3p expression [138]. Finally, miR-133b attenuates osteosarcoma cell proliferation and, when overexpressed, suppresses Sirt1 levels. Forced expression of Sirt1 can partly rescue the inhibitory effect of miR-133b in osteosarcoma cells [149]. Taken together, these examples represent the number of ways in which HDACi can influence cellular behavior in osteosarcoma. As demonstrated in other cancers, the implied mechanisms are far reaching and consistent with the broad activity of HDACi.

Clinical Evidence for Histone Deacetylase Inhibitors in Osteosarcoma

HDACs as Biomarkers in Osteosarcoma

Epigenetic biomarkers are increasingly commercially available and translatable in many cancers to inform the risk of disease, diagnosis, prognosis, or response to treatment [5, 123]. Available evidence in primary human osteosarcoma tissues suggests that HDAC expression is increased and can be prognostic. In one study, patient osteosarcoma samples expressed HDAC1 and HDAC2 at high levels, but low HDAC3 compared to osteoblasts. Interestingly, while HDAC expression was increased overall, low levels were associated with advanced disease and poor survival [17]. In a related genomic database analysis, HDAC2 levels were also noted to be high in osteosarcoma, but increased expression was associated with poor survival [65]. Other studies have also noted overexpression of HDACs in patient tissues, including class I HDACs and SIRT1, but did not correlate levels with clinical outcomes [27, 154]. This evidence, though limited to small patient cohorts and specific HDACs, is largely consistent with the emerging role of epigenetics in osteosarcoma. Future research could identify HDAC-related biomarkers that could aid in prognostication and personalize the prescription of HDACi to minimize toxicity.

HDACi in Clinical Trials for Osteosarcoma

No clinical trials to date have evaluated HDACi in an osteosarcoma-specific study. Two completed phase I clinical trials treated patients with the HDACi belinostat for primary or metastatic solid tumors refractory to standard therapy, including osteosarcoma [91, 92]. However, both trials were industry-run and have no posted or published results at the time of this publication. A phase I study of the HDACi etinostat in patients with refractory solid tumors and lymphoid malignancies did include osteosarcoma in the inclusion criteria, though only two "sarcoma" patients were recruited without further histologic specification [60, 90]. Etinostat was well tolerated in this study but had poor efficacy, as no patients had a partial or complete response to therapy. Another phase I study of the HDACi vorinostat in combination with bortezomib for children with refractory or recurrent solid tumors showed good tolerability but no response in the one osteosarcoma patient included [88]. Considering the lack of clinical experience using HDACi in osteosarcoma, its efficacy in osteosarcoma patients is uncertain.

Combination Therapy with Histone Deacetylase Inhibitors in Osteosarcoma

The combination of HDACi with other anticancer agents has led to promising preclinical results and provides opportunities to mitigate drug resistance and minimize toxicity [121]. For osteosarcoma, the most likely clinical application for HDACi is in combination with standard of care therapy, consisting of methotrexate, doxorubicin, and cisplatin (MAP). This longstanding regimen causes cellular stress by reducing nucleoside pools and inducing DNA damage, resulting in apoptosis. HDACi may enhance tumor suppressor activity in the setting of cellular stress and further cell death. Not surprisingly, HDAC inhibitors show additive effects when combined with the antimetabolite gemcitabine and the topoisomerase inhibitor doxorubicin in patient-derived osteosarcoma cultures [73]. Similarly, pretreatment of osteosarcoma cell lines with the HDACi valproic acid followed by doxorubicin is superior to either agent alone [135]. The HDACi AR-42 synergizes with doxorubicin in osteosarcoma cell lines [87]. However, synergy for vorinostat and doxorubicin was cell line dependent in one study, while cisplatin was uniformly synergistic with vorinostat [104].

DNA damage can also be generated by radiation. Though infrequently used in osteosarcoma, a series of preclinical studies demonstrates that radiation combined with HDACi can improve efficacy, compared to radiation alone. In osteosarcoma cell lines, pretreatment with either vorinostat or valproic acid increases radiosensitization by attenuating radiation-induced DNA repair protein expression, including Rad51 and Ku80 [8]. These findings were confirmed in a murine osteosarcoma model where radiation and HDACi combination therapy reduces osteosarcoma proliferation, angiogenesis, and increases apoptosis with elevated expression of p53 and p21 [9, 10].

Perhaps the most rational and successful combination therapy involving HDACi includes proteasome inhibitors. The degradation of ubiquitinated proteins in the proteasome is essential in tumor cells, and impairment can lead to cellular stress and apoptosis [121]. HDACi are thought to inhibit the proteasome pathway in osteosarcoma, and proteasome inhibition decreases HDAC levels in multiple myeloma [55, 137]. Indeed, biologic synergy has been proposed in multiple myeloma where clinical trials demonstrate efficacy with romidepsin and bortezomib [40, 41]. In osteosarcoma, HDACi and proteasome inhibitors demonstrated excellent activity together in a preclinical screen [151]. As already discussed, a pediatric phase I clinical trial reported that vorinostat and bortezomib were well tolerated in combination, though the one osteosarcoma patient included did not have a response [88].

Dual epigenetic therapy with HDACi and DNMTi has been successful in hematologic malignancies [35]. Treatment of osteosarcoma cell lines with the HDACi 4-phenylbutyrate and the DNMTi

5-azacytidine is associated with genome-wide gene expression changes, including induction of proapoptotic genes and pronounced cytotoxicity [124]. Similar results are seen with valproic acid and the DNMTi hydralazine, acting through the extrinsic apoptotic pathway and by reducing angiogenesis [59, 145]. Combination of trichostatin A and 5-azacytidine results in more efficient reexpression of key apoptotic and differentiation pathways and can overcome multidrug resistance in osteosarcoma cell lines [14].

Many additional agents have been studied in combination with HDACi for other cancers with limited or no evidence in osteosarcoma, including drugs that target DNA repair pathways, hormonal therapy, immunotherapy, and tyrosine kinase inhibitors [121]. Future preclinical research on HDACi for the treatment of osteosarcoma should consider the effects of HDACi in combination with other agents. Not only is combination therapy with MAP the most likely clinical scenario, but this approach also provides the greatest potential to address the challenges associated with HDACi therapy discussed next.

Challenges in the Development of Histone Deacetylase Inhibitor Therapy for Osteosarcoma

Lack of efficacy to date for HDACi as monotherapies against solid tumors is well documented, despite promising results in many hematologic malignancies [36, 69]. Although it is not clear why this discrepancy exists, it may be due to ineffectively low drug concentrations since HDACi appear potent in many preclinical studies, where there are fewer barriers to drug delivery. Ongoing efforts to improve HDACi delivery are therefore focused on local administration and targeted drug design. Local delivery of HDACi may be enhanced by topical application in the setting of cutaneous malignancies, intratumoral injection, and surgically placed biodegradable polymers [36]. None of these techniques have been applied in osteosarcoma. Targeted drug design can incorporate ligands known to selectively accumulate within tumors, such as conjugating HDACi to folic acid derivatives to engage folate receptors overexpressed on the cell surface of solid tumors [118]. Alternatively, combination therapy may bypass limitations in drug delivery by decreasing the effective concentration required for HDACi through synergistic interactions with other agents.

Approaches to improve tissue-specific delivery may mitigate the dose-limiting adverse events associated with HDACi therapy, including cardiotoxicity, myelosuppression, nausea, vomiting, and hepatic dysfunction. This can also be accomplished by developing greater isoform selectivity with newer HDACi to minimize off-target effects. Another described strategy is to weaken HDACi binding to hERG, the critical pathway thought to be involved in HDACi-mediated cardiotoxicity [36]. Finally, with the emergence of HDAC levels as biomarkers, patient selection for HDAC therapy can be improved and the risk-benefit ratio optimized.

Despite the broad effects of HDACi, mechanisms of tumor cell resistance are described for many cancers, including osteosarcoma. Drug efflux, overexpression of pro-survival BCL-2 family proteins, JAK-STAT signaling, and antioxidants have all been implicated in HDACi resistance [30]. In osteosarcoma, doxorubicin-resistant osteosarcoma cell lines are resistant to romidepsin through upregulation of the multidrug resistance transporters P-glycoprotein (P-gp) and multidrug resistance-associated protein 1 (MRP1) [79, 98]. Inhibition of P-gp and MRP1 reverses romidepsin resistance. These findings underscore the importance of combination therapy to combat drug resistance and the utility of biomarkers and personalized approaches to assess HDACi sensitivity prior to administration. Three-dimensional osteosarcoma models, for example, including osteosarcoma spheroids (sarcospheres), offer a potential pathway to effectively screen patient biopsy samples and tailor chemotherapy accordingly [23, 24].

The final and perhaps most significant barrier to the development of HDACi therapies in osteosarcoma is the scarcity of disease-specific clinical trials, particularly those that contain patients with early or localized disease that is not refractory to standard therapies. To overcome this barrier, improvements in HDACi will have to be coupled with substantial preclinical evidence and early-phase clinical trials demonstrating both safety and efficacy in osteosarcoma.

Conclusions

The emergence of genomic therapies has been met with great enthusiasm by those studying osteosarcoma, owing to the relatively limited activity of other newer agents and the significant epigenetic dysregulation observed in osteosarcoma tumors. The resulting research to date is largely limited to preclinical studies, which deliver a positive outlook for HDACi in osteosarcoma. However, these results must be considered in the context of high attrition rates in the drug discovery pipeline. Future research should aim to address the challenges facing HDACi for osteosarcoma, including drug delivery, toxicity, resistance, and the availability of clinical trials. Critical to these efforts will be the development of more selective HDACi and combination therapy.

References

1. Allfrey VG, Mirsky AE (1964) Structural modifications of histones and their possible role in the regulation of RNA synthesis. Science 144(3618):559–559
2. Baker EK, Taylor S, Gupte A, Sharp PP, Walia M, Walsh NC, Zannettino AC, Chalk AM, Burns CJ, Walkley CR (2015) BET inhibitors induce apoptosis through a MYC independent mechanism and synergise with CDK inhibitors to kill osteosarcoma cells. Sci Rep 5:10120. https://doi.org/10.1038/srep10120
3. Bayles I, Krajewska M, Pontius WD, Saiakhova A, Morrow JJ, Bartels C, Lu J, Faber ZJ, Fedorov Y, Hong ES, Karnuta JM, Rubin B, Adams DJ, George RE, Scacheri PC (2019) Ex vivo screen identifies CDK12 as a metastatic vulnerability in osteosarcoma. J Clin Invest. https://doi.org/10.1172/JCI127718
4. Bennani-Baiti IM, Machado I, Llombart-Bosch A, Kovar H (2012) Lysine-specific demethylase 1 (LSD1/KDM1A/AOF2/BHC110) is expressed and is an epigenetic drug target in chondrosarcoma, Ewing's sarcoma, osteosarcoma, and rhabdomyosarcoma. Hum Pathol 43(8):1300–1307
5. Berdasco M, Esteller M (2019) Clinical epigenetics: seizing opportunities for translation. Nat Rev Genet 20(2):109–127. https://doi.org/10.1038/s41576-018-0074-2
6. Bhadury J, Nilsson LM, Muralidharan SV, Green LC, Li Z, Gesner EM, Hansen HC, Keller UB, McLure KG, Nilsson JA (2014) BET and HDAC inhibitors induce similar genes and biological effects and synergize to kill in Myc-induced murine lymphoma. Proc Natl Acad Sci U S A 111(26):E2721–E2730. https://doi.org/10.1073/pnas.1406722111
7. Bhaskara S, Knutson SK, Jiang G, Chandrasekharan MB, Wilson AJ, Zheng S, Yenamandra A, Locke K, Yuan J-L, Bonine-Summers AR (2010) Hdac3 is essential for the maintenance of chromatin structure and genome stability. Cancer Cell 18(5):436–447
8. Blattmann C, Oertel S, Ehemann V, Thiemann M, Huber PE, Bischof M, Witt O, Deubzer HE, Kulozik AE, Debus J, Weber KJ (2010) Enhancement of radiation response in osteosarcoma and rhabdomyosarcoma cell lines by histone deacetylase inhibition. Int J Radiat Oncol Biol Phys 78(1):237–245. https://doi.org/10.1016/j.ijrobp.2010.03.010
9. Blattmann C, Oertel S, Thiemann M, Dittmar A, Roth E, Kulozik AE, Ehemann V, Weichert W, Huber PE, Stenzinger A, Debus J (2015) Histone deacetylase inhibition sensitizes osteosarcoma to heavy ion radiotherapy. Radiat Oncol (London, England) 10:146. https://doi.org/10.1186/s13014-015-0455-z
10. Blattmann C, Thiemann M, Stenzinger A, Christmann A, Roth E, Ehemann V, Debus J, Kulozik AE, Weichert W, Huber PE, Oertel S, Abdollahi A (2013) Radiosensitization by histone deacetylase inhibition in an osteosarcoma mouse model. Strahlentherapie und Onkologie: Organ der Deutschen Rontgengesellschaft [et al] 189(11):957–966. https://doi.org/10.1007/s00066-013-0372-8
11. Byles V, Zhu L, Lovaas J, Chmilewski L, Wang J, Faller D, Dai Y (2012) SIRT1 induces EMT by cooperating with EMT transcription factors and enhances prostate cancer cell migration and metastasis. Oncogene 31(43):4619
12. Cain JE, McCaw A, Jayasekara WS, Rossello FJ, Marini KD, Irving AT, Kansara M, Thomas DM, Ashley DM, Watkins DN (2013) Sustained low-dose treatment with the histone deacetylase inhibitor LBH589 induces terminal differentiation of osteosarcoma cells. Sarcoma 2013:608964. https://doi.org/10.1155/2013/608964
13. Capdeville R, Buchdunger E, Zimmermann J, Matter A (2002) Glivec (STI571, imatinib), a rationally developed, targeted anticancer drug. Nat Rev Drug Discov 1(7):493–502. https://doi.org/10.1038/nrd839
14. Capobianco E, Mora A, La Sala D, Roberti A, Zaki N, Badidi E, Taranta M, Cinti C (2014) Separate and combined effects of DNMT and HDAC inhibitors in treating human multi-drug resistant osteosarcoma HosDXR150 cell line. PLoS One 9(4):e95596. https://doi.org/10.1371/journal.pone.0095596
15. Carol H, Gorlick R, Kolb EA, Morton CL, Manesh DM, Keir ST, Reynolds CP, Kang MH, Maris JM, Wozniak A, Hickson I, Lyalin D, Kurmasheva RT,

Houghton PJ, Smith MA, Lock R (2014) Initial testing (stage 1) of the histone deacetylase inhibitor, quisinostat (JNJ-26481585), by the Pediatric Preclinical Testing Program. Pediatr Blood Cancer 61(2):245–252. https://doi.org/10.1002/pbc.24724
16. Cha T-L, Chuang M-J, Wu S-T, Sun G-H, Chang S-Y, Yu D-S, Huang S-M, Huan SK-H, Cheng T-C, Chen T-T (2009) Dual degradation of aurora A and B kinases by the histone deacetylase inhibitor LBH589 induces G2-M arrest and apoptosis of renal cancer cells. Clin Cancer Res 15(3):840–850
17. Chaiyawat P, Pruksakorn D, Phanphaisarn A, Teeyakasem P, Klangjorhor J, Settakorn J (2018) Expression patterns of class I histone deacetylases in osteosarcoma: a novel prognostic marker with potential therapeutic implications. Mod Pathol 31(2):264–274. https://doi.org/10.1038/modpathol.2017.125
18. Chan HM, Krstic-Demonacos M, Smith L, Demonacos C, La Thangue NB (2001) Acetylation control of the retinoblastoma tumour-suppressor protein. Nat Cell Biol 3(7):667–674. https://doi.org/10.1038/35083062
19. Chen J, Xia J, Yu YL, Wang SQ, Wei YB, Chen FY, Huang GY, Shi JS (2014) HDAC5 promotes osteosarcoma progression by upregulation of Twist 1 expression. Tumour Biol 35(2):1383–1387. https://doi.org/10.1007/s13277-013-1189-x
20. Cheng DD, Yang QC, Zhang ZC, Yang CX, Liu YW (2012) Antitumor activity of histone deacetylase inhibitor trichostatin A in osteosarcoma cells. Asian Pac J Cancer Prev 13(4):1395–1399. https://doi.org/10.7314/apjcp.2012.13.4.1395
21. Cheng M, Cai W, Huang W, Chen Y, Wu Z, Luo P, Yan W (2018) Histone deacetylase 6 regulated expression of IL-8 is involved in the doxorubicin (Dox) resistance of osteosarcoma cells via modulating ABCB1 transcription. Eur J Pharmacol 840:1–8. https://doi.org/10.1016/j.ejphar.2018.09.032
22. Choudhary C, Weinert BT, Nishida Y, Verdin E, Mann M (2014) The growing landscape of lysine acetylation links metabolism and cell signalling. Nat Rev Mol Cell Biol 15(8):536
23. Collier CD, Buschbach JD, Gandhi DA, Getty PJ, Greenfield EM (2016) Opportunities for drug repurposing in osteosarcoma: a screen of FDA-approved oncology drugs in a micrometastatic model of disease. Paper presented at the Orthopaedic Research Society annual meeting, Orlando
24. Collier CD, Wirtz EC, Knafler GJ, Morris WZ, Getty PJ, Greenfield EM (2018) Micrometastatic drug screening platform shows heterogeneous response to MAP chemotherapy in osteosarcoma cell lines. Clin Orthop Relat Res 476(7):1400–1411. https://doi.org/10.1007/s11999.0000000000000059
25. Dawson MA, Kouzarides T (2012) Cancer epigenetics: from mechanism to therapy. Cell 150(1):12–27. https://doi.org/10.1016/j.cell.2012.06.013
26. Demetriadou C, Kirmizis A (2017) Histone acetyltransferases in cancer: guardians or hazards? Crit Rev Oncog 22(3–4):195–218. https://doi.org/10.1615/CritRevOncog.2017024506
27. Deng Z, Liu X, Jin J, Xu H, Gao Q, Wang Y, Zhao J (2016) Histone deacetylase inhibitor trichostatin a promotes the apoptosis of osteosarcoma cells through p53 signaling pathway activation. Int J Biol Sci 12(11):1298–1308. https://doi.org/10.7150/ijbs.16569
28. Dovey OM, Foster CT, Conte N, Edwards SA, Edwards JM, Singh R, Vassiliou G, Bradley A, Cowley SM (2013) Histone deacetylase 1 and 2 are essential for normal T-cell development and genomic stability in mice. Blood 121(8):1335–1344
29. Duo J, Ma Y, Wang G, Han X, Zhang C (2013) Metformin synergistically enhances antitumor activity of histone deacetylase inhibitor trichostatin a against osteosarcoma cell line. DNA Cell Biol 32(4):156–164. https://doi.org/10.1089/dna.2012.1926
30. Falkenberg KJ, Johnstone RW (2014) Histone deacetylases and their inhibitors in cancer, neurological diseases and immune disorders. Nat Rev Drug Discov 13(9):673–691. https://doi.org/10.1038/nrd4360
31. Fan J, Lou B, Chen W, Zhang J, Lin S, Lv F-F, Chen Y (2014) Down-regulation of HDAC5 inhibits growth of human hepatocellular carcinoma by induction of apoptosis and cell cycle arrest. Tumor Biol 35(11):11523–11532
32. Geng H, Harvey CT, Pittsenbarger J, Liu Q, Beer TM, Xue C, Qian DZ (2011) HDAC4 protein regulates HIF1α protein lysine acetylation and cancer cell response to hypoxia. J Biol Chem 286(44):38095–38102
33. Genta S, Pirosa MC, Stathis A (2019) BET and EZH2 inhibitors: novel approaches for targeting cancer. Curr Oncol Rep 21(2):13. https://doi.org/10.1007/s11912-019-0762-x
34. Glozak MA, Sengupta N, Zhang X, Seto E (2005) Acetylation and deacetylation of non-histone proteins. Gene 363:15–23. https://doi.org/10.1016/j.gene.2005.09.010
35. Gore SD, Baylin S, Sugar E, Carraway H, Miller CB, Carducci M, Grever M, Galm O, Dauses T, Karp JE, Rudek MA, Zhao M, Smith BD, Manning J, Jiemjit A, Dover G, Mays A, Zwiebel J, Murgo A, Weng LJ, Herman JG (2006) Combined DNA methyltransferase and histone deacetylase inhibition in the treatment of myeloid neoplasms. Cancer Res 66(12):6361–6369. https://doi.org/10.1158/0008-5472.CAN-06-0080
36. Gryder BE, Sodji QH, Oyelere AK (2012) Targeted cancer therapy: giving histone deacetylase inhibitors all they need to succeed. Future Med Chem 4(4):505–524. https://doi.org/10.4155/fmc.12.3
37. Gupta P, Reid RC, Iyer A, Sweet MJ, Fairlie DP (2012) Towards isozyme-selective HDAC inhibitors for interrogating disease. Curr Top Med Chem 12(14):1479–1499

38. Haberland M, Montgomery RL, Olson EN (2009) The many roles of histone deacetylases in development and physiology: implications for disease and therapy. Nat Rev Genet 10(1):32–42. https://doi.org/10.1038/nrg2485
39. Harrison DJ, Geller DS, Gill JD, Lewis VO, Gorlick R (2018) Current and future therapeutic approaches for osteosarcoma. Expert Rev Anticancer Ther 18(1):39–50. https://doi.org/10.1080/14737140.2018.1413939
40. Harrison SJ, Quach H, Link E, Seymour JF, Ritchie DS, Ruell S, Dean J, Januszewicz H, Johnstone R, Neeson P, Dickinson M, Nichols J, Prince HM (2011) A high rate of durable responses with romidepsin, bortezomib, and dexamethasone in relapsed or refractory multiple myeloma. Blood 118(24):6274–6283. https://doi.org/10.1182/blood-2011-03-339879
41. Hideshima T, Richardson PG, Anderson KC (2011) Mechanism of action of proteasome inhibitors and deacetylase inhibitors and the biological basis of synergy in multiple myeloma. Mol Cancer Ther 10(11):2034–2042. https://doi.org/10.1158/1535-7163.MCT-11-0433
42. Houtkooper RH, Pirinen E, Auwerx J (2012) Sirtuins as regulators of metabolism and healthspan. Nat Rev Mol Cell Biol 13(4):225–238. https://doi.org/10.1038/nrm3293
43. Huang R, Xu Y, Wan W, Shou X, Qian J, You Z, Liu B, Chang C, Zhou T, Lippincott-Schwartz J (2015) Deacetylation of nuclear LC3 drives autophagy initiation under starvation. Mol Cell 57(3):456–466
44. Igarashi K, Yamamoto N, Hayashi K, Takeuchi A, Kimura H, Miwa S, Hoffman RM, Tsuchiya H (2016) Non-toxic efficacy of the combination of caffeine and valproic acid on human osteosarcoma cells in vitro and in orthotopic nude-mouse models. Anticancer Res 36(9):4477–4482. https://doi.org/10.21873/anticanres.10992
45. Iyer NG, Ozdag H, Caldas C (2004) p300/CBP and cancer. Oncogene 23(24):4225–4231. https://doi.org/10.1038/sj.onc.1207118
46. Jin Z, Jiang W, Jiao F, Guo Z, Hu H, Wang L, Wang L (2014) Decreased expression of histone deacetylase 10 predicts poor prognosis of gastric cancer patients. Int J Clin Exp Pathol 7(9):5872
47. Jones PA, Issa JP, Baylin S (2016) Targeting the cancer epigenome for therapy. Nat Rev Genet 17(10):630–641. https://doi.org/10.1038/nrg.2016.93
48. Jung KH, Noh JH, Kim JK, Eun JW, Bae HJ, Xie HJ, Chang YG, Kim MG, Park H, Lee JY (2012) HDAC2 overexpression confers oncogenic potential to human lung cancer cells by deregulating expression of apoptosis and cell cycle proteins. J Cell Biochem 113(6):2167–2177
49. Kaluza D, Kroll J, Gesierich S, Manavski Y, Boeckel J-N, Doebele C, Zelent A, Rössig L, Zeiher AM, Augustin HG (2013) Histone deacetylase 9 promotes angiogenesis by targeting the antiangiogenic microRNA-17–92 cluster in endothelial cells. Arterioscler Thromb Vasc Biol 33(3):533–543
50. Kang Y, Nian H, Rajendran P, Kim E, Dashwood W, Pinto J, Boardman L, Thibodeau S, Limburg P, Löhr C (2014) HDAC8 and STAT3 repress BMF gene activity in colon cancer cells. Cell Death Dis 5(10):e1476
51. Kao GD, McKenna WG, Guenther MG, Muschel RJ, Lazar MA, Yen TJ (2003) Histone deacetylase 4 interacts with 53BP1 to mediate the DNA damage response. J Cell Biol 160(7):1017–1027
52. Kato H, Tamamizu-Kato S, Shibasaki F (2004) Histone deacetylase 7 associates with hypoxia-inducible factor 1α and increases transcriptional activity. J Biol Chem 279(40):41966–41974
53. Keremu A, Aimaiti A, Liang Z, Zou X (2019) Role of the HDAC6/STAT3 pathway in regulating PD-L1 expression in osteosarcoma cell lines. Cancer Chemother Pharmacol 83(2):255–264. https://doi.org/10.1007/s00280-018-3721-6
54. Keshelava N, Houghton PJ, Morton CL, Lock RB, Carol H, Keir ST, Maris JM, Reynolds CP, Gorlick R, Kolb EA, Wu J, Smith MA (2009) Initial testing (stage 1) of vorinostat (SAHA) by the pediatric preclinical testing program. Pediatr Blood Cancer 53(3):505–508. https://doi.org/10.1002/pbc.21988
55. Kikuchi J, Wada T, Shimizu R, Izumi T, Akutsu M, Mitsunaga K, Noborio-Hatano K, Nobuyoshi M, Ozawa K, Kano Y, Furukawa Y (2010) Histone deacetylases are critical targets of bortezomib-induced cytotoxicity in multiple myeloma. Blood 116(3):406–417. https://doi.org/10.1182/blood-2009-07-235663
56. Kisseberth WC, Murahari S, London CA, Kulp SK, Chen CS (2008) Evaluation of the effects of histone deacetylase inhibitors on cells from canine cancer cell lines. Am J Vet Res 69(7):938–945. https://doi.org/10.2460/ajvr.69.7.938
57. Kopljar I, Gallacher DJ, De Bondt A, Cougnaud L, Vlaminckx E, Van den Wyngaert I, Lu HR (2016) Functional and transcriptional characterization of histone deacetylase inhibitor-mediated cardiac adverse effects in human induced pluripotent stem cell-derived cardiomyocytes. Stem Cells Transl Med 5(5):602–612. https://doi.org/10.5966/sctm.2015-0279
58. Koshkina NV, Rao-Bindal K, Kleinerman ES (2011) Effect of the histone deacetylase inhibitor SNDX-275 on Fas signaling in osteosarcoma cells and the feasibility of its topical application for the treatment of osteosarcoma lung metastases. Cancer 117(15):3457–3467. https://doi.org/10.1002/cncr.25884
59. Kumanishi S, Yamanegi K, Nishiura H, Fujihara Y, Kobayashi K, Nakasho K, Futani H, Yoshiya S (2019) Epigenetic modulators hydralazine and sodium valproate act synergistically in VEGI-mediated anti-angiogenesis and VEGF interference in human osteosarcoma and vascular endothe-

lial cells. Int J Oncol 55(1):167–178. https://doi.org/10.3892/ijo.2019.4811
60. Kummar S, Gutierrez M, Gardner ER, Donovan E, Hwang K, Chung EJ, Lee MJ, Maynard K, Kalnitskiy M, Chen A, Melillo G, Ryan QC, Conley B, Figg WD, Trepel JB, Zwiebel J, Doroshow JH, Murgo AJ (2007) Phase I trial of MS-275, a histone deacetylase inhibitor, administered weekly in refractory solid tumors and lymphoid malignancies. https://www.ncbi.nlm.nih.gov/pubmed/17875771. Accessed 18 Pt 1 13
61. Lamoureux F, Baud'huin M, Rodriguez Calleja L, Jacques C, Berreur M, Redini F, Lecanda F, Bradner JE, Heymann D, Ory B (2014) Selective inhibition of BET bromodomain epigenetic signalling interferes with the bone-associated tumour vicious cycle. Nat Commun 5:3511. https://doi.org/10.1038/ncomms4511
62. Laporte AN, Barrott JJ, Yao RJ, Poulin NM, Brodin BA, Jones KB, Underhill TM, Nielsen TO (2017) HDAC and proteasome inhibitors synergize to activate pro-apoptotic factors in synovial sarcoma. PLoS One 12(1):e0169407. https://doi.org/10.1371/journal.pone.0169407
63. Lasko LM, Jakob CG, Edalji RP, Qiu W, Montgomery D, Digiammarino EL, Hansen TM, Risi RM, Frey R, Manaves V, Shaw B, Algire M, Hessler P, Lam LT, Uziel T, Faivre E, Ferguson D, Buchanan FG, Martin RL, Torrent M, Chiang GG, Karukurichi K, Langston JW, Weinert BT, Choudhary C, de Vries P, Van Drie JH, McElligott D, Kesicki E, Marmorstein R, Sun C, Cole PA, Rosenberg SH, Michaelides MR, Lai A, Bromberg KD (2017) Discovery of a selective catalytic p300/CBP inhibitor that targets lineage-specific tumours. Nature 550(7674):128–132. https://doi.org/10.1038/nature24028
64. Lee JY, Koga H, Kawaguchi Y, Tang W, Wong E, Gao YS, Pandey UB, Kaushik S, Tresse E, Lu J (2010) HDAC6 controls autophagosome maturation essential for ubiquitin-selective quality-control autophagy. EMBO J 29(5):969–980
65. Li J, Yan X, Tang J, Wang Y, Tang J, Wu W, Liu M (2019) HDAC2-mediated upregulation of IL-6 triggers the migration of osteosarcoma cells. Cell Biol Toxicol. https://doi.org/10.1007/s10565-019-09459-7
66. Li M, Luo J, Brooks CL, Gu W (2002) Acetylation of p53 inhibits its ubiquitination by Mdm2. J Biol Chem 277(52):50607–50611. https://doi.org/10.1074/jbc.C200578200
67. Li Y, Kao GD, Garcia BA, Shabanowitz J, Hunt DF, Qin J, Phelan C, Lazar MA (2006) A novel histone deacetylase pathway regulates mitosis by modulating Aurora B kinase activity. Genes Dev 20(18):2566–2579
68. Li Y, Peng L, Seto E (2015) Histone deacetylase 10 regulates the cell cycle G2/M phase transition via a novel Let-7–HMGA2–cyclin A2 pathway. Mol Cell Biol 35(20):3547–3565
69. Li Y, Seto E (2016) HDACs and HDAC inhibitors in cancer development and therapy. Cold Spring Harb Perspect Med 6(10). https://doi.org/10.1101/cshperspect.a026831
70. Licht JD (2001) AML1 and the AML1-ETO fusion protein in the pathogenesis of t (8; 21) AML. Oncogene 20(40):5660
71. Lim J-H, Lee Y-M, Chun Y-S, Chen J, Kim J-E, Park J-W (2010) Sirtuin 1 modulates cellular responses to hypoxia by deacetylating hypoxia-inducible factor 1α. Mol Cell 38(6):864–878
72. Liu Y, Peng L, Seto E, Huang S, Qiu Y (2012) Modulation of histone deacetylase 6 (HDAC6) nuclear import and tubulin deacetylase activity through acetylation. J Biol Chem 287(34):29168–29174. https://doi.org/10.1074/jbc.M112.371120
73. Loh AHP, Stewart E, Bradley CL, Chen X, Daryani V, Stewart CF, Calabrese C, Funk A, Miller G, Karlstrom A, Krafcik F, Goshorn DR, Vogel P, Bahrami A, Shelat A, Dyer MA (2019) Combinatorial screening using orthotopic patient derived xenograft-expanded early phase cultures of osteosarcoma identify novel therapeutic drug combinations. Cancer Lett 442:262–270. https://doi.org/10.1016/j.canlet.2018.10.033
74. Lu C, Jain SU, Hoelper D, Bechet D, Molden RC, Ran L, Murphy D, Venneti S, Hameed M, Pawel BR, Wunder JS, Dickson BC, Lundgren SM, Jani KS, De Jay N, Papillon-Cavanagh S, Andrulis IL, Sawyer SL, Grynspan D, Turcotte RE, Nadaf J, Fahiminiyah S, Muir TW, Majewski J, Thompson CB, Chi P, Garcia BA, Allis CD, Jabado N, Lewis PW (2016) Histone H3K36 mutations promote sarcomagenesis through altered histone methylation landscape. Science 352(6287):844–849. https://doi.org/10.1126/science.aac7272
75. Lv Z, Weng X, Du C, Zhang C, Xiao H, Cai X, Ye S, Cheng J, Ding C, Xie H (2016) Downregulation of HDAC6 promotes angiogenesis in hepatocellular carcinoma cells and predicts poor prognosis in liver transplantation patients. Mol Carcinog 55(5):1024–1033
76. Ma X, Liu Y, Liu Y, Alexandrov LB, Edmonson MN, Gawad C, Zhou X, Li Y, Rusch MC, Easton J, Huether R, Gonzalez-Pena V, Wilkinson MR, Hermida LC, Davis S, Sioson E, Pounds S, Cao X, Ries RE, Wang Z, Chen X, Dong L, Diskin SJ, Smith MA, Guidry Auvil JM, Meltzer PS, Lau CC, Perlman EJ, Maris JM, Meshinchi S, Hunger SP, Gerhard DS, Zhang J (2018) Pan-cancer genome and transcriptome analyses of 1,699 paediatric leukaemias and solid tumours. Nature 555(7696):371–376. https://doi.org/10.1038/nature25795
77. Mack SC, Witt H, Piro RM, Gu L, Zuyderduyn S, Stutz AM, Wang X, Gallo M, Garzia L, Zayne K, Zhang X, Ramaswamy V, Jager N, Jones DT, Sill M, Pugh TJ, Ryzhova M, Wani KM, Shih DJ, Head R, Remke M, Bailey SD, Zichner T, Faria CC, Barszczyk M, Stark S, Seker-Cin H, Hutter S, Johann P, Bender S, Hovestadt V, Tzaridis T, Dubuc AM,

Northcott PA, Peacock J, Bertrand KC, Agnihotri S, Cavalli FM, Clarke I, Nethery-Brokx K, Creasy CL, Verma SK, Koster J, Wu X, Yao Y, Milde T, Sin-Chan P, Zuccaro J, Lau L, Pereira S, Castelo-Branco P, Hirst M, Marra MA, Roberts SS, Fults D, Massimi L, Cho YJ, Van Meter T, Grajkowska W, Lach B, Kulozik AE, von Deimling A, Witt O, Scherer SW, Fan X, Muraszko KM, Kool M, Pomeroy SL, Gupta N, Phillips J, Huang A, Tabori U, Hawkins C, Malkin D, Kongkham PN, Weiss WA, Jabado N, Rutka JT, Bouffet E, Korbel JO, Lupien M, Aldape KD, Bader GD, Eils R, Lichter P, Dirks PB, Pfister SM, Korshunov A, Taylor MD (2014) Epigenomic alterations define lethal CIMP-positive ependymomas of infancy. Nature 506(7489):445–450. https://doi.org/10.1038/nature13108

78. Manara MC, Valente S, Cristalli C, Nicoletti G, Landuzzi L, Zwergel C, Mazzone R, Stazi G, Arimondo PB, Pasello M, Guerzoni C, Picci P, Nanni P, Lollini PL, Mai A, Scotlandi K (2018) A quinoline-based DNA methyltransferase inhibitor as a possible adjuvant in osteosarcoma therapy. Mol Cancer Ther 17(9):1881–1892. https://doi.org/10.1158/1535-7163.MCT-17-0818
79. Matsubara H, Watanabe M, Imai T, Yui Y, Mizushima Y, Hiraumi Y, Kamitsuji Y, Watanabe K, Nishijo K, Toguchida J, Nakahata T, Adachi S (2009) Involvement of extracellular signal-regulated kinase activation in human osteosarcoma cell resistance to the histone deacetylase inhibitor FK228 [(1S,4S,7Z,10S,16E,21R)-7-ethylidene-4,21-bis(propan-2-yl)-2-oxa-12,13-dithia-5,8,20,23-tetraazabicyclo[8.7.6]tricos-16-ene-3,6,9,19,22-pentone]. J Pharmacol Exp Therap 328(3):839–848. https://doi.org/10.1124/jpet.108.147462
80. McClure JJ, Li X, Chou CJ (2018) Advances and challenges of HDAC inhibitors in cancer therapeutics. Adv Cancer Res 138:183–211. https://doi.org/10.1016/bs.acr.2018.02.006
81. Miller KM, Tjeertes JV, Coates J, Legube G, Polo SE, Britton S, Jackson SP (2010) Human HDAC1 and HDAC2 function in the DNA-damage response to promote DNA nonhomologous end-joining. Nat Struct Mol Biol 17(9):1144
82. Moresi V, Carrer M, Grueter CE, Rifki OF, Shelton JM, Richardson JA, Bassel-Duby R, Olson EN (2012) Histone deacetylases 1 and 2 regulate autophagy flux and skeletal muscle homeostasis in mice. Proc Natl Acad Sci 109(5):1649–1654
83. Morin RD, Johnson NA, Severson TM, Mungall AJ, An J, Goya R, Paul JE, Boyle M, Woolcock BW, Kuchenbauer F, Yap D, Humphries RK, Griffith OL, Shah S, Zhu H, Kimbara M, Shashkin P, Charlot JF, Tcherpakov M, Corbett R, Tam A, Varhol R, Smailus D, Moksa M, Zhao Y, Delaney A, Qian H, Birol I, Schein J, Moore R, Holt R, Horsman DE, Connors JM, Jones S, Aparicio S, Hirst M, Gascoyne RD, Marra MA (2010) Somatic mutations altering EZH2 (Tyr641) in follicular and diffuse large B-cell lymphomas of germinal-center origin. Nat Genet 42(2):181–185. https://doi.org/10.1038/ng.518
84. Morrow JJ, Bayles I, Funnell APW, Miller TE, Saiakhova A, Lizardo MM, Bartels CF, Kapteijn MY, Hung S, Mendoza A, Dhillon G, Chee DR, Myers JT, Allen F, Gambarotti M, Righi A, DiFeo A, Rubin BP, Huang AY, Meltzer PS, Helman LJ, Picci P, Versteeg HH, Stamatoyannopoulos JA, Khanna C, Scacheri PC (2018) Positively selected enhancer elements endow osteosarcoma cells with metastatic competence. Nat Med 24(2):176–185. https://doi.org/10.1038/nm.4475
85. Morrow JJ, Khanna C (2015) Osteosarcoma genetics and epigenetics: emerging biology and candidate therapies. Crit Rev Oncog 20(3–4):173–197. https://doi.org/10.1615/CritRevOncog.2015013713
86. Mu X, Brynien D, Weiss KR (2015) The HDAC inhibitor Vorinostat diminishes the in vitro metastatic behavior of Osteosarcoma cells. Biomed Res Int 2015:290368. https://doi.org/10.1155/2015/290368
87. Murahari S, Jalkanen AL, Kulp SK, Chen CS, Modiano JF, London CA, Kisseberth WC (2017) Sensitivity of osteosarcoma cells to HDAC inhibitor AR-42 mediated apoptosis. BMC Cancer 17(1):67. https://doi.org/10.1186/s12885-017-3046-6
88. Muscal JA, Thompson PA, Horton TM, Ingle AM, Ahern CH, McGovern RM, Reid JM, Ames MM, Espinoza-Delgado I, Weigel BJ, Blaney SM (2013) A phase I trial of vorinostat and bortezomib in children with refractory or recurrent solid tumors: a Children's Oncology Group phase I consortium study (ADVL0916). Pediatr Blood Cancer 60(3):390–395. https://doi.org/10.1002/pbc.24271
89. Narita T, Weinert BT, Choudhary C (2019) Functions and mechanisms of non-histone protein acetylation. Nat Rev Mol Cell Biol 20(3):156–174. https://doi.org/10.1038/s41580-018-0081-3
90. NCT00020579 MS-275 in treating patients with advanced solid tumors or lymphoma. https://ClinicalTrials.gov/show/NCT00020579
91. NCT00413075 Study of oral PXD101 in patients with advanced solid tumors or lymphoma. https://ClinicalTrials.gov/show/NCT00413075
92. NCT00413322 Study of PXD101 alone and in combination with 5-fluorouracil (5-FU) in patients with advanced solid tumors. https://ClinicalTrials.gov/show/NCT00413322
93. NCT01241162 Decitabine followed by a cancer antigen vaccine for patients with neuroblastoma and sarcoma. https://ClinicalTrials.gov/show/NCT01241162
94. NCT03213665 Tazemetostat in treating patients with relapsed or refractory advanced solid tumors, Non-Hodgkin lymphoma, or histiocytic disorders with EZH2, SMARCB1, or SMARCA4 gene mutations (A Pediatric MATCH Treatment Trial)
95. NCT03600649 Clinical trial of SP-2577 (Seclidemstat) in patients with relapsed or refractory

Ewing sarcoma. https://ClinicalTrials.gov/show/NCT03600649
96. NCT03628209 Nivolumab or nivolumab and azacitidine in patients with recurrent, resectable osteosarcoma. https://ClinicalTrials.gov/show/NCT03628209
97. Oehme I, Linke J-P, Böck BC, Milde T, Lodrini M, Hartenstein B, Wiegand I, Eckert C, Roth W, Kool M (2013) Histone deacetylase 10 promotes autophagy-mediated cell survival. Proc Natl Acad Sci 110(28):E2592–E2601
98. Okada T, Tanaka K, Nakatani F, Sakimura R, Matsunobu T, Li X, Hanada M, Nakamura T, Oda Y, Tsuneyoshi M, Iwamoto Y (2006) Involvement of P-glycoprotein and MRP1 in resistance to cyclic tetrapeptide subfamily of histone deacetylase inhibitors in the drug-resistant osteosarcoma and Ewing's sarcoma cells. Int J Cancer 118(1):90–97. https://doi.org/10.1002/ijc.21297
99. Osada H, Tatematsu Y, Saito H, Yatabe Y, Mitsudomi T, Takahashi T (2004) Reduced expression of class II histone deacetylase genes is associated with poor prognosis in lung cancer patients. Int J Cancer 112(1):26–32
100. Papa L, Germain D (2014) SirT3 regulates the mitochondrial unfolded protein response. Mol Cell Biol 34(4):699–710
101. Park PJ (2009) ChIP-seq: advantages and challenges of a maturing technology. Nat Rev Genet 10(10):669–680. https://doi.org/10.1038/nrg2641
102. Pattenden SG, Simon JM, Wali A, Jayakody CN, Troutman J, McFadden AW, Wooten J, Wood CC, Frye SV, Janzen WP, Davis IJ (2016) High-throughput small molecule screen identifies inhibitors of aberrant chromatin accessibility. Proc Natl Acad Sci U S A 113(11):3018–3023. https://doi.org/10.1073/pnas.1521827113
103. Peng L, Yuan Z, Li Y, Ling H, Izumi V, Fang B, Fukasawa K, Koomen J, Chen J, Seto E (2015) Ubiquitinated sirtuin 1 (SIRT1) function is modulated during DNA damage-induced cell death and survival. J Biol Chem 290(14):8904–8912
104. Pettke A, Hotfilder M, Clemens D, Klco-Brosius S, Schaefer C, Potratz J, Dirksen U (2016) Suberanilohydroxamic acid (vorinostat) synergistically enhances the cytotoxicity of doxorubicin and cisplatin in osteosarcoma cell lines. Anti-Cancer Drugs 27(10):1001–1010. https://doi.org/10.1097/CAD.0000000000000418
105. Radhakrishnan R, Li Y, Xiang S, Yuan F, Yuan Z, Telles E, Fang J, Coppola D, Shibata D, Lane WS (2015) Histone deacetylase 10 regulates DNA mismatch repair and may involve the deacetylation of MutS homolog 2. J Biol Chem 290(37):22795–22804
106. Rao-Bindal K, Koshkina NV, Stewart J, Kleinerman ES (2013) The histone deacetylase inhibitor, MS-275 (entinostat), downregulates c-FLIP, sensitizes osteosarcoma cells to FasL, and induces the regression of osteosarcoma lung metastases. Curr Cancer Drug Targets 13(4):411–422
107. Ray A, Alalem M, Ray BK (2013) Loss of epigenetic Kruppel-like factor 4 histone deacetylase (KLF-4-HDAC)-mediated transcriptional suppression is crucial in increasing vascular endothelial growth factor (VEGF) expression in breast cancer. J Biol Chem 288(38):27232–27242
108. Rettig I, Koeneke E, Trippel F, Mueller W, Burhenne J, Kopp-Schneider A, Fabian J, Schober A, Fernekorn U, von Deimling A (2015) Selective inhibition of HDAC8 decreases neuroblastoma growth in vitro and in vivo and enhances retinoic acid-mediated differentiation. Cell Death Dis 6(2):e1657
109. Riley J, Hutchinson R, McArt D, Crawford N, Holohan C, Paul I, Van Schaeybroeck S, Salto-Tellez M, Johnston P, Fennell DA (2013) Prognostic and therapeutic relevance of FLIP and procaspase-8 overexpression in non-small cell lung cancer. Cell Death Dis 4(12):e951
110. Roadmap Epigenomics Consortium, Kundaje A, Meuleman W, Ernst J, Bilenky M, Yen A, Heravi-Moussavi A, Kheradpour P, Zhang Z, Wang J, Ziller MJ, Amin V, Whitaker JW, Schultz MD, Ward LD, Sarkar A, Quon G, Sandstrom RS, Eaton ML, Wu YC, Pfenning AR, Wang X, Claussnitzer M, Liu Y, Coarfa C, Harris RA, Shoresh N, Epstein CB, Gjoneska E, Leung D, Xie W, Hawkins RD, Lister R, Hong C, Gascard P, Mungall AJ, Moore R, Chuah E, Tam A, Canfield TK, Hansen RS, Kaul R, Sabo PJ, Bansal MS, Carles A, Dixon JR, Farh KH, Feizi S, Karlic R, Kim AR, Kulkarni A, Li D, Lowdon R, Elliott G, Mercer TR, Neph SJ, Onuchic V, Polak P, Rajagopal N, Ray P, Sallari RC, Siebenthall KT, Sinnott-Armstrong NA, Stevens M, Thurman RE, Wu J, Zhang B, Zhou X, Beaudet AE, Boyer LA, De Jager PL, Farnham PJ, Fisher SJ, Haussler D, Jones SJ, Li W, Marra MA, McManus MT, Sunyaev S, Thomson JA, Tlsty TD, Tsai LH, Wang W, Waterland RA, Zhang MQ, Chadwick LH, Bernstein BE, Costello JF, Ecker JR, Hirst M, Meissner A, Milosavljevic A, Ren B, Stamatoyannopoulos JA, Wang T, Kellis M (2015) Integrative analysis of 111 reference human epigenomes. Nature 518(7539):317–330. https://doi.org/10.1038/nature14248
111. Roberts RD, Lizardo MM, Reed DR, Hingorani P, Glover J, Allen-Rhoades W, Fan T, Khanna C, Sweet-Cordero EA, Cash T, Bishop MW, Hegde M, Sertil AR, Koelsche C, Mirabello L, Malkin D, Sorensen PH, Meltzer PS, Janeway KA, Gorlick R, Crompton BD (2019) Provocative questions in osteosarcoma basic and translational biology: a report from the Children's Oncology Group. Cancer. https://doi.org/10.1002/cncr.32351
112. Roh MS, Kim CW, Park BS, Kim GC, Jeong JH, Kwon HC, Suh DJ, Cho KH, Yee SB, Yoo YH (2004) Mechanism of histone deacetylase inhibitor Trichostatin A induced apoptosis in human osteo-

sarcoma cells. Apoptosis 9(5):583–589. https://doi.org/10.1023/B:APPT.0000038037.68908.6e
113. Sabari BR, Zhang D, Allis CD, Zhao Y (2017) Metabolic regulation of gene expression through histone acylations. Nat Rev Mol Cell Biol 18(2):90
114. Schiattarella GG, Sannino A, Toscano E, Cattaneo F, Trimarco B, Esposito G, Perrino C (2016) Cardiovascular effects of histone deacetylase inhibitors epigenetic therapies: systematic review of 62 studies and new hypotheses for future research. Int J Cardiol 219:396–403. https://doi.org/10.1016/j.ijcard.2016.06.012
115. Seo H-W, Kim E-J, Na H, Lee M-O (2009) Transcriptional activation of hypoxia-inducible factor-1α by HDAC4 and HDAC5 involves differential recruitment of p300 and FIH-1. FEBS Lett 583(1):55–60
116. Shah RR (2019) Safety and tolerability of histone deacetylase (HDAC) inhibitors in oncology. Drug Saf 42(2):235–245. https://doi.org/10.1007/s40264-018-0773-9
117. Shultz MD, Cao X, Chen CH, Cho YS, Davis NR, Eckman J, Fan J, Fekete A, Firestone B, Flynn J, Green J, Growney JD, Holmqvist M, Hsu M, Jansson D, Jiang L, Kwon P, Liu G, Lombardo F, Lu Q, Majumdar D, Meta C, Perez L, Pu M, Ramsey T, Remiszewski S, Skolnik S, Traebert M, Urban L, Uttamsingh V, Wang P, Whitebread S, Whitehead L, Yan-Neale Y, Yao YM, Zhou L, Atadja P (2011) Optimization of the in vitro cardiac safety of hydroxamate-based histone deacetylase inhibitors. J Med Chem 54(13):4752–4772. https://doi.org/10.1021/jm200388e
118. Sodji QH, Kornacki JR, McDonald JF, Mrksich M, Oyelere AK (2015) Design and structure activity relationship of tumor-homing histone deacetylase inhibitors conjugated to folic and pteroic acids. Eur J Med Chem 96:340–359. https://doi.org/10.1016/j.ejmech.2015.04.014
119. Sonnemann J, Trommer N, Becker S, Wittig S, Grauel D, Palani CD, Beck JF (2012) Histone deacetylase inhibitor-mediated sensitization to TRAIL-induced apoptosis in childhood malignancies is not associated with upregulation of TRAIL receptor expression, but with potentiated caspase-8 activation. Cancer Biol Ther 13(6):417–424. https://doi.org/10.4161/cbt.19293
120. Sun R, Shen J, Gao Y, Zhou Y, Yu Z, Hornicek F, Kan Q, Duan Z (2016) Overexpression of EZH2 is associated with the poor prognosis in osteosarcoma and function analysis indicates a therapeutic potential. Oncotarget 7(25):38333
121. Suraweera A, O'Byrne KJ, Richard DJ (2018) Combination therapy with histone deacetylase inhibitors (HDACi) for the treatment of cancer: achieving the full therapeutic potential of HDACi. Front Oncol 8:92. https://doi.org/10.3389/fonc.2018.00092
122. Takasaka N, Araya J, Hara H, Ito S, Kobayashi K, Kurita Y, Wakui H, Yoshii Y, Yumino Y, Fujii S (2014) Autophagy induction by SIRT6 through attenuation of insulin-like growth factor signaling is involved in the regulation of human bronchial epithelial cell senescence. J Immunol 192(3):958–968
123. Taylor EM, Koss B, Davis LE, Tackett AJ (2020) Histone modifications as biomarkers for immunotherapy. In: Biomarkers for immunotherapy of cancer. Springer, New York, pp 213–228
124. Thayanithy V, Park C, Sarver AL, Kartha RV, Korpela DM, Graef AJ, Steer CJ, Modiano JF, Subramanian S (2012) Combinatorial treatment of DNA and chromatin-modifying drugs cause cell death in human and canine osteosarcoma cell lines. PLoS One 7(9):e43720. https://doi.org/10.1371/journal.pone.0043720
125. Toiber D, Erdel F, Bouazoune K, Silberman DM, Zhong L, Mulligan P, Sebastian C, Cosentino C, Martinez-Pastor B, Giacosa S (2013) SIRT6 recruits SNF2H to DNA break sites, preventing genomic instability through chromatin remodeling. Mol Cell 51(4):454–468
126. Tonak M, Becker M, Graf C, Eckhard L, Theobald M, Rommens PM, Wehler TC, Proschek D (2014) HDAC inhibitor-loaded bone cement for advanced local treatment of osteosarcoma and chondrosarcoma. Anticancer Res 34(11):6459–6466
127. Tseng AH, Shieh S-S, Wang DL (2013) SIRT3 deacetylates FOXO3 to protect mitochondria against oxidative damage. Free Radic Biol Med 63:222–234
128. Turtoi A, Mottet D, Matheus N, Dumont B, Peixoto P, Hennequiere V, Deroanne C, Colige A, De Pauw E, Bellahcene A (2012) The angiogenesis suppressor gene AKAP12 is under the epigenetic control of HDAC7 in endothelial cells. Angiogenesis 15(4):543–554
129. Verdin E, Ott M (2015) 50 years of protein acetylation: from gene regulation to epigenetics, metabolism and beyond. Nat Rev Mol Cell Biol 16(4):258
130. von Burstin J, Eser S, Paul MC, Seidler B, Brandl M, Messer M, von Werder A, Schmidt A, Mages J, Pagel P (2009) E-cadherin regulates metastasis of pancreatic cancer in vivo and is suppressed by a SNAIL/HDAC1/HDAC2 repressor complex. Gastroenterology 137(1):361–371. e365
131. Wang CK, Yu XD, Li Q, Xie G, Teng Y (2013) Chloroquine and valproic acid combined treatment in vitro has enhanced cytotoxicity in an osteosarcoma cell line. Asian Pac J Cancer Prev 14(8):4651–4654. https://doi.org/10.7314/apjcp.2013.14.8.4651
132. Watanabe K, Okamoto K, Yonehara S (2005) Sensitization of osteosarcoma cells to death receptor-mediated apoptosis by HDAC inhibitors through downregulation of cellular FLIP. Cell Death Differ 12(1):10–18. https://doi.org/10.1038/sj.cdd.4401507
133. Weinstein IB (2002) Addiction to oncogenes--the Achilles heal of cancer. Science 297(5578):63–64
134. Wirries A, Jabari S, Jansen EP, Roth S, Figueroa-Juarez E, Wissniowski TT, Neureiter D, Klieser E, Lechler P, Ruchholtz S, Bartsch DK, Boese CK, Di Fazio P (2018) Panobinostat mediated cell

death: a novel therapeutic approach for osteosarcoma. Oncotarget 9(68):32997–33010. https://doi.org/10.18632/oncotarget.26038
135. Wittenburg LA, Bisson L, Rose BJ, Korch C, Thamm DH (2011) The histone deacetylase inhibitor valproic acid sensitizes human and canine osteosarcoma to doxorubicin. Cancer Chemother Pharmacol 67(1):83–92. https://doi.org/10.1007/s00280-010-1287-z
136. Wittenburg LA, Gustafson DL, Thamm DH (2010) Phase I pharmacokinetic and pharmacodynamic evaluation of combined valproic acid/doxorubicin treatment in dogs with spontaneous cancer. Clin Cancer Res 16(19):4832–4842. https://doi.org/10.1158/1078-0432.CCR-10-1238
137. Wittenburg LA, Ptitsyn AA, Thamm DH (2012) A systems biology approach to identify molecular pathways altered by HDAC inhibition in osteosarcoma. J Cell Biochem 113(3):773–783. https://doi.org/10.1002/jcb.23403
138. Wu G, Yu W, Zhang M, Yin R, Wu Y, Liu Q (2018) MicroRNA-145-3p suppresses proliferation and promotes apotosis and autophagy of osteosarcoma cell by targeting HDAC4. Artif Cells Nanomed Biotechnol 46(sup2):579–586. https://doi.org/10.1080/21691401.2018.1464459
139. Wu Z, Ma C, Shan Z, Ju Y, Li S, Zhao Q (2013) Histone deacetylase inhibitors suppress the growth of human osteosarcomas in vitro and in vivo. J BUON 18(4):1032–1037
140. Xiao Q, Huang L, Zhang Z, Chen X, Luo J, Zhang Z, Chen S, Shu Y, Han Z, Cao K (2017) Overexpression of miR-140 inhibits proliferation of osteosarcoma cells via suppression of histone deacetylase 4. Oncol Res 25(2):267–275. https://doi.org/10.3727/096504016X14732510786564
141. Xie C, Wu B, Chen B, Shi Q, Guo J, Fan Z, Huang Y (2016) Histone deacetylase inhibitor sodium butyrate suppresses proliferation and promotes apoptosis in osteosarcoma cells by regulation of the MDM2-p53 signaling. Onco Targets Ther 9:4005–4013. https://doi.org/10.2147/OTT.S105418
142. Yamaguchi T, Cubizolles F, Zhang Y, Reichert N, Kohler H, Seiser C, Matthias P (2010) Histone deacetylases 1 and 2 act in concert to promote the G1-to-S progression. Genes Dev 24(5):455–469
143. Yamanegi K, Kawabe M, Futani H, Nishiura H, Yamada N, Kato-Kogoe N, Kishimoto H, Yoshiya S, Nakasho K (2015) Sodium valproate, a histone deacetylase inhibitor, modulates the vascular endothelial growth inhibitor-mediated cell death in human osteosarcoma and vascular endothelial cells. Int J Oncol 46(5):1994–2002. https://doi.org/10.3892/ijo.2015.2924
144. Yamanegi K, Yamane J, Hata M, Ohyama H, Yamada N, Kato-Kogoe N, Futani H, Nakasho K, Okamura H, Terada N (2009) Sodium valproate, a histone deacetylase inhibitor, decreases the secretion of soluble Fas by human osteosarcoma cells and increases their sensitivity to Fas-mediated cell death. J Cancer Res Clin Oncol 135(7):879–889. https://doi.org/10.1007/s00432-008-0522-z
145. Yamanegi K, Yamane J, Kobayashi K, Kato-Kogoe N, Ohyama H, Nakasho K, Yamada N, Hata M, Fukunaga S, Futani H, Okamura H, Terada N (2012) Valproic acid cooperates with hydralazine to augment the susceptibility of human osteosarcoma cells to Fas- and NK cell-mediated cell death. Int J Oncol 41(1):83–91. https://doi.org/10.3892/ijo.2012.1438
146. Yamanegi K, Yamane J, Kobayashi K, Kato-Kogoe N, Ohyama H, Nakasho K, Yamada N, Hata M, Nishioka T, Fukunaga S, Futani H, Okamura H, Terada N (2010) Sodium valproate, a histone deacetylase inhibitor, augments the expression of cell-surface NKG2D ligands, MICA/B, without increasing their soluble forms to enhance susceptibility of human osteosarcoma cells to NK cell-mediated cytotoxicity. Oncol Rep 24(6):1621–1627. https://doi.org/10.3892/or_00001026
147. Yang XJ, Seto E (2008) The Rpd3/Hda1 family of lysine deacetylases: from bacteria and yeast to mice and men. Nat Rev Mol Cell Biol 9(3):206–218. https://doi.org/10.1038/nrm2346
148. Yang Y, Huang Y, Wang Z, Wang HT, Duan B, Ye D, Wang C, Jing R, Leng Y, Xi J, Chen W, Wang G, Jia W, Zhu S, Kang J (2016) HDAC10 promotes lung cancer proliferation via AKT phosphorylation. Oncotarget 7(37):59388–59401. https://doi.org/10.18632/oncotarget.10673
149. Ying S, Jianjun H, Xue Y, Shuwei Y, Liyuan Z, Jie W, Lixian C (2017) MicroRNA-133b inhibits cell proliferation and invasion in osteosarcoma by targeting Sirt1. Oncol Res 25(9):1421–1430. https://doi.org/10.3727/096504016X14826089198805
150. Yoo YG, Kong G, Lee MO (2006) Metastasis-associated protein 1 enhances stability of hypoxia-inducible factor-1α protein by recruiting histone deacetylase 1. EMBO J 25(6):1231–1241
151. Yu D, Kahen E, Cubitt CL, McGuire J, Kreahling J, Lee J, Altiok S, Lynch CC, Sullivan DM, Reed DR (2015) Identification of synergistic, clinically achievable, combination therapies for osteosarcoma. Sci Rep 5:16991
152. Zhang C, Yang C, Feldman MJ, Wang H, Pang Y, Maggio DM, Zhu D, Nesvick CL, Dmitriev P, Bullova P, Chittiboina P, Brady RO, Pacak K, Zhuang Z (2017) Vorinostat suppresses hypoxia signaling by modulating nuclear translocation of hypoxia inducible factor 1 alpha. Oncotarget 8(34):56110–56125. https://doi.org/10.18632/oncotarget.18125
153. Zhang M, Xiang S, Joo H-Y, Wang L, Williams KA, Liu W, Hu C, Tong D, Haakenson J, Wang C (2014) HDAC6 deacetylates and ubiquitinates MSH2 to maintain proper levels of MutSα. Mol Cell 55(1):31–46
154. Zhang N, Xie T, Xian M, Wang YJ, Li HY, Ying MD, Ye ZM (2016) SIRT1 promotes metastasis of human osteosarcoma cells. Oncotarget 7(48):79654–79669. https://doi.org/10.18632/oncotarget.12916
155. Zhang ZL, Zhang AH, Feng ZQ, An L, Zhao QH (2013) Histone deacetylase inhibits the growth and migration of human osteosarcoma cells. Zhonghua Yi Xue Za Zhi 93(33):2680–2685

156. Zhu S, Denman CJ, Cobanoglu ZS, Kiany S, Lau CC, Gottschalk SM, Hughes DP, Kleinerman ES, Lee DA (2015) The narrow-spectrum HDAC inhibitor entinostat enhances NKG2D expression without NK cell toxicity, leading to enhanced recognition of cancer cells. Pharm Res 32(3):779–792. https://doi.org/10.1007/s11095-013-1231-0

157. Zimmermann S, Kiefer F, Prudenziati M, Spiller C, Hansen J, Floss T, Wurst W, Minucci S, Göttlicher M (2007) Reduced body size and decreased intestinal tumor rates in HDAC2-mutant mice. Cancer Res 67(19):9047–9054

Oncolytic Viruses and Their Potential as a Therapeutic Opportunity in Osteosarcoma

5

Mary Frances Wedekind and Timothy P. Cripe

Abstract

Osteosarcoma remains an unmet medical need. Oncolytic viruses are gaining traction as novel cancer therapeutics. These viruses are either naturally nonpathogenic or engineered to be safe by specific genetic deletions yet retain the ability to infect and kill human cancer cells and elicit anticancer immunity. Some versions are being specifically designed and tested in patients with osteosarcoma, though due to their generalized mechanism of action most are being tested in patients across a broad range of cancer types. The activity of these viruses is impacted not only by the susceptibility of tumor cells to infection but also by the tumor microenvironment (TME) and by tumor immunogenicity. Here we review the field of oncolytic viruses with a particular emphasis on highlighting any available data in preclinical osteosarcoma models or in patients with osteosarcoma. While in general the viruses have been shown safe to administer to patients by a variety of routes, their therapeutic efficacy to date has been limited. Given the low rate of adverse events and the likely absence of long-term side effects, the utility of oncolytic viruses will most likely be realized when used in combination with other agents.

Keywords

Osteosarcoma · Oncolytic virus · Virotherapy · Viroimmunotherapy

Introduction

Oncolytic viroimmunotherapy represents a novel class of biologic "drugs" wherein modified viruses are utilized for the treatment of many types of cancers. More specifically, oncolytic viruses are live viruses that are used to (i) infect and lyse tumor cells, (ii) induce an antitumor immune response, and (iii) express foreign transgenes, with the ability to self-propagate through tumors. In 1904, George Dock reported complete remission for a patient with leukemia after an influenza infection [27]. This report and other similar clinical observations lead to several clinical trials utilizing virotherapy in the treatment of cancer in the 1950s [48]. Several patients experienced tumor reduction or clinical improvements; however, due to side effects (infections) and the emergence of chemotherapy, oncolytic virother-

M. F. Wedekind
National Institutes of Health, Bethesda, MD, USA

T. P. Cripe (✉)
Nationwide Children's Hospital, The Ohio State University, Columbus, OH, USA
e-mail: timothy.cripe@nationwidechildrens.org

E. S. Kleinerman, R. Gorlick (eds.), *Current Advances in the Science of Osteosarcoma*, Advances in Experimental Medicine and Biology 1258, https://doi.org/10.1007/978-3-030-43085-6_5

apy did not flourish in that era [44, 48, 94]. It was not until 1991, after successful treatment with genetically attenuated and altered herpes simplex virus (HSV) in a murine glioma model, did oncolytic virotherapy intrigue begin a resurgence [68]. Since that time, with continued advances in genetic manipulation, oncolytic virotherapy has gained significant traction with numerous RNA and DNA virus options being developed and tested via intravenous or intratumoral injection. Talimogene laherparepvec (T-VEC, also known as Imlygic), a genetically modified herpes simplex virus (HSV) type 1 with expression of a human GM-CSF transgene, was FDA approved in 2015 for the treatment of metastatic melanoma after observations of durable primary and abscopal tumor site regressions [4]. With continued adequate safety profiles of administration of genetically altered oncolytic virotherapy in adults and the pediatric population, the future of virotherapy for the treatment of numerous kinds of cancers is promising [4, 76, 92, 95].

Osteosarcoma is the most common bone tumor in the pediatric, adolescent, and young adult populations. Current standard of care includes chemotherapy and surgery, with overall 5-year survival of 70% in localized disease [65]. However, for patients with metastatic osteosarcoma, overall 5-year survival is <20% with no improvements in therapy options in the last 50 years despite numerous attempts. Novel therapeutic approaches are much needed to help improve the outcomes for these patients. In this chapter, we will review the novel therapy of oncolytic viroimmunotherapy for the treatment of osteosarcoma as well the tumor microenvironment and its influences on oncolytic viroimmunotherapy.

Tumor Microenvironment

The immune system's interaction with a tumor is a complex and sophisticated network of connections. It is the communication of stimulation or exhaustion that tips the balance of tumor suppression and growth in the body.

Immunologic Landscape and Immunoediting

The innate immune system consisting of dendritic cells, macrophages, natural killer cells, neutrophils, basophils, and eosinophils is the frontline defense against foreign antigens. The adoptive immune system consisting of B lymphocytes, CD4+ helper T lymphocytes, and CD8+ cytotoxic T lymphocytes is directly activated by the antigen presenting cells of the innate immune system [69]. So-called inflamed tumors are generally composed of high levels of tumor-infiltrating lymphocytes (TIL), including CD8+ cytotoxic T cells, high expression of PD-1 on tumor-infiltrating immune cells, genomic instability generating numerous tumor antigens, and the presence of preexisting antitumor immune response. In contrast, a "noninflamed" tumor exhibits a low mutational burden, low expression of major histocompatibility complex I (MHC-I), and high expression of immunosuppressive cytokines, such as TGFβ [40] (Fig. 5.1).

Cancer immunoediting is the theory that described the interplay of the immune system and the tumor in three different phases [90]. "Elimination" occurs when the immune system recognizes tumor-specific antigens, whether from the products of mutated genes or overexpressed normal genes or genes encoding viral proteins, which activate the innate and adaptive immune systems to eradicate the tumor. Tumor mutational burden (TMB) has been shown to increase tumor-specific antigens leading to a more "inflamed" TME. Numerous adult solid tumors, melanoma, renal cell carcinoma, and non-small-cell lung cancer have shown an association with increased TMB and improved survival leading to its consideration as a predictive biomarker for some immunotherapeutic agents [16, 82, 87]. If all the cancer cells are not destroyed, some may have the ability to lay quiescent, leading to the "equilibrium" phase. During this phase, the tumor is neither growing nor being completely eliminated. Finally, some tumor cells have the ability to evade the immune system and "escape"

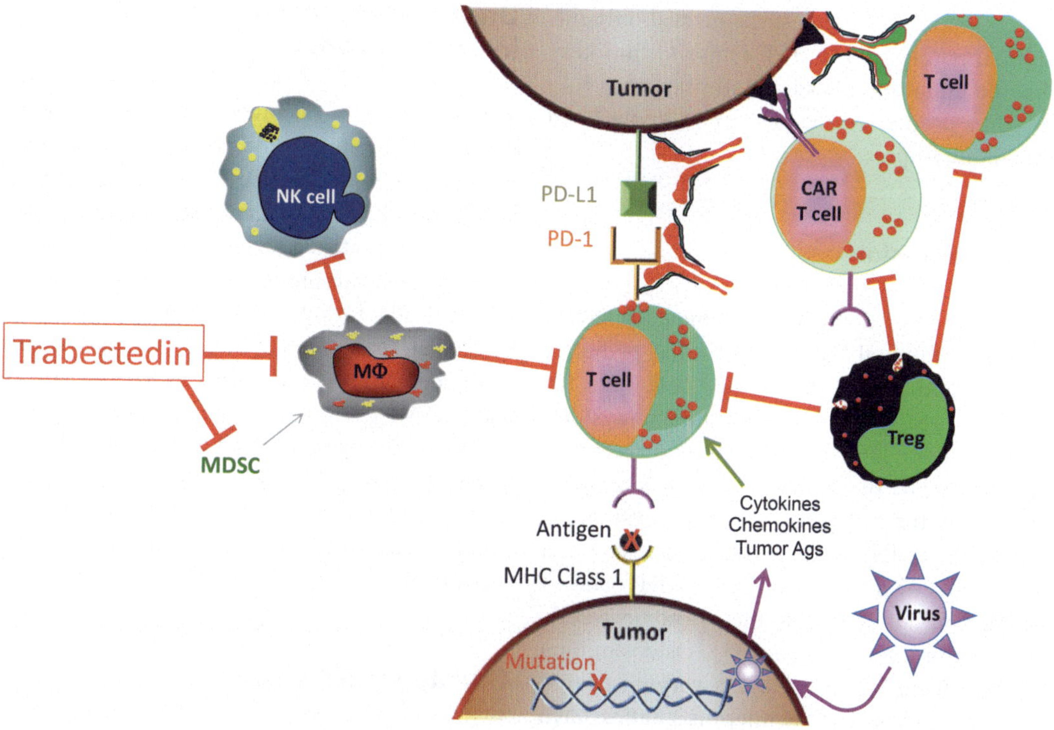

Fig. 5.1 Tumor microenvironment and immunomodulatory therapies. The tumor microenvironment in general, and there is much evidence for osteosarcoma in particular, is replete with immune cells that act to either promote or repress anti-cancer immune responses. The relative importance of each molecule or cell likely varies from patient to patient, and possibly even from region to region within the same tumor. Thus, in the future it may be important to personalize immunotherapy for each patient based on detailed analyses of the microenvironment on biopsy. A variety of small molecule, antibody, cellular and viral therapies have been established or are under investigation to promote innate and adaptive anticancer immunity and/or reverse immune suppression, as described in the text. Green arrows represent immunologic stimulation; red stops represent immunologic suppression; red Y represent antibody therapy. NK cell, natural killer cell; Treg, T-regulatory cell; MHC, major histocompatibility complex; PD-1, programmed cell death-1; PD-L1, programmed cell death ligand-1; CAR-T, chimeric antigen receptor T cell; MDSC, myeloid-derived suppressor cells; MΦ, macrophage

the immune system to proliferate [83, 98]. Numerous mechanisms exist for tumor evasion including recruitment of T-regulatory cells, myeloid-derived stem cells, or macrophages, downregulation of MHC-I, and upregulation of inhibitor ligands on tumor cells including TIM-3, PD-L1, and LAG-3 [42, 69]. Over the recent years, the TME of many different cancers has been found to be an important aspect of prognosis and possible indications for therapeutic options. In numerous adult solid tumors, the higher the immunoscore and immunoediting, or the percentage of tumor-infiltrating lymphocytes, the lower the metastatic rate and the longer the survival [6, 71].

Osteosarcoma TME

Tumor Mutational Burden

Within the pediatric population, it has been found that the majority of the tumors have a low tumor mutational burden [15, 55, 93]. Osteosarcoma is a genetically chaotic tumor with numerous deletions, somatic copy-number alterations, chromothripsis, and a few point mutations leading it to have a relatively high mutational burden among pediatric cancers [42, 89, 91]. However, when comparing to melanoma with a median frequency of somatic mutations per megabase pair of approximately 13, osteosarcoma still only has approximately 1.2 [16, 19]. Thus, it is unlikely that the mutational burden will play a big role in the predictive value of response to immunotherapies.

Immunoevasion

Despite the fact that osteosarcoma does have a higher mutational burden than other pediatric tumors and thus has some potential for increased recognition by the immune system, osteosarcoma is still utilizing other mechanisms to evade the immune system. A requirement for the adaptive immune system is utilization of the MHC class I. Osteosarcoma has been shown to downregulate MHC class I which would hide it from the cytotoxic T cells and thus promote growth [38]. Osteosarcoma has also been found to have a varying degree of T cell exhaustion. Numerous studies have found an increased PD-L1 expression, while others have not confirmed these findings [36, 52, 64, 70, 75]. A consistent finding, however, is that expression of PD-L1 has an association with lower survival [52, 75]. In regard to the immune cell infiltrates of osteosarcoma, it has been found that a higher CD8+ cytotoxic T cell infiltration is associated with a superior survival, and more specifically, a low CD8/Foxp3 ratio has been identified as a poor prognostic feature that is independent from metastatic status or response to chemotherapy [70, 75] [31]. Finally, osteosarcoma has been found to have dysregulation with the TME innate immune system. Macrophages, specifically M-2 polarized, have been correlated with higher TIM-3 and PD-1 expression [36]. TAMs have been shown to inhibit the T cell response, recruit immunosuppressive cell populations, inhibit dendritic cells, and lead to vascular dysfunction with limited T cell migration [25]. Interestingly, inhibition of M-2-polarized macrophages has been found to enhance proliferation of T cells within the osteosarcoma TME and also prevent metastasis in a murine model [36, 104].

Oncolytic Virotherapy in Osteosarcoma

The main goal of oncolytic viroimmunotherapy is to utilize naturally occurring or genetically modified viruses which replicate selectively within a tumor cell without damage to healthy tissues. The first step includes direct lysis of the cancer cells causing immunogenic cell death which leads to release of damage-associated molecular patterns (DAMPs) and pathogen-associated molecular patterns (PAMPs). DAMPs and PAMPs lead to the release of pro-inflammatory cytokines and activation of innate immune cells [1]. Lastly, antigen stimulation from the interaction of the innate and adaptive immune cells leads to T cell priming and TIL recruitment.

Oncolytic DNA Viruses

Adenovirus

Adenovirus (AdV) is a non-enveloped, dsDNA virus that results most commonly as a mild upper respiratory tract infection. Over 40 serotypes have been identified in AdV, but the most common strains utilized in oncolytic viroimmunotherapy are serotypes 2 and 5. During infection, adenovirus binds to cellular receptors such as coxsackie-adenovirus receptor (for attachment) and integrins (for entry). The virus will then be endocytosed and brought to the nucleus while disassembling its viral capsid. Viral transcription occurs in three phases. In the early phase, the viral genome early-region 1A (E1A) or early-region 1B (E1B) is transcribed. E1A protein activates the cellular repair or apoptotic pathway mediated by p53, while E1B protein prevents early death by binding to p53 and inducing degradation or binding the antiapoptotic factor BCL2. The first oncolytic adenovirus called dl1520 (also known as ONYX-015 and Cl-1042) contained a deletion of the E1B gene and thus no formation of the E1B-55kD protein leading to growth arrest [20]. Initial preclinical studies showed that dl1520 was effective in decreasing tumor size and thus was the first AdV to go to clinical trial [12, 84]. In the clinical trial, AdV was combined with systemic chemotherapy in patients with advanced sarcoma. Only one partial response was seen in a patient with malignant peripheral nerve sheath tumor [73]. Concerns for dl1520 arose when its specificity was found to be independent of p53 status [20]. Another site for gene inactivation within osteosarcoma is the Rb mutation. The AdV, Ad5-Δ24, removes the binding site for the Rb protein with a 24-base pair deletion in the E1A region [32]. This deletion results in virus replication that is selective for cells that have a defective Rb mutation. A study utilizing AdΔ24 alone in human osteosarcoma cell line in vitro and in vivo demonstrated antitumor efficacy and persistence of viral particles within the tumors. Unfortunately, no complete cures were observed in vivo [102]. Therefore, in another in vitro and in vivo study, standard chemotherapy agents to treat osteosarcoma were used in combination with virus. AdΔ24 alone again showed some efficacy; however, when cisplatin was added, the antitumor efficacy was enhanced [66]. AdΔ24, renamed DNX-2401, was utilized in a Phase I clinical trial with recurrent malignant glioma with 12% demonstrating a ≥95% reduction in tumor size and 20% of patients survived >3 years after treatment. No current clinical trials are open for osteosarcoma.

Another approach to adenovirotherapy includes placing genes essential for replication under the control of a tumor-specific promotor. Osteocalcin is a protein hormone found in the bone, limited to osteoblasts in healthy humans, but has high activity in osteosarcoma [51]. Ad-OC-E1A is a constructed AdV with the osteocalcin promoter. It has been tested in canine osteosarcoma cells with enhanced killing in vitro and a therapeutic benefit in vivo [41]. Ad-OC-E1A also showed efficacy in human osteosarcoma cells in vivo and in vitro with significant tumor reduction in a pulmonary metastatic model [58]. A Phase I/II trial was planned for AD-OC-E1A in osteosarcoma patients, but the study has yet to be published [9].

Telomerase activity is another target for regulated expression as 44–81% of bone and soft-tissue sarcomas have been detected to have activity [37]. Human telomerase reverse transcriptase (hTERT) gene expression stabilizes telomere lengths and is highly expressed on cancer cells but not normal tissues [50]. OB-301, telomelysin, is an AdV that utilizes the hTERT promoter to restrict replication to cells with high telomerase activity [37]. OBP-301 was tested on osteosarcoma cell lines and demonstrated insensitivity to the antitumor effects unlike other bone and soft-tissue sarcomas [57, 88]. Thus, enhanced OBP-301, with added expression of wild-type p53 (OBP-702), or combination with a p53-expressing replication-deficient adenovirus was utilized in vitro and in vivo osteosarcoma cell lines. OBP-702 was found to induce more profound tumor delays and cell killing. Through a Phase I clinical trial, OBP-301 was well tolerated in patients with advanced solid tumors; however, no patients with osteosarcoma were enrolled [72].

Finally, AdV has been manipulated to alter the TME. IL-24 has exhibited antitumor activity in numerous types of cancer like osteosarcoma (58). In one study, an hTERT promoter AdV was equipped with IL-24 (OA-IL-24), thus resulting in high levels of IL-24 in the TME. Utilizing osteosarcoma cells, OA-IL-24 demonstrated higher killing in vitro and suppressed tumor growth in vivo. More importantly, OA-IL-24 increased sensitivity to doxorubicin [61]. Lastly, an oncolytic AdV named VCN-01 utilized the defective pRB restrictive pathway with an addition of an expression cassette of human PH20 gene, which expressed hyaluronidase leading to degradation of extracellular matrix hyaluronic acid [67]. An association between high hyaluronic acid levels and low survival rates and development of chemoresistance is seen in some tumors [67]. VCN-01 demonstrated potent antitumor effect in vivo and in vitro including in a metastatic model. VCN-01 is currently being utilized in two clinical trials (NCT02045602 and NCT2045589) in adults with refractory solid tumors.

Herpes Simplex Virus

Herpes simplex virus (HSV) is an enveloped, double-stranded linear DNA virus with a 152 kb genome. Due to its large genome and nonessential joint regions, a large portion can be removed without affecting viral potency, resulting in space for the insertion of transgenes [34]. HSV is part of the Herpesviridae family with characteristic oral or genital ulcerations manifesting with infection. HSV attaches to host cell membrane through binding of surface glycoproteins to several different cellular receptors as well as heparan sulfate proteoglycans followed by a conformational change that triggers the fusion of the viral envelope to the host cellular membrane. The virion is then released into the cytosol to be degraded and delivered to the nucleus where replication will occur followed by death of the host cell [22]. Most oncolytic HSV-1 have been engineered to have a mutation of the gene encoding ICP34.5, which is the neurovirulence gene which results in marked attenuation in normal cells but not most tumor cells. Oncolytic HSV continues to have an intact thymidine kinase (tk) to allow for treatment with antiherpetic agents in the event of a viral outbreak [22]. Some oncolytic HSVs have inactivated UL39 that encodes ICP6 protein, which is required for viral DNA replication and highly expressed in rapidly dividing cells [59].

Several oncolytic HSVs (talimogene laherparepvec (T-VEC), HSV1716, NV1020, G207, M032, rRp450, etc.) are being utilized in preclinical and clinical studies and have been efficacious in numerous tumor types including sarcomas, melanomas, and colon, breast, lung, and hepatic tumors [11]. NV1020 and G207 have been utilized preclinically in osteosarcoma with only modest sensitivity [11]. HSV1716 has been utilized in preclinical and clinical studies. Preclinical data suggest the favorable responses observed in various solid tumor types are based on not only a direct lytic effect but also an antitumor immune response [56].

HSV1716 was also utilized in a Phase I clinical trial evaluating intratumoral injection in children and young adults with non-central nervous system tumors including osteosarcoma. Despite no durable responses, a patient with osteosarcoma experienced an immunologic flare-up as a result of viral injection [95]. T-VEC is similar to HSV1716 except with the deletion of ICP47, which normally blocks antigen presentation to MHC classes I and II, and contains the coding sequence for human granulocyte-macrophage colony-stimulating factor in place of ICP34.5 [21]. A Phase I trial of T-VEC in children with non-central nervous system solid tumors is underway (NCT02756845).

Vaccinia Virus

Vaccinia virus is an enveloped, double-stranded linear DNA virus with a genome of 190 kb and is a member of the poxvirus family. The original use of vaccinia virus eradicated smallpox in 1979. With its large genome, as with HSV, the ability to delete nonessential genes makes vaccinia virus attractive as an oncolytic virotherapy option. After utilizing numerous proteins for viral entry into cells, vaccinia virus is unique among the other DNA viruses as viral replication occurs independently in the cytoplasm [24]. The utilization of vaccinia growth factor (VGF) via epidermal growth factor receptor allows for viral spread to uninfected tissues [24].

To engineer vaccinia virus toward oncolytic properties, deletion of VGF restricts viral infection to cells with epidermal growth factor receptors, which is often observed in cancer cells [14]. For further attenuation, deletion of the J2R gene encoding for viral thymidine kinase (tk) leads vaccinia virus to be dependent on cells with overexpression of cellular tk, which is also often observed in cancer cells [78]. With both gene deletions, the resulting oncolytic virus was named double-deleted vaccinia virus and is very specific for cancer cells and thus increasing the safety profile [96].

In addition to a double-deleted vaccinia virus, a single-deleted vaccinia virus with additional gene insertions has also been developed. JX-594, also called Pexa-Vec, includes thymidine kinase gene deletions, granulocyte-macrophage colony-stimulating factor insertion to induce a systemic antitumor immune response, and lac-Z gene insertion under control of the p7.5 promoter [23]. JX-963 is a double-deleted vaccinia virus with the insertion of granulocyte-macrophage colony-stimulating factor.

Preclinical studies have shown high cytotoxicity of Pexa-Vec in vitro against osteosarcoma [39, 63]. In a metastatic osteosarcoma model, double-deleted vaccinia vaccine demonstrated reduced lung metastatic lesions with significantly prolonged survival [63]. Numerous other solid tumors in the pediatric and adult population have shown antitumor activity with numerous vaccinia virus including rhabdomyosarcoma, fibrosarcoma, and fibrohistiocytoma [39]. Several clinical trials in adult patients have proven the safety of oncolytic vaccinia virus with one Phase II trial demonstrating improvement in survival in patients with advanced liver cancer [23]. A Phase I clinical trial utilizing Pexa-Vec including pediatric patients with unresectable refractory solid tumors demonstrated that Pexa-Vec was safe in this population [23]. As far as we are aware, no clinical trials to date have included osteosarcoma patients.

Protoparvovirus

Protoparvovirus H-1 (H-1PV) is a wild-type virus occurring naturally in rats and is a single-stranded, non-enveloped DNA virus with a genome of 5.1 kb. The genome contains only two main transcription units, nonstructural proteins (essential in the replication and cytotoxicity of the virus) and viral capsid proteins [99]. There have been numerous preclinical data suggesting H-1PV oncolytic properties and preference toward tumor cells; however, no one mechanism has been found likely due to numerous mechanisms combined [5]. Wild-type H-1PV or a mutant and deletion of 114 in-frame nucleotides called Del H-1PV have been studied with osteosarcoma. In the pioneering clinical trial utilizing H-1PV in the 1960s, two adolescent patients with osteosarcoma were injected with wild-type H-1PV with no adverse side effects noted. However, both succumbed to their disease shortly thereafter [97]. We found only one other report of utilizing H-1PV in osteosarcoma cells. A preclinical in vitro study utilizing wild-type H-1PV and Del H-1PV demonstrated more effective cytotoxicity of osteosarcoma cells with wild-type versus Del. This result is in contrast to all other preclinical studies reported that suggest Del H-1PV is more effective [33]. More preclinical in vitro and in vivo studies need to be performed utilizing H-1PV.

Oncolytic RNA Viruses

Reovirus

Reovirus is a member of the Reoviridae family and is a non-enveloped double-stranded RNA virus that has low pathogenicity in adults with no exhibited clinical symptoms [45]. Reovirus shows selectivity toward infection of malignant cells with an activated Ras-signaling pathway, which is typically present in osteosarcoma cells [47]. In a preclinical study of canine solid tumors, osteosarcoma was susceptible to reovirus but not to the degree as other solid canine solid tumors [45]. Reolysin is a formulation of reovirus type 3 Dearing strain developed by Oncolytics Biotech and is the only clinically utilized reovirus. This formulation was utilized in vitro and in vivo with significant results in sarcoma cell lines, including osteosarcoma. Stable disease was seen in the osteosarcoma murine models with partial responses when combined with cisplatin [43]. Numerous clinical trials utilizing Reolysin have been performed for many different solid tumors in the adult population with a very tolerable toxicity profile and minimal side effects. A Phase I trial in pediatric patients of Reolysin alone or in combination with cyclophosphamide included three osteosarcoma patients. Both Reolysin alone and in combination were well tolerated; however, no objective responses were seen [53]. A Phase II trial with Reolysin given intravenously to patients with bone and soft-tissue sarcomas metastatic to the lungs has been completed, but no completed published results have been reported to our knowledge.

Semliki Forest Virus

Semliki Forest virus is an enveloped positive-stranded RNA virus of the Togaviridae family [49]. Wild-type Semliki Forest Virus is pathogenic to small rodents and mice but is nonpathogenic in humans [8]. A mutated version of Semliki Forest Virus, SFVA7, has been mutated at its opal codon and amino acids in the nonstructural genome, and rodents, mice, or humans remain asymptomatic with injection. SFVA7 was further altered to express green fluorescent protein, VA7-EGFP, and compared against adenovirus, Ad5Δ24, in osteosarcoma xenografts. In vitro, VA7-EGFP was superior to Ad5Δ24 with more extensive cell death in a shorter period of time with a lower multiplicity of infection. In vivo, VA7-EGFP demonstrated significantly improved survival compared to Ad5Δ24 [49].

Vesicular Stomatitis Virus

Vesicular stomatitis virus (VSV) is a negative-sense RNA virus of the rhabdovirus family. Infection with vesicular stomatitis virus is typically seen in cattle, horses, and swine, with rare, insignificant disease in humans. Oncolytic properties of VSV are due to the sensitivity to interferon with replication only in cells with defective interferon responses, mostly malignant cells [46]. In vitro study of VSV compared to reovirus demonstrated that osteosarcoma was highly susceptible to VSV infection and had more effective cytotoxicity at similar multiplicity of infection [74]. Another study utilized isolated limb perfusion to reduce the viral amount dispersed to the rest of the body. This technique leads to viral gene expression largely limited to osteosarcoma cells without infection of other tissues local or distant, while the primary tumor site demonstrated significant tumor growth delays [54]. A recombinant VSV with near-infrared fluorescent protein, Katushka (rVSV-K), inserted into the genome has been studied in two in vitro studies. First, rVSV-K was utilized in a metastatic osteosarcoma model where it was determined that metastatic lesions had slower growth and prolonged survival compared to control mice, but also the metastatic cells were easily visualized in whole blood samples [46]. Following this study, the same group demonstrated that rVSV-K was also useful in surgical resections with margins significantly larger in the rVSV-K group than in the nonfluorescent group [86].

Measles Virus

Measles virus is a negative single-stranded RNA virus belonging to the Paramyxoviridae family. Measles virus is highly contagious with high morbidity and mortality without a prior exposure via vaccination. Measles virus enters a cell through interactions of its H protein, CD46, and signaling lymphocyte-activating molecule. Oncolytic selectivity is achieved via overexpression of CD46 on most tumor cells. In one in vitro study, an osteosarcoma cell line was resistant to measles virus due to inhibition of viral replication and entry [10]. However, in another in vitro and in vivo study, there were numerous osteosarcoma cell lines sensitive in vitro and antitumor activity with one highly metastatic model in vivo. They found that all six osteosarcoma cell lines were susceptible to measles virus and it was highly cytotoxic. In one metastatic xenograft model, they found reduction of metastatic lesions, prolonged tumor growth at the primary site, and prolonged survival with utilization of measles virus [28].

Poliovirus

Poliovirus is a non-enveloped, positive-stranded RNA virus belonging to the Picornaviridae family which causes paralytic poliomyelitis in humans. Poliovirus attachment and entry are mediated by a single molecule, CD155, which is upregulated in most solid tumors [13]. After confirmation that several osteosarcoma cell lines expressed CD155, poliovirus was shown to induce apoptosis in vitro [7]. As far as we are aware, no in vivo or clinical trials have been performed utilizing osteosarcoma and poliovirus.

Newcastle Disease Virus

Newcastle disease virus is a negative, single-stranded RNA virus belonging to the Paramyxoviridae family and results in a contagious viral bird disease with humans unaffected. PV701 is an attenuated strain with oncoselectivity based on defective interferon signaling. In vitro, PV701 shows a cytotoxic effect on osteosarcoma cells [80]. Newcastle disease virus has been utilized in clinical trials for patients with solid tumors and well tolerated [62]. No clinical trials to our knowledge have included osteosarcoma patients.

Maraba Virus

Maraba virus is a negative single-stranded RNA virus belonging to the Rhabdoviridae family and is found in sand flies with no detection outside of South America currently [77]. The most likely mechanism for malignancy precedence is the utilization of the ubiquitous low-density lipoprotein receptor, which is expressed on a wide range of malignancies [77]. In a comparative in vitro analysis of Maraba virus compared against HSV-1, adenovirus, reovirus, and VSV, Maraba virus exhibited more cytotoxicity. Currently, as far as we are aware, there are no active clinical trials utilizing Maraba virus for osteosarcoma patients.

Combination Therapy

Oncolytic virotherapy has shown numerous promising results in preclinical and clinical studies; however, it has become apparent that single-agent therapy will unlikely eradicate disease. Combination therapy is attractive as a way to overlap immunogenic cell death and enhance responses to therapy. These combination therapies with oncolytic viruses are currently underway. Investigators have begun to study T-VEC in combination with checkpoint inhibition into the clinical trials with good results in melanoma and greater efficacy and also tolerability [29]. Numerous other oncolytic viruses are being combined with checkpoint inhibition preclinically and clinically [17, 60, 81]. In one preclinical study utilizing osteosarcoma murine models, combination of HSV-1 and anti-programmed death ligand-1 demonstrated prolonged survival in an “inflamed” TME murine model compared to a “noninflamed” TME model (Wedekind and Cripe, unpublished results). To our knowledge, no other checkpoint inhibition with oncolytic virus therapy combinations is currently published for osteosarcoma.

Another intriguing combination is two viruses in combination. Newcastle disease virus in combination with adenovirus demonstrated enhanced cytotoxicity in vitro and superior antitumor efficacy in vivo compared to either virus alone in an osteosarcoma model [18]. Another group utilized combination of fowl pox viruses, an enveloped linear double-stranded DNA virus, and Newcastle disease virus. They determined the combination of these two viruses had increased cytotoxicity to osteosarcoma cell line in vitro with prolonged survival and tumor growth in vivo [103]. Despite a limited number of current combination therapies for osteosarcoma, there are numerous potentials that warranted exploration in the future and will be discussed in the future directions section.

Challenges to Oncolytic Virotherapy

Despite much preclinical and clinical data supporting the utilization of oncolytic virotherapy, there are challenges to overcome. One of the main limitations to oncolytic virotherapy is the mode of delivery. The ideal situation would be for intravenous administration to reach metastatic sites; however, there are concerns about the amount of viral particles that reach tumor sites as the liver can sequester much of the particles [3]. To avoid this limitation, virus may be administered via intratumoral injection. For tumors that are very superficial and accessible, intratumoral injection would not be a limitation; however, numerous patients have inaccessible tumors or tumors that are very large. For osteosarcoma in particular, with pulmonary metastatic lesions, intratumoral injections raise more complications such as the risk of pneumothorax. More data need to be collected to determine the extent of an abscopal effect, as melanoma treated with T-VEC demonstrated abscopal effects that may render injection of distant sites unnecessary [30].

Another complicating factor is that neutralizing antibodies can dampen the efficacy of oncolytic viruses [35]. It has been demonstrated that 50–80% of humans have antibodies against HSV-1 and 90% against reovirus [101]. In a Phase I clinical trial utilizing oncolytic measles virus, the dose level was very high before any efficacy was seen due to neutralizing antibodies [85]. In another Phase I trial, maximum neutralizing antibodies were reached by one-third of patients in 1 week and by almost two-thirds by 2 weeks. Thus, the recommendation for the systemic treatment of reovirus is to administer numerous high doses in the first week of treatment before the neutralizing antibodies can diminish its effect. This limitation reconfirms the need for combination therapy [35, 100].

Future Directions

The future of oncolytic virotherapy for osteosarcoma has numerous possibilities. As there has been limited data in regard to the use of oncolytic virotherapy for osteosarcoma, there is much that needs be learned. The use of combination therapy needs to be explored preclinically with osteosarcoma to determine the best applications for clinical trials. In a preclinical in vivo and in vitro study, trabectedin was observed to inhibit osteosarcoma tumor growth and metastatic growth [79]. Trabectedin in combination with HSV-1 oncolytic virus has shown complete regression of tumors in a Ewing sarcoma mouse model [26]. There is potential that this combination may show benefit in osteosarcoma. Another future combination therapy that has great potential but will require thoughtfulness with its execution is CAR-T cell therapy and oncolytic virotherapy [2]. There are numerous options and pitfalls for combining these two therapies, but through this combination, there may be the potential to counteract immunosuppression within the TME, reverse tumor immunosuppression, or improve the CAR-T homing and activation [2]. Continuing to gain knowledge of the effects of oncolytic virotherapy within the TME and the utilization of combination therapies to enhance the body's immune response are necessary to propel oncolytic virotherapy as a future treatment option for patients with osteosarcoma.

References

1. Achard C, Surendran A, Wedge ME et al (2018) Lighting a fire in the tumor microenvironment using oncolytic immunotherapy. EBioMedicine 31:17–24
2. Ajina A, Maher J (2017) Prospects for combined use of oncolytic viruses and CAR T-cells. J Immunother Cancer 5:90
3. Alemany R, Curiel DT (2001) CAR-binding ablation does not change biodistribution and toxicity of adenoviral vectors. Gene Ther 8:1347–1353
4. Andtbacka RH, Kaufman HL, Collichio F et al (2015) Talimogene Laherparepvec improves durable response rate in patients with advanced melanoma. J Clin Oncol 33:2780–2788
5. Angelova AL, Geletneky K, Nuesch JP et al (2015) Tumor selectivity of oncolytic parvoviruses: from in vitro and animal models to cancer patients. Front Bioeng Biotechnol 3:55
6. Angelova M, Mlecnik B, Vasaturo A et al (2018) Evolution of metastases in space and time under immune selection. Cell 175:751–765.e16
7. Atsumi S, Matsumine A, Toyoda H et al (2012) Oncolytic virotherapy for human bone and soft tissue sarcomas using live attenuated poliovirus. Int J Oncol 41:893–902
8. Barrett PN, Sheahan BJ, Atkins GJ (1980) Isolation and preliminary characterization of Semliki Forest virus mutants with altered virulence. J Gen Virol 49:141–147
9. Benjamin R, Helman L, Meyers P et al (2001) A phase I/II dose escalation and activity study of intravenous injections of OCaP1 for subjects with refractory osteosarcoma metastatic to lung. Hum Gene Ther 12:1591–1593
10. Berchtold S, Lampe J, Weiland T et al (2013) Innate immune defense defines susceptibility of sarcoma cells to measles vaccine virus-based oncolysis. J Virol 87:3484–3501
11. Bharatan NS, Currier MA, Cripe TP (2002) Differential susceptibility of pediatric sarcoma cells to oncolysis by conditionally replication-competent herpes simplex viruses. J Pediatr Hematol Oncol 24:447–453
12. Bischoff JR, Kirn DH, Williams A et al (1996) An adenovirus mutant that replicates selectively in p53-deficient human tumor cells. Science 274:373–376

13. Brown MC, Gromeier M (2015) Cytotoxic and immunogenic mechanisms of recombinant oncolytic poliovirus. Curr Opin Virol 13:81–85
14. Buijs PR, Verhagen JH, Van Eijck CH et al (2015) Oncolytic viruses: from bench to bedside with a focus on safety. Hum Vaccin Immunother 11:1573–1584
15. Chalmers ZR, Connelly CF, Fabrizio D et al (2017) Analysis of 100,000 human cancer genomes reveals the landscape of tumor mutational burden. Genome Med 9:34
16. Champiat S, Ferte C, Lebel-Binay S et al (2014) Exomics and immunogenics: bridging mutational load and immune checkpoints efficacy. Onco Targets Ther 3:e27817
17. Chen CY, Hutzen B, Wedekind MF et al (2018a) Oncolytic virus and PD-1/PD-L1 blockade combination therapy. Oncolytic Virother 7:65–77
18. Chen S, Zhang Q, Xu D et al (2018b) Antitumor effect of the Newcastle disease viral hemagglutinin-neuraminidase gene is expressed through an oncolytic adenovirus effect in osteosarcoma cells. Anti-Cancer Drugs 29:197–207
19. Chen X, Bahrami A, Pappo A et al (2014) Recurrent somatic structural variations contribute to tumorigenesis in pediatric osteosarcoma. Cell Rep 7:104–112
20. Chu RL, Post DE, Khuri FR et al (2004) Use of replicating oncolytic adenoviruses in combination therapy for cancer. Clin Cancer Res 10:5299–5312
21. Conry RM, Westbrook B, Mckee S et al (2018) Talimogene laherparepvec: first in class oncolytic virotherapy. Hum Vaccin Immunother 14:839–846
22. Cripe TP, Chen CY, Denton NL et al (2015a) Pediatric cancer gone viral. Part I: strategies for utilizing oncolytic herpes simplex virus-1 in children. Mol Ther Oncolytics 2:1–9
23. Cripe TP, Ngo MC, Geller JI et al (2015b) Phase 1 study of intratumoral Pexa-Vec (JX-594), an oncolytic and immunotherapeutic vaccinia virus, in pediatric cancer patients. Mol Ther 23:602–608
24. De Magalhaes JC, Andrade AA, Silva PN et al (2001) A mitogenic signal triggered at an early stage of vaccinia virus infection: implication of MEK/ERK and protein kinase A in virus multiplication. J Biol Chem 276:38353–38360
25. Denardo DG, Ruffell B (2019) Macrophages as regulators of tumour immunity and immunotherapy. Nat Rev Immunol 19:369–382
26. Denton NL, Chen CY, Scott TR et al (2016) Tumor-associated macrophages in oncolytic virotherapy: friend or foe? Biomedicines 4:1–9
27. Dock G (1904) The influence of complicating diseases upon leukaemia. Am J Med Sci 127:561–592
28. Domingo-Musibay E, Allen C, Kurokawa C et al (2014) Measles Edmonston vaccine strain derivatives have potent oncolytic activity against osteosarcoma. Cancer Gene Ther 21:483–490
29. Dummer R, Hoeller C, Gruter IP et al (2017) Combining talimogene laherparepvec with immunotherapies in melanoma and other solid tumors. Cancer Immunol Immunother 66:683–695
30. Fountzilas C, Patel S, Mahalingam D (2017) Review: oncolytic virotherapy, updates and future directions. Oncotarget 8:102617–102639
31. Fritzsching B, Fellenberg J, Moskovszky L et al (2015) CD8(+)/FOXP3(+)-ratio in osteosarcoma microenvironment separates survivors from non-survivors: a multicenter validated retrospective study. Onco Targets Ther 4:e990800
32. Fueyo J, Gomez-Manzano C, Alemany R et al (2000) A mutant oncolytic adenovirus targeting the Rb pathway produces anti-glioma effect in vivo. Oncogene 19:2–12
33. Geiss C, Kis Z, Leuchs B et al (2017) Preclinical testing of an oncolytic parvovirus: standard Protoparvovirus H-1PV efficiently induces osteosarcoma cell lysis in vitro. Viruses 9:1–18
34. Glorioso JC (2014) Herpes simplex viral vectors: late bloomers with big potential. Hum Gene Ther 25:83–91
35. Gong J, Sachdev E, Mita AC et al (2016) Clinical development of reovirus for cancer therapy: an oncolytic virus with immune-mediated antitumor activity. World J Methodol 6:25–42
36. Han Q, Shi H, Liu F (2016) CD163(+) M2-type tumor-associated macrophage support the suppression of tumor-infiltrating T cells in osteosarcoma. Int Immunopharmacol 34:101–106
37. Hasei J, Sasaki T, Tazawa H et al (2013) Dual programmed cell death pathways induced by p53 transactivation overcome resistance to oncolytic adenovirus in human osteosarcoma cells. Mol Cancer Ther 12:314–325
38. Haworth KB, Arnold MA, Pierson CR et al (2016) Characterization of MHC class I and beta-2-microglobulin expression in pediatric solid malignancies to guide selection of immune-based therapeutic trials. Pediatr Blood Cancer 63:618–626
39. He S, Li P, Chen CH et al (2012) Effective oncolytic vaccinia therapy for human sarcomas. J Surg Res 175:e53–e60
40. Hegde PS, Karanikas V, Evers S (2016) The where, the when, and the how of immune monitoring for cancer immunotherapies in the era of checkpoint inhibition. Clin Cancer Res 22:1865–1874
41. Hemminki A, Kanerva A, Kremer EJ et al (2003) A canine conditionally replicating adenovirus for evaluating oncolytic virotherapy in a syngeneic animal model. Mol Ther 7:163–173
42. Heymann MF, Lezot F, Heymann D (2017) The contribution of immune infiltrates and the local microenvironment in the pathogenesis of osteosarcoma. Cell Immunol 343:103711
43. Hingorani P, Zhang W, Lin J et al (2011) Systemic administration of reovirus (Reolysin) inhibits growth of human sarcoma xenografts. Cancer 117:1764–1774
44. Hoster HA, Zanes RP Jr, Von Haam E (1949) Studies in Hodgkin's syndrome; the association of viral hepatitis and Hodgkin's disease; a preliminary report. Cancer Res 9:473–480

45. Igase M, Hwang CC, Coffey M et al (2015) The oncolytic effects of reovirus in canine solid tumor cell lines. J Vet Med Sci 77:541–548
46. Johan MP, Kubo T, Furuta T et al (2018) Metastatic tumor cells detection and anti-metastatic potential with vesicular stomatitis virus in immunocompetent murine model of osteosarcoma. J Orthop Res 36:2562–2569
47. Kasel JA, Rosen L, Evans HE (1963) Infection of human volunteers with a reovirus of bovine origin. Proc Soc Exp Biol Med 112:979–981
48. Kelly E, Russell SJ (2007) History of oncolytic viruses: genesis to genetic engineering. Mol Ther 15:651–659
49. Ketola A, Hinkkanen A, Yongabi F et al (2008) Oncolytic Semliki forest virus vector as a novel candidate against unresectable osteosarcoma. Cancer Res 68:8342–8350
50. Kim NW, Piatyszek MA, Prowse KR et al (1994) Specific association of human telomerase activity with immortal cells and cancer. Science 266:2011–2015
51. Ko SC, Cheon J, Kao C et al (1996) Osteocalcin promoter-based toxic gene therapy for the treatment of osteosarcoma in experimental models. Cancer Res 56:4614–4619
52. Koirala P, Roth ME, Gill J et al (2016) Immune infiltration and PD-L1 expression in the tumor microenvironment are prognostic in osteosarcoma. Sci Rep 6:30093
53. Kolb EA, Sampson V, Stabley D et al (2015) A phase I trial and viral clearance study of reovirus (Reolysin) in children with relapsed or refractory extra-cranial solid tumors: a Children's Oncology Group Phase I Consortium report. Pediatr Blood Cancer 62:751–758
54. Kubo T, Shimose S, Matsuo T et al (2011) Oncolytic vesicular stomatitis virus administered by isolated limb perfusion suppresses osteosarcoma growth. J Orthop Res 29:795–800
55. Lawrence MS, Stojanov P, Polak P et al (2013) Mutational heterogeneity in cancer and the search for new cancer-associated genes. Nature 499:214–218
56. Leddon JL, Chen CY, Currier MA et al (2015) Oncolytic HSV virotherapy in murine sarcomas differentially triggers an antitumor T-cell response in the absence of virus permissivity. Mol Ther Oncolytics 1:14010
57. Li G, Kawashima H, Ogose A et al (2011) Efficient virotherapy for osteosarcoma by telomerase-specific oncolytic adenovirus. J Cancer Res Clin Oncol 137:1037–1051
58. Li X, Jung C, Liu YH et al (2006) Anti-tumor efficacy of a transcriptional replication-competent adenovirus, Ad-OC-E1a, for osteosarcoma pulmonary metastasis. J Gene Med 8:679–689
59. Liu TC, Kirn D (2005) Viruses with deletions in antiapoptotic genes as potential oncolytic agents. Oncogene 24:6069–6079
60. Liu Z, Ravindranathan R, Kalinski P et al (2017) Rational combination of oncolytic vaccinia virus and PD-L1 blockade works synergistically to enhance therapeutic efficacy. Nat Commun 8:14754
61. Liu Z, Xu L, Yuan H et al (2015) Oncolytic adenovirus-mediated mda7/IL24 expression suppresses osteosarcoma growth and enhances sensitivity to doxorubicin. Mol Med Rep 12:6358–6364
62. Lorence RM, Pecora AL, Major PP et al (2003) Overview of phase I studies of intravenous administration of PV701, an oncolytic virus. Curr Opin Mol Ther 5:618–624
63. Lun X, Ruan Y, Jayanthan A et al (2013) Double-deleted vaccinia virus in virotherapy for refractory and metastatic pediatric solid tumors. Mol Oncol 7:944–954
64. Majzner RG, Simon JS, Grosso JF et al (2017) Assessment of programmed death-ligand 1 expression and tumor-associated immune cells in pediatric cancer tissues. Cancer 123(19):3807–3815
65. Martin A, Leavey P (2015) Signs and symptoms. In: Cripe TP, Yeager N (eds) Malignant pediatric bone tumors – treatment & malignant. Springer International Publishing Switzerland
66. Martinez-Velez N, Xipell E, Jauregui P et al (2014) The oncolytic adenovirus Delta24-RGD in combination with cisplatin exerts a potent anti-osteosarcoma activity. J Bone Miner Res 29:2287–2296
67. Martinez-Velez N, Xipell E, Vera B et al (2016) The oncolytic adenovirus VCN-01 as therapeutic approach against pediatric osteosarcoma. Clin Cancer Res 22:2217–2225
68. Martuza RL, Malick A, Markert JM et al (1991) Experimental therapy of human glioma by means of a genetically engineered virus mutant. Science 252:854–856
69. Miwa S, Shirai T, Yamamoto N et al (2019) Current and emerging targets in immunotherapy for osteosarcoma. J Oncol 2019:7035045
70. Mochizuki K, Kawana S, Yamada S et al (2019) Various checkpoint molecules, and tumor-infiltrating lymphocytes in common pediatric solid tumors: possibilities for novel immunotherapy. Pediatr Hematol Oncol 36:17–27
71. Monjazeb AM, Zamora AE, Grossenbacher SK et al (2013) Immunoediting and antigen loss: overcoming the achilles heel of immunotherapy with antigen non-specific therapies. Front Oncol 3:197
72. Nemunaitis J, Tong AW, Nemunaitis M et al (2010) A phase I study of telomerase-specific replication competent oncolytic adenovirus (telomelysin) for various solid tumors. Mol Ther 18:429–434
73. Opyrchal M, Aderca I, Galanis E (2009) Phase I clinical trial of locoregional administration of the oncolytic adenovirus ONYX-015 in combination with mitomycin-C, doxorubicin, and cisplatin chemotherapy in patients with advanced sarcomas. Methods Mol Biol 542:705–717
74. Paglino JC, Van Den Pol AN (2011) Vesicular stomatitis virus has extensive oncolytic activity against human sarcomas: rare resistance is overcome by blocking interferon pathways. J Virol 85:9346–9358
75. Palmerini E, Agostinelli C, Picci P et al (2017) Tumoral immune-infiltrate (IF), PD-L1 expression

and role of CD8/TIA-1 lymphocytes in localized osteosarcoma patients treated within protocol ISG-OS1. Oncotarget 8:111836–111846
76. Park BH, Hwang T, Liu TC et al (2008) Use of a targeted oncolytic poxvirus, JX-594, in patients with refractory primary or metastatic liver cancer: a phase I trial. Lancet Oncol 9:533–542
77. Pol JG, Atherton MJ, Bridle BW et al (2018) Development and applications of oncolytic Maraba virus vaccines. Oncolytic Virother 7:117–128
78. Puhlmann M, Brown CK, Gnant M et al (2000) Vaccinia as a vector for tumor-directed gene therapy: biodistribution of a thymidine kinase-deleted mutant. Cancer Gene Ther 7:66–73
79. Ratti C, Botti L, Cancila V et al (2017) Trabectedin overrides osteosarcoma differentiative block and reprograms the tumor immune environment enabling effective combination with immune checkpoint inhibitors. Clin Cancer Res 23:5149–5161
80. Reichard KW, Lorence RM, Cascino CJ et al (1992) Newcastle disease virus selectively kills human tumor cells. J Surg Res 52:448–453
81. Ribas A, Dummer R, Puzanov I et al (2017) Oncolytic virotherapy promotes intratumoral T cell infiltration and improves anti-PD-1 immunotherapy. Cell 170(1109–1119):e10
82. Rizvi NA, Hellmann MD, Snyder A et al (2015) Cancer immunology. Mutational landscape determines sensitivity to PD-1 blockade in non-small cell lung cancer. Science 348:124–128
83. Roberts SS, Chou AJ, Cheung NK (2015) Immunotherapy of childhood sarcomas. Front Oncol 5:181
84. Rogulski KR, Wing MS, Paielli DL et al (2000) Double suicide gene therapy augments the antitumor activity of a replication-competent lytic adenovirus through enhanced cytotoxicity and radiosensitization. Hum Gene Ther 11:67–76
85. Russell SJ, Federspiel MJ, Peng KW et al (2014) Remission of disseminated cancer after systemic oncolytic virotherapy. Mayo Clin Proc 89:926–933
86. Sakuda T, Kubo T, Johan MP et al (2019) Novel near-infrared fluorescence-guided surgery with vesicular stomatitis virus for complete surgical resection of osteosarcomas in mice. J Orthop Res 37:1192–1201
87. Samstein RM, Lee CH, Shoushtari AN et al (2019) Tumor mutational load predicts survival after immunotherapy across multiple cancer types. Nat Genet 51:202–206
88. Sasaki T, Tazawa H, Hasei J et al (2011) Preclinical evaluation of telomerase-specific oncolytic virotherapy for human bone and soft tissue sarcomas. Clin Cancer Res 17:1828–1838
89. Sayles LC, Breese MR, Koehne AL et al (2019) Genome-informed targeted therapy for osteosarcoma. Cancer Discov 9:46–63
90. Schreiber RD, Old LJ, Smyth MJ (2011) Cancer immunoediting: integrating immunity's roles in cancer suppression and promotion. Science 331:1565–1570
91. Scott MC, Temiz NA, Sarver AE et al (2018) Comparative transcriptome analysis quantifies immune cell transcript levels, metastatic progression, and survival in osteosarcoma. Cancer Res 78:326–337
92. Senzer NN, Kaufman HL, Amatruda T et al (2009) Phase II clinical trial of a granulocyte-macrophage colony-stimulating factor-encoding, second-generation oncolytic herpesvirus in patients with unresectable metastatic melanoma. J Clin Oncol 27:5763–5771
93. Shlien A, Campbell BB, De Borja R et al (2015) Combined hereditary and somatic mutations of replication error repair genes result in rapid onset of ultra-hypermutated cancers. Nat Genet 47:257–262
94. Southam CM, Moore AE (1952) Clinical studies of viruses as antineoplastic agents with particular reference to Egypt 101 virus. Cancer 5:1025–1034
95. Streby KA, Geller JI, Currier MA et al (2017) Intratumoral injection of HSV1716, an oncolytic herpes virus, is safe and shows evidence of immune response and viral replication in young cancer patients. Clin Cancer Res 23:3566–3574
96. Thorne SH, Hwang TH, O'gorman WE et al (2007) Rational strain selection and engineering creates a broad-spectrum, systemically effective oncolytic poxvirus, JX-963. J Clin Invest 117:3350–3358
97. Toolan HW, Saunders EL, Southam CM et al (1965) H-1 virus viremia in the human. Proc Soc Exp Biol Med 119:711–715
98. Wedekind MF, Denton NL, Chen CY, Cripe TP (2018) Immunotherapy in pediatric oncology. Pediatr Drugs (in press) 20:395–408
99. Weiss N, Stroh-Dege A, Rommelaere J et al (2012) An in-frame deletion in the NS protein-coding sequence of parvovirus H-1PV efficiently stimulates export and infectivity of progeny virions. J Virol 86:7554–7564
100. White CL, Twigger KR, Vidal L et al (2008) Characterization of the adaptive and innate immune response to intravenous oncolytic reovirus (Dearing type 3) during a phase I clinical trial. Gene Ther 15:911–920
101. Whitley RJ, Kimberlin DW, Roizman B (1998) Herpes simplex viruses. Clin Infect Dis 26:541–553. quiz 554–5
102. Witlox AM, Van Beusechem VW, Molenaar B et al (2004) Conditionally replicative adenovirus with tropism expanded towards integrins inhibits osteosarcoma tumor growth in vitro and in vivo. Clin Cancer Res 10:61–67
103. Zhang ZY, Wang LQ, Fu CF et al (2013) Combination of targeting gene-viro therapy with recombinant Fowl-pox viruses with HN and VP3 genes on mouse osteosarcoma. Eur Rev Med Pharmacol Sci 17:767–776
104. Zhou Q, Xian M, Xiang S et al (2017) All-trans retinoic acid prevents osteosarcoma metastasis by inhibiting M2 polarization of tumor-associated macrophages. Cancer Immunol Res 5:547–559

Applying Osteosarcoma Immunology to Understand Disease Progression and Assess Immunotherapeutic Response

6

Hillary G. Pratt, E. Markel Justin, and Brock A. Lindsey

Abstract

Osteosarcoma, the most common malignant bone tumor in children and adolescents, remains a complicated disease to treat; no new treatments have been developed in more than three decades. Due to the importance of the immune system in osteosarcoma disease progression, immunotherapeutic strategies have been explored to potentially improve long-term survival. However, most immunotherapeutics have not reached the level of success hoped would occur in this disease. Understanding the immune system in osteosarcoma will be key to optimizing treatments and improving patient outcomes. Therefore, immunophenotyping can be used as a very powerful tool to help better understand the complexity of the immune response seen in osteosarcoma and in the use of immunotherapy in this malignancy. This book chapter will provide an overview of the known immune responses seen in this disease and potential developments for the future of immunophenotyping. Indeed, it appears that being able to track the immune system throughout the disease and treatment of patients with osteosarcoma could allow for a personalized approach to immunotherapy.

H. G. Pratt · E. M. Justin · B. A. Lindsey (✉)
West Virginia University School of Medicine, Morgantown, WV, USA
e-mail: blindsey@hsc.wvu.edu

Keywords

Immunology · Immunopathology · Immunotherapy · Myeloid · Granulocyte · Lymphoid · Lymphocyte · Cytokine

Introduction

Despite being the most common malignant bone tumor in children and adolescents, osteosarcoma remains a difficult tumor to treat. Over the past three decades, no new treatments for osteosarcoma have been developed; thus, improvements in long-term survival have remained stagnant since the first addition of chemotherapy for the disease. Nonetheless, although osteosarcoma has been deemed a highly pleomorphic, immunologically cold tumor historically speaking, more recent data has suggested immunotherapy may be successful [1]. Recent studies have shown osteosarcoma to be an immunologically "hot" cancer as both the innate and adaptive immune systems are affected not only in the local tumor microenvironment but also systemically. In order to overcome nonresponse and immunoresistance, both of which have been problematic in immunotherapy, it logically follows that we should also attempt to understand the impact of osteosarcoma on the immune system. One mechanism by which these alterations can be assessed involves

E. S. Kleinerman, R. Gorlick (eds.), *Current Advances in the Science of Osteosarcoma*, Advances in Experimental Medicine and Biology 1258, https://doi.org/10.1007/978-3-030-43085-6_6

immunophenotyping various tissue samples from the patient, including tumor biopsies, metastatic tumor biopsies, splenic biopsies, and serum samples. Immunophenotyping, commonly performed by immunohistochemistry (IHC) or flow cytometry, allows us to quantify particular subsets of white blood cells using highly specific fluorescent antibodies. Furthermore, the cell surface epitopes marked by these antibodies can also be quantified to assess the status of the immune cell in question. Thus, this chapter outlines the known immunological alterations that occur with osteosarcoma disease progression and subsequently discusses the immunotherapies that have been explored to combat this malignancy.

Assessing Osteosarcoma Disease Progression

Osteosarcoma Immunology: Innate Immune System

Myeloid Lineage

Granulocytes

The most prominent granulocyte in osteosarcoma immunopathology is the neutrophil, which is one of the main players in the acute inflammatory response. Following injury or infection, the affected tissue releases interleukin (IL)-8, which is a neutrophil chemoattractant and degranulation agent. Upon arrival to the site of the insult, they release inflammatory mediators and phagocytose cellular debris. As neutrophils are one of the first immune cells to arrive in infected tissues, it is likely that they contribute to the tumor microenvironment although the details of this are not well defined. Indeed, little work has been done to examine neutrophils in the context of osteosarcoma disease progression; however, there are a few studies with some indirect evidence to support a role for neutrophils in tumor progression. Postoperative infections have been associated with significantly increased event-free survival and 5-year overall survival (100% and 100% with infection versus 54% and 43% without, respectively; $P = 0.01$ [2]), while high neutrophil-lymphocyte ratios (NLR) are associated with poor prognoses ($P < 0.05$); furthermore, multivariate Cox regression has shown that NLR is a top risk factor associated with death in osteosarcoma patients [3]. Osteosarcoma patients have also been found to have high serum levels of IL-8, as well as polymorphisms in its coding gene that are linked to increased likelihood of metastasis [4, 5]. High expression levels of another neutrophil chemoattractant, CXCL5, have also been linked to advanced clinical stage and metastasis [6]. Interestingly, both IL-8 and CXCL5 have been found to have direct effects on osteosarcoma tumor cells themselves by inducing migration and invasion. Anther cell type, related to neutrophils, polymorphonuclear myeloid-derived suppressor cells (PMN-MDSCs) will be discussed further in the MDSC section below.

Mononuclear Phagocytic System: Monocytes, Macrophages, and Dendritic Cells

Monocytes

Monocytes are antigen-presenting cells (APCs) that are one of the main means of communication between the innate and adaptive arms of the immune system. They are produced in the bone marrow and migrate to areas of inflammation where they differentiate into macrophages or dendritic cells. Early in vitro studies found that monocytes were cytotoxic to osteosarcoma tumor cells leading to the belief that they contributed to tumor immunosurveillance [7, 8]; however, multiple in vivo studies have suggested a more immunosuppressive role for monocytes in the context of a developing malignancy. Notably, it was discovered that in osteosarcoma patients, circulating T cell immunoglobulin and mucin-domain containing-3 (Tim-3)-positive monocytes suppress antitumor T-helper type 1 (T_H1) responses by interacting with Tim-3+ T cells, naïve CD4+ T cells, and galectin-9 (Gal-9)-expressing T regulatory cells (T_{reg}). Additionally, a group in 2017 found clinical relevance in the ratio of absolute peripheral monocyte count after initial treatment divided by absolute monocyte count before initial treatment (monocyte ratio); when a patient's monocyte ratio was greater than 1, he or she was over five times more likely to develop metastases compared to patients with ratios less than or equal to 1 (OR 5.367; 95% CI, 3.083–9.343) [9]. Osteosarcoma cells release the cytokine IL-34 which recruits inflammatory monocytes to areas of tumorigenesis; upon arrival, they mature into tumor-associated macrophages (TAM) discussed in the subsequent section.

Macrophages

Macrophages, derived from circulating monocytes, reside in many tissues throughout the body and phagocytose cellular debris, tumor cells, and pathogens. Tissue-resident macrophages can produce a diverse response to inflammation, infection, healing, and cancer based on the cytokines in the local microenvironment. Furthermore, macrophages can be delineated into a number of subsets based on their cell surface expression, including classically activated (M1) macrophages, alternatively activated (M2) macrophages, TAMs, and myeloid-derived suppressor cells (MDSCs). Specifically, in tumorigenesis, TAMs of the M1 subtype have a protective effect, while the M2 subtype and MDSCs suppress antitumor immune responses [10]. The M1/M2 TAM subpopulations will be discussed here, while MDSCs will be discussed in a subsequent section.

In osteosarcoma, immunohistochemical stains of resected primary tumors have shown macrophages to be the predominant infiltrating cell type [11]. A number of studies have shown a correlation between TAMs and metastases in osteosarcoma [10–12]. Similar to their activity in other solid tumors, M1 macrophages have been shown to suppress tumorigenesis in osteosarcoma. One in vitro study demonstrated that IFNγ-activated $CD86^+HLA\text{-}DR^+$ M1 macrophages can inhibit osteosarcoma cell growth [13]. In patients, the quantity of M1 macrophages, defined by iNOS positivity, in the primary tumor was negatively correlated with metastatic disease (p = 0.001) [11]. This observation may be related to the ability of the M1 macrophage to promote a T_H1 antitumor response [10]. Although M1 macrophages may be the first to invade the primary tumor, in vivo studies have shown that early $F4/80^+CD163^-$ M1 macrophages are replaced by $F4/80^+CD163^+$ M2 macrophages between the first and third weeks of tumor establishment [14]. According to one group's in vivo study, M2 TAMs are likely recruited to the primary tumor due to IL-34 overexpression by osteosarcoma cells [15]. In contrast to M1 macrophages, $CD163^+$M2 macrophages generate a T_H2 response [16]. Furthermore, in vitro studies of M2 macrophages have shown these cells enhance the migratory capacity of osteosarcoma cells [17]. M2 TAMs likely promote tumorigenesis via immunosuppression, production of matrix-degrading enzymes, and stimulation of angiogenesis. Notably, $CD163^+$ M2 macrophages have been shown to correlate with increased density of $CD31^+$ blood vessels and $CD146^+$ vascular cells in vivo [11]. Furthermore, patients with a higher proportion of $CD163^+$ M2 macrophages present with metastases [18]. In addition to stimulating neoangiogenesis, M2 macrophages overexpress epidermal growth factor (EGF), which can bind epidermal growth factor receptor (EGFR) on osteosarcoma cells to promote growth [14].

Although many studies attempt to stratify M1 and M2 populations, cancer is a complex disease process that may stimulate a more intermediate TAM phenotype. For example, Buddingh et al. found that a higher percentage of TAMs, irrespective of M1 or M2 phenotype, was correlated with a higher overall survival rate in patients [12]. Similarly, a phase III clinical trial found a higher overall survival rate and higher rate of metastasis-free survival in patients whose primary tumor samples showed >50% $CD163^+$ M2-like macrophages with a trend toward significance for $CD68^+$ M2 macrophages [16]. Furthermore, one group's in vivo study mimicked these observations; both M1 markers, Cxcl9 and iNOS, and M2 marker, Tgm2, were significantly elevated in splenic biopsies of all tumor-bearing mice. Taken together, these findings suggest that macrophage polarization in malignancy is complex and requires more than one cell marker to accurately capture polarization [19].

Dendritic Cells

Like macrophages, conventional dendritic cells (cDCs) are derived from circulating monocytes and serve to prime and activate a T cell response against neoantigens produced by tumor cells [20, 21]. CDCs can be divided into cDC1 cells and cDC2 cells with the CD103$^+$ cDC1s performing the roles of antigen presentation and cytotoxic T cell (T_c) activation. In early in vitro studies, CD14$^+$ DCs, isolated cDC1s, and isolated cDC2s respectively cultured with osteosarcoma cell lines showed significantly decreased expression of maturation markers, including CD80, CD83, CD86, CD40, HLA-DR, and CCR7 [20, 22]. Furthermore, in contrast to control cDC1 cells, which express IL-12, cDC1s cultured with osteosarcoma cells increased expression of IL-10 but did not express IL-12. Isolated cDC2s cultured with osteosarcoma cells also significantly increased IL-10 production. Elevated levels of IL-10 prime the T_H2 response and suppress the T_H1 response, which may inhibit an adequate immune response to the primary tumor [20]. Although in vitro studies of cDCs have shown a possible role for DCs in osteosarcoma pathogenesis, it should be noted that a lower level of DCs has been found to infiltrate pediatric tumors, including osteosarcoma, when compared to adult onset tumors. The relative lack of infiltrating DCs, taken with the observation that osteosarcoma can suppress a cDC-induced T_H1 antitumor response, may suggest a role for vaccination with tumor-antigen-loaded dendritic cells in treatment of osteosarcoma, which will be discussed later [23].

Myeloid-Derived Suppressor Cells

Myeloid-derived suppressor cells (MDSCs) constitute a population of myeloid cells that have failed to achieve maturation to dendritic cells, monocytes, or granulocytes. Associated with pathology, especially chronic inflammation and cancer, MDSCs can be categorized as CD11b$^+$HLA-DR$^-$CD14$^+$CD15$^-$ monocytic MDSCs (M-MDSCs) or CD66b$^+$CD15$^+$CD14$^{-/dim}$CD33^{dim}HLA-DR$^-$ polymorphonuclear MDSCs (PMN-MDSCs) in humans [24]. While M-MDSCs suppress T cell activity via both antigen-dependent and antigen-independent mechanisms, including production of reactive oxygen species through arginase (Arg) and nitric oxide synthase (iNOS), the PMN-MDSCs suppress T cell activity mainly by an antigen-dependent pathway [25]. For patients with solid tumors, increased levels of MDSCs have been correlated with lower overall survival [26]. In vivo studies of MDSCs in murine osteosarcoma models show elevated levels of both PMN-MDSCs and M-MDSCs in peripheral blood and spleen (Unpublished data) [27]. Although elevated early in the disease process, the M-MDSC cell levels trended downward as the disease progressed, while the PMN-MDSC population remained stable as shown in Fig. 6.1 (Unpublished data). Similarly, in pediatric sarcoma patients, CD14$^+$HLA-DR$^{lo/-}$ M-MDSCs were found to be increased compared to healthy controls [28]. Regarding HLA-DR, a marker for antigen-presenting cells of the myeloid lineage, a decrease has been shown on monocytes of patients with aggressive sarcoma, which is postulated to be one of the initial steps in development of M-MDSCs from monocytes [28, 29]. Studies done in vivo have shown that HLA-DR expression decreases on myeloid-type cells as PD-L1 expression increases. Specifically, the ratio of PD-L1 to HLA-DR (P [1]/M[II]) in peripheral blood increased with osteosarcoma and trended upward in both peripheral blood and spleen as the disease progressed. Moreover, the systemic M-MDSC P [1]/M[II], including blood, bone marrow, and spleen tissues, showed a positive correlation with the lung M-MDSC ratio that was significant in blood (Fig. 6.2: $R = 0.972$; $p < 0.01$) and the spleen (Fig. 6.2: $R = 0.998$; $p < 0.01$) and trending toward significance in bone marrow (Unpublished data).

IL-18 expression has also been shown to be elevated in vivo on MDSCs with osteosarcoma disease progression. As inhibition of IL-18 decreased levels of MDSCs in both peripheral blood and the primary tumor, this cytokine may play a role in the development and recruitment of MDSCs [27]. A more immature population of MDSCs has also been identified as Lin (including CD3, CD14, CD15, CD19, and

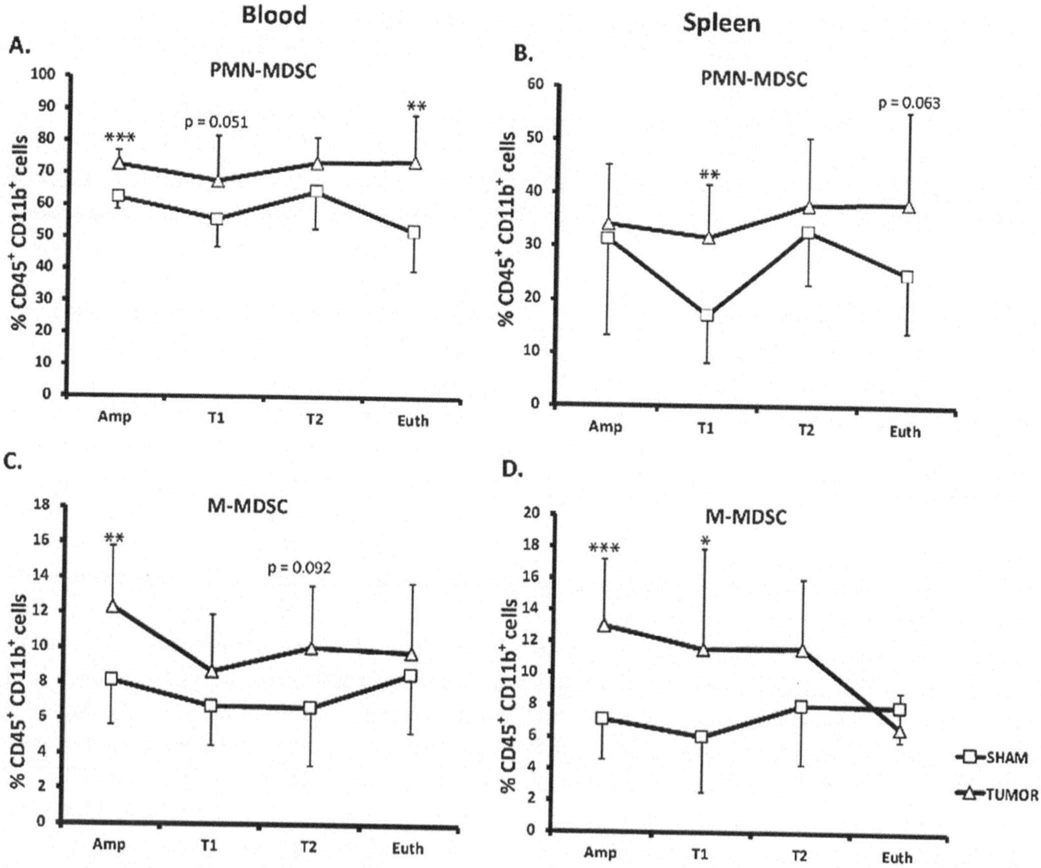

Fig. 6.1 Systemic PMN-MDSCs remain relatively constant, while M-MDSCs decrease during disease progression in vivo. $***p = 0.001$, $**p = 0.01$, $*p = 0.05$

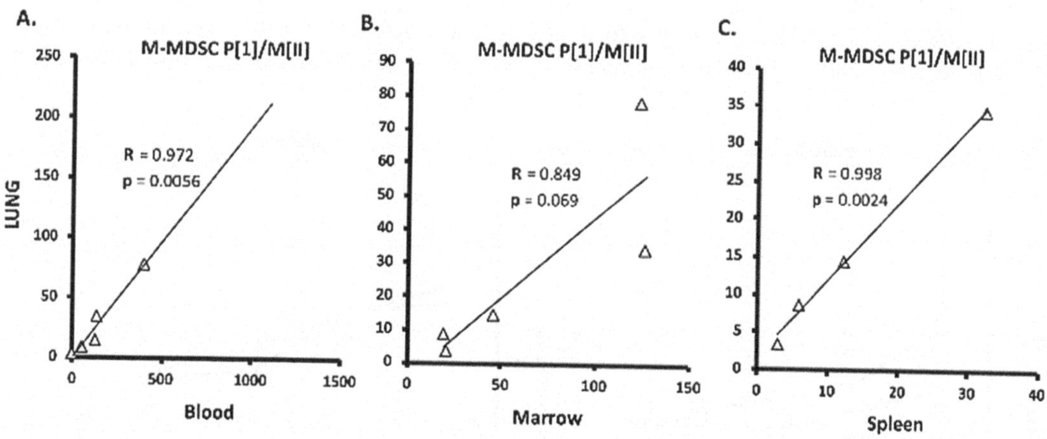

Fig. 6.2 Ratio of PD-L1 to HLA-DR (P [1]/M[II]) on M-MDSCs correlated with ratio of M-MDSCs in the lung in vivo

CD56)-HLA-DR-CD33^{+} but was not found to be elevated in pediatric sarcoma and has not been studied further [25, 28]. Certainly, due to the apparent role of MDSCs in malignancy, these cell populations may be targeted to improve treatment strategies, as will be discussed with treatment strategies.

Osteosarcoma Immunology: Adaptive Immune System

Lymphoid Lineage

T Lymphocytes

Helper T Cells

Helper T (T_H) cells, commonly defined as being CD4^{+}, can generate several different types of immune response: T_H1 or T_H2 being the most common. T_H1 cells primarily secrete IFNγ and TNFα cytokines enhancing both the effector and memory functions of CD8^{+} cytotoxic T cells (CTLs) [30–31]. CD4^{+} T_H cells can also enhance the antitumor response by optimizing the cDC response, which in turn stimulates the CTLs [32]. Regarding tumor rejection in particular, T_H1 cells can recruit NK cells and macrophages to form an antitumor response. Often considered to be in opposition of T_H1 cells, T_H2 response involves IL-4, IL-5, and IL-13, which together activate a humoral response [30]. Although IL-4 recruits macrophages and eosinophils in murine models, both IL-4 and IL-13 have been shown to activate macrophages toward an M2-like immunosuppressive phenotype [30, 33, 34]. The role of another distinct CD4^{+} T_H cell type, T_H17, has been debated for its potential antitumor function [35]. One in vitro study found that IL-22, commonly produced by T_H17 cells in vivo, stimulated the proliferation and invasion of two different osteosarcoma cell lines [36]. Thus, an abundance of CD4^{+} T_H17 cells in the primary tumor might suggest a poor prognosis.

While numerous studies have shown that CD4^{+} T_H can improve the antitumor response, one group has shown that CD8^{+} CTLs can have an antitumor effect in osteosarcoma without CD4^{+} T_H aid [31]. Nonetheless, for adult osteosarcoma, CD4^{+} T_H cells likely play a role in the immune process as this cell type represented the predominant TIL and demonstrated lytic activity against several osteosarcoma cell lines [37]. Furthermore, an in vivo study of the immunologic changes during osteosarcoma pathogenesis showed that percentages of CD45^{+}CD4^{+}CD8^{-} T_H trended downward in the peripheral blood and spleen as the disease progressed, which is shown in Fig. 6.3 (Unpublished data). Similarly, for peripheral blood samples of patients with osteosarcoma, there were decreased overall numbers of CD4^{+} T cells [28]. In addition to the overall decrease in CD4^{+} T cells, patients with osteosarcoma exhibited increased expression of PD-1

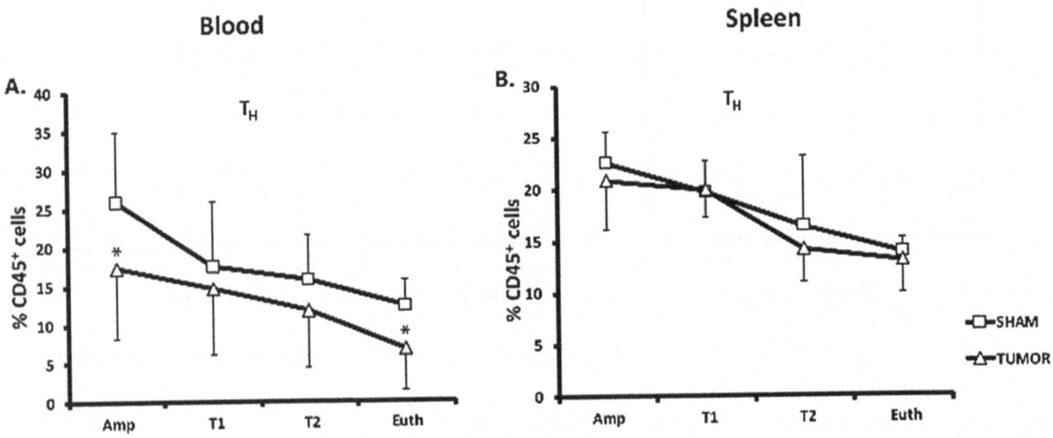

Fig. 6.3 Helper T cells decrease systemically during disease progression in vivo. $*p = 0.05$

and Tim-3, markers of T cell dysfunction and exhaustion, respectively, in $CD4^+$ T_H cells compared to patients without osteosarcoma [38–40]. Furthermore, elevated levels of $Tim\text{-}3^+CD4^+$ T cells in the peripheral blood were shown to be correlated with later stage of disease and increased expression of IFNγ and IL-2, which stimulate a T cell response. Elevated levels of peripheral $Tim\text{-}3^+CD4^+$ T cells also correlated with decreased overall survival in osteosarcoma patients [41].

In addition to traditional T_H cells, $CD4^+CXCR5^+$ follicular T-helper (T_{FH}) cells, which promote a B cell response and maintain germinal centers in secondary lymphoid organs, have been found to be of importance in osteosarcoma [42–43]. Indeed, levels of $CXCR3^-CCR6^-$ T_H2-type cells and $CXCR3^+CCR6^+$ T_H17-type cells were found to be nearly two times greater in osteosarcoma patients, while levels of $CXCR3^+CCR6^-$ T_H1-type cells were lower. Furthermore, T_{FH} levels were found to be elevated peripherally in patients with both metastatic and high-grade osteosarcoma, specifically the T_H1 and T_H17 types but not the T_H2 subtypes [42]. Although $CD4^+$ T_H and T_{FH} have been shown to play a role in the disease process of osteosarcoma, these cells have received less attention than CTLs.

Cytotoxic T Cells

Data accumulated from a number of human cancers have shown that the number, type, and location of tumor-infiltrating lymphocytes (TILs) in the primary tumor can be prognostic for overall survival. In particular, an increased number of $CD8^+$ TILs has been shown to correlate with increased survival rates and decreased rates of recurrence [44]. $CD8^+$ TILs are also prominent in osteosarcoma as they represent the major infiltrating cell type in pediatric osteosarcoma and make up 15.5% of infiltrating cells in adult osteosarcoma [45, 37]. An early study of primary tumor samples from patients with high-grade osteosarcoma found $CD3^+CD8^+$ TILs to be predominate infiltrating inflammatory cells. As the pattern of $CD3^+CD8^+$ CTL infiltration did not differ significantly from benign bone tumor samples, $CD8^+$ TILs in osteosarcoma were not thought to have any prognostic significance [46]. Later, in a retrospective study, it was found that higher $CD8^+$ TIL density correlated with improved survival in osteosarcoma. Additionally, in this study, a positive relationship between the CD8/Foxp3 TIL ratio and overall survival was observed [47]. This relationship between $CD8^+$ cells and $Foxp3^+$ cells has been mirrored in other cancer types as CTLs have been shown to upregulate T_{reg} cells [47, 48]. In addition to demonstrating a relationship with overall survival, levels of $CD8^+$ TILs have also been found to correlate with pretreatment metastases; patients presenting with metastases had lower levels of $CD8^+$ TILs [16]. Further studies of primary tumors in patients with poor disease outcome showed that higher density of $CD3^+CD8^+$ TILs correlated with increased expression of PD-L1 on tumor cells and immune cells, which may be a mechanism by which the tumor evades the immune system and causes T cell exhaustion [19, 49, 50]. Furthermore, patients with high expression of PD-L1 had a median overall survival of 28 months, while median survival for patients with low PD-L1 expression was 89 months [50]. PD-1 has also been shown to be upregulated on $CD8^+$ T cells in patients with osteosarcoma [37]. Thus, the PD-1/PD-L1 axis has become an attractive therapeutic target in osteosarcoma. Overall, $CD8^+$ TILs have been shown to improve outcomes in osteosarcoma, as they have cytolytic activity, but can be subject to T cell exhaustion in patients with severe disease [37].

Peripheral $CD8^+$ CTLs have also been shown to have significant role in the pathogenesis of osteosarcoma. One in vivo study demonstrated that the percentages of $CD45^+CD4^-CD8^+$ CTLs decreased in the peripheral blood and spleen as the disease progressed as demonstrated in Fig. 6.4 (Unpublished data). Nevertheless, the amount of circulating $CD8^+$ CTLs was found to be higher than levels of $CD8^+$ T cells within the primary tumor in patients [37]. Some of these cells were found to express Tim-3, a marker of T cell exhaustion, and were particularly prevalent in patients at later stages of disease. Furthermore, increased levels of $Tim\text{-}3^+CD8^+$ T cells in the periphery negatively correlated with serum levels

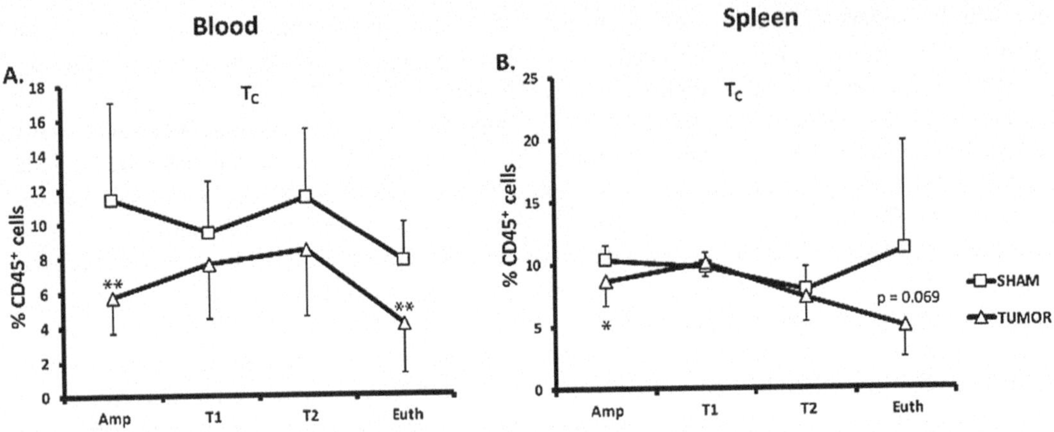

Fig. 6.4 In vivo osteosarcoma progression induces a decrease in percent population of systemic cytotoxic T cells. $**p = 0.0.1$, $*p = 0.05$

of IFNγ and TNFα. Since these cytokines activate the T_H1 response, as discussed previously, it follows that overall survival was decreased in patients with high levels of circulating Tim-3⁺CD8⁺ T cells [39, 41]. For pediatric patients with osteosarcoma, cytotoxic T lymphocyte-associated protein 4 (CTLA-4), a T cell inhibitory receptor, was found to be upregulated on $CD8^+$ cells circulating in the peripheral blood [28]. This receptor has been theorized to be another immunosuppressive immune escape mechanism that can create resistance in response to solitary immune checkpoint blockade. Therefore, since CTLA-4 has been shown to play a role in osteosarcoma progression, this receptor has become a target for immune blockade, which will be discussed in a subsequent section.

Regulatory T Cells

Regulatory T cells (T_{regs}) play a role in peripheral self-tolerance. While these $CD4^+CD25^+$ T cells have been shown to prevent autoimmunity, particularly in vivo, an abundance of these cells can also suppress antitumor immunity to tumors overexpressing self-antigens. The T_{reg} population has been identified as expressing CD25 (IL-2αR), glucocorticoid-induced TNF receptor (GITR), CTLA-4, and Foxp3 transcription factor. In vivo studies have shown that the presence of T_{reg} cells promotes tumor growth and inhibits $CD8^+$ CTL immunogenic response to the tumor [51]. For example, when $CD4^+CD25^+$ T_{reg} cells are cocultured with $CD8^+$ T cells, the $CD8^+$ T cells show significant suppression of proliferation and production of IFNγ, IL-2, and CD25 [52]. Interestingly, splenic T_{reg} cells in vivo are increased later in disease progression (Fig. 6.5) [19]. Similarly, a retrospective study of human osteosarcoma samples demonstrated that increased $Foxp3^+$ T_{regs} infiltrating the tumor indicated a worse prognosis. In fact, as $Foxp3^+$ T_{regs} increased relative to $CD8^+$ CTLs, survival decreased independent of primary metastasis and chemotherapy response. This relationship is likely due to T_{reg}-mediated inhibition of CTLs [47]. Furthermore, an increase in $CD3^+CD8^-Foxp3^+$ T cells has been found in local relapses of osteosarcoma, potentially suggesting

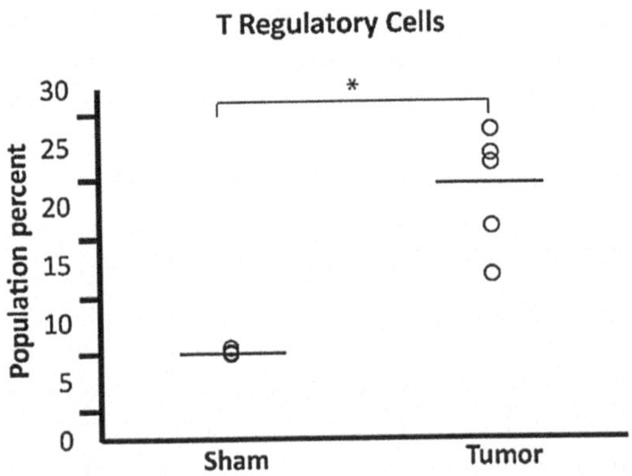

Fig. 6.5 Regulatory T cell populations (CD45 + D4 + Foxp3+) are increased in the spleens of tumor-bearing mice

a tumor escape mechanism [49]. Importantly, while Foxp3$^+$ cells have classically been classified as T_{reg} cells, it should be noted that recent evidence suggests that Foxp3 can be upregulated in conventional TILs, which suggests that Foxp3 alone cannot be utilized to delineate T_{reg} populations [47]. Also, despite the apparent role of T_{regs} in the local tumor microenvironment, these cells have not been shown to be elevated systemically in pediatric osteosarcoma [28].

Lymphocytes

In addition to tumor-infiltrating T lymphocytes, B lymphocytes can also be found in the tumor microenvironment; these cells, however, have been studied less than their T cell counterparts [53]. Activated B cells can generate autoantibodies that promote an antitumor response in malignancy [53, 54]. Furthermore, B cells may enhance the T cell response via antigen presentation and costimulation [53–55]. For several malignancies, the presence of CD20$^+$ TILs in addition to CD4$^+$ and CD8$^+$ TILs has been correlated with better overall outcomes when compared to independent CD20$^+$ B cell or CD8$^+$ T cell tumor infiltration, respectively [53, 56, 57]. In particular, for soft tissue non-gastrointestinal stromal tumor sarcomas, increased CD20$^+$ TILs correlated with improved disease-specific survival [58]. Additionally, a study of pediatric patients with soft tissue sarcomas found increased memory class switched B cells [28]. Despite these findings, utilizing CD20$^+$ TILs as a prognostic indicator of disease outcome and survival may be difficult as some, including immature and those producing IL-10 and TGF-β, CD20$^+$ B cells may suppress the antitumor response [53]. Further study into the effects of CD20$^+$ B cells should be done to elucidate the role of these cells in osteosarcoma.

Natural Killer Cells

As lymphocyte members of the innate immune system, natural killer (NK) cells play a role in immunosurveillance and are defined by CD3 negativity as well as CD56 or CD16 positivity [59]. In cancer specifically, these granulocytic NK cells can release their cytotoxic granules to directly kill malignantly transformed cells. Released tumor cell antigens are then presented to B and T lymphocytes via APCs to generate an adaptive immune response [60]. NK cell-mediated killing is negatively correlated to expression of HLA class I molecules, which serves to maintain self-tolerance. This mechanism, however, can be exploited by malignant cells in order to evade immunosurveillance [60, 61]. In osteosarcoma, analysis of HLA class I expression in primary and metastatic tumors showed downregulation in 52% and 88% of samples, respectively [62]. To inhibit NK cells, HLA class I molecules on tumor cells interact with killer immunoglobulin-like receptor (KIR), which is expressed on the NK cell surface and commonly upregulated in malignancy [60, 63]. For patients with osteosarcoma, NK cells can express combinations of all three KIR molecules, including KIR2DL1, KIR2DL2/2DL3, and KIR3DL1, and thus display decreased activity [63].

Early in osteosarcoma disease pathogenesis, peripheral blood NKG2D$^+$ NK cells do not differ in number, phenotype, or functionality [64]. In general, however, circulating NK cells is commonly lower in osteosarcoma patients compared to healthy patients [65, 66]. These findings may be explained by in vivo studies which suggest that CD45$^+$NKp46$^+$ NK cells are significantly increased systemically with osteosarcoma [19]. As the disease progresses, though, NK cell populations in the peripheral blood decrease. These in vivo findings may be explained by the relative increase in M-MDSCs and PMN-MDSCs in osteosarcoma, which inhibit NK cells (Unpublished data). Due to the potent antitumor activity of NK cells but dysregulation in osteosarcoma, these cells have become an important focus for future clinical trials [66].

Assessing Immunotherapeutic Response in Osteosarcoma

Checkpoint Blockade

Immune checkpoints serve to preserve self-tolerance and prevent autoimmunity and excess immune stimulation; however, these pathways can also be used by tumor cells as a means of escaping immune system detection. Due to their role in tumor immune tolerance, these T cell checkpoint molecules, programmed death receptor 1 (PD-1) and CTLA-4, are therapeutic targets for advanced cancers [67]. Since the PD-1/PD-L1 axis has been shown to negatively impact overall survival in patients with osteosarcoma, these molecules are of particular interest [49, 50, 68]. In vivo anti-PD-1 treatment facilitated macrophage-mediated decrease in both micro- and macrometastases but not primary tumor growth [68, 69]. Specifically, Ki-67, a marker of tumor cell proliferation, expression decreased, while TUNEL, a marker of cell apoptosis, expression increased in tumor tissue [69]. Anti-PD-1 also increased $CD4^+$ T cell, $CD8^+$ T cell, $NKp46^+$ NK cell, and $F4/80^+CD11b^+$ macrophage infiltration into lung metastases [68, 69]. Furthermore, anti-PD-1 shifted the macrophage polarization from $CD163^+$ M2 to $CD86^+$ M1 macrophages in lung metastases, which facilitates an antitumor response [69]. Importantly, treatment with anti-PD-1 downregulated tumor cell expression of PD-L1, which suggests utilization of another mechanism for immune suppression [68].

Therapeutic strategies have also targeted PD-L1 in an attempt to interrupt the PD-1/PD-L1 axis. Similar to anti-PD-1 treatment, anti-PD-L1 treatment downregulated expression of PD-L1 by primary tumor and metastatic osteosarcoma cells. Increased length of treatment in vivo did not enhance response to anti-PD-L1, which further suggests another pathway for immune suppression. Interestingly, following treatment with anti-PD-L1, $CD8^+$ TILs showed decreased PD-1 expression and increased CTLA-4 expression [19, 70]. CTLA-4 is a membrane glycoprotein expressed by activated effector T cells and tumor cells which represses T cell proliferation, cell cycle progression, and cytokine production [71]. Furthermore, in response to anti-PD-L1, the tumor cells upregulated CD80 and CD86, which bind to CTLA-4 to inhibit T cell activation [71, 72]. Thus, blockade of both PD-L1 and CTLA-4 in vivo completely controlled metastases and increased long-term survival to 60% from 0% shown with anti-PD-L1 treatment alone. Combination of anti-PD-L1 treatment with doxorubicin did not similarly enhance long-term survival [70]. In addition to stimulating changes in tumor cells, in vivo anti-PD-L1 treatment also has an effect on systemic immune cells. Specifically, total percentage of monocytes/macrophages increased in response to anti-PD-L1. Additionally, the expression of M1 markers, Nos2 and Cxcl9, and M2 marker, Tgm2, decreased resulting in a near normal M1/M2 macrophage ratio and returned the immune phenotype to that of sham status (Fig. 6.6) [19].

Regarding the efficacy of immune blockade in humans, anti-PD-L1 has shown a durable response and prolonged disease stabilization across a variety of tumor types [72]. A phase II trial for pembrolizumab monotherapy, an anti-PD-1 antibody, showed a confirmed partial response to treatment in 5% of patients with osteosarcoma. This response, however, resulted in greater than 50% tumor shrinkage for 6 months. Use of muramyl tripeptide phosphatidylethanolamine (mifamurtide), which will be discussed later, has been suggested to enhance immune cell infiltration into metastases, thereby potentially improving the efficacy of pembrolizumab in osteosarcoma [73]. Efficacy of anti-PD-1 treatment may also be enhanced by combination therapy. For example, trabectedin, an antitumor medication approved in Europe for treatment of soft tissue sarcomas, decreases osteosarcoma tumor cell proliferation and lung metastases in vivo. Following this treatment, $CD8^+$ TILs exhibited upregulation of PD-1 suggesting a potential role for trabectedin in immune blockade. Indeed, treatment with both trabectedin and anti-PD-1 inhibited tumor growth and enhanced T cell activation to a greater degree than either treatment alone [74].

Other Immunotherapeutics

Cytokine Stimulation

Cytokines play an important role in the coordination of immune response against tumor antigens. Due to their unique ability to modulate the immune system, these molecules have been

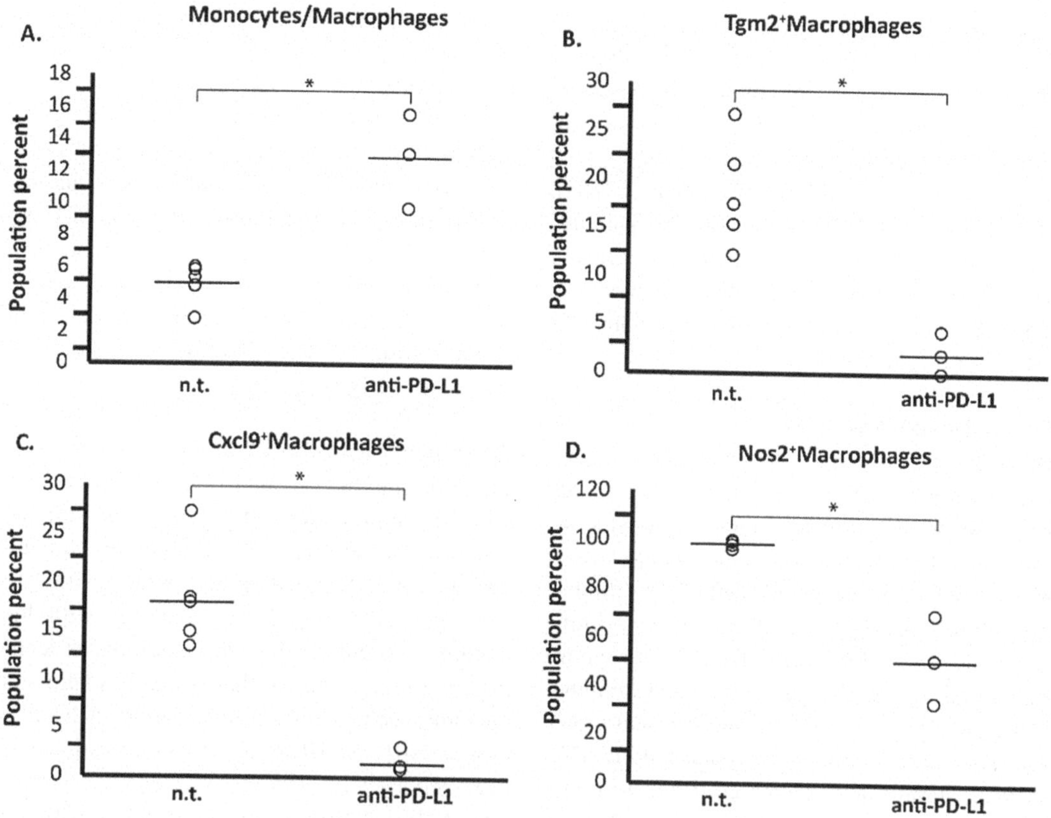

Fig. 6.6 In vivo treatment with anti-PD-L1 reduces macrophage markers to near normal

of interest for immunotherapy and have been implemented as treatment in a number of malignancies. One such cytokine, interferon alpha (IFNα), has demonstrated the ability to inhibit proliferation and induce apoptosis in tumor cells as well as inhibit angiogenesis [67]. Early studies of IFNα showed the ability to inhibit growth of osteosarcoma cell lines in vitro and xenograft osteosarcoma tumor growth in vivo [67, 75–76]. Relatively little clinical data is available regarding IFNα treatment in osteosarcoma. Two early clinical studies showed 5-year recurrence-free survival as well as a 10-year metastases-free survival rate and a disease-specific survival rate of 39% and 43%, respectively [77–80]. Furthermore, with elevated dose, the disease-free survival rate increased to 63% with higher treatment dose, indicating a possible dose-response relationship that mirrors in vivo studies [77, 78, 80]. Combination therapy including methotrexate, doxorubicin, and cisplatin with pegylated IFNα-2b demonstrated improved event-free survival and overall survival; however, these differences were not significant [81].

IL-2 is another cytokine that has been implemented as immunotherapy for malignancy, specifically osteosarcoma. Produced mostly by antigen-stimulated $CD4^+$ T cells with some production by NK cells, $CD8^+$ T cell, and activated DCs, IL-2 stimulates growth and differentiation in $CD8^+$ T cells, maintains $CD4^+$ T_{reg} populations, and drives differentiation of $CD4^+$ T cells into effector cells [82]. Early in vitro studies of IL-2 suggest it indirectly stimulates NK cell activity by increasing IFNγ [83]. Furthermore, IL-2 has been shown to stimulate lymphokine-activated killer cells (LAK), which can be primed to form an antitumor response; thus, IL-2 has received attention as a potential therapeutic agent [67, 82, 84–86]. IL-2 combined with autologous LAK cells first demonstrated the ability to mediate tumor regression in a study of 25 metastatic patients that had failed standard chemotherapy [82, 84]. In patients with metastatic osteosarcoma at diagnosis, the 3-year event-free survival rate and overall survival rate were 34.4% and 45.0%, respectively, with IL-2 treatment [67, 85]. Despite the positive tumor response in patients, IL-2 may simultaneously stimulate high levels of T_{reg} cells, thereby stimulating an environment tolerant of tumorigenesis rather than stimulating an antitumor response [51].

Dendritic Cell Vaccines

Dendritic cells have the ability to prime the adaptive immune system, specifically a T cell response, in response to tumor antigens; however, in osteosarcoma and other malignancies, these DCs fail to mature and adequately stimulate a T cell response [20, 21, 23, 87]. Thus, priming a patient's own dendritic cells ex vivo against tumor antigens has become a potential immunotherapy for patients with advanced disease [87]. In vivo studies of tumor-lysate-pulsed DCs showed significantly increased CTL response and IFNγ levels, while levels of IL-4 were decreased. Increased expression of costimulatory CD86, CD83, CD205, and CD209 in the DCs correlated with the allogenic T cell response [88]. Furthermore, combining the tumor-lysate DC treatment with anti-CTLA-4 antibody treatment increased the number $CD8^+$ TILs, decreased the amount of $Foxp3^+$ T_{reg} cells in metastatic tumor and the spleen, and increased serum levels of IFNγ. This combined treatment reduced the metastatic tumor burden and prolonged overall survival [79]. Similarly, in vivo studies combining tumor-lysate-pulsed DCs with agonist anti-GITR increased the number of $CD8^+$ TILs, increased serum IFNγ, decreased serum IL-10, and decreased numbers of $CD4^+Foxp3^+$ T_{reg} cells in the spleen [90].

In response to treatment with tumor-lysate-pulsed DCs, patients with metastatic pediatric malignancy generated a specific T cell response allowing for tumor regression. Levels of IFNγ were increased ten-fold in response to tumor antigen in these patients [91]. In patients with pediatric sarcoma treated with tumor-lysate-pulsed DCs, patients having an immune response had a 73% 5-year overall survival rate, while those without a response had a 37% 5-year overall survival rate [92]. In contrast, patients with advanced soft tissue sarcoma treated with DCs pulsed with autologous tumor lysate had increased levels of IFNγ and IL-12, which stimulate a T_H1 response; nonetheless, only 1 of 35 patients achieved a partial response. These findings suggest that immune stimulation is not enough to overcome advanced malignancy [93].

Genetically Modified T Cells

Chimeric antigen receptor T cells consist of a specific antibody-derived single-chain variable fragment (scFv) with a T cell signaling domain such that the effector function is released when the target cell is bound. One challenge with this therapy, particularly for osteosarcoma, is identifying a tumor antigen that is highly expressed on tumor cells with low expression in healthy tissues [94]. As HER2 gene amplification has been reported in osteosarcoma and is particularly linked with poor outcome, administration of HER2-CAR T cells in patients with osteosarcoma resulted in stable disease for 24% of patients [94–95]. Three of these patients had the tumor removed with one having more than 90% necrosis. IL-8 was found to be significantly elevated 1 week after infusion and remained elevated for 4 weeks, while levels of HER2-CAR-T cells lasted for approximately 6 weeks without toxicity [95]. Pediatric sarcoma has also been shown to strongly express GD2, a glycosphingolipid [94]. Although GD2-CAR T cells demonstrated lytic activity against $GD2^+$ tumor cells in vitro, the in vivo efficacy of treatment was poor. Poor in vivo efficacy was attributed to the presence of MDSCs. However, addition of all-trans retinoic acid (ATRA) improved efficacy, likely by depleting of M-MDSCs and diminishing the suppressive capacity of PMN-MDSCs [96]. Additionally, CD166, a protein that mediates communication between adjacent cells and has been implicated in tumorigenesis, has been found to be exclusively expressed in four human osteosarcoma cell lines. CD166-CAR-T cells efficiently suppressed tumor growth in tumor xenografts without significant toxicity to healthy organs [97].

Similar to CAR-T cells, T cells can be genetically engineered to express T cell receptors against particular tumor antigens; however, unlike CAR-T cells, the response is dependent upon HLA haplotype [98]. Genetically engineered lymphocytes reactive with NY-ESO-1, a cancer/testis antigen expressed in 80% of synovial cell sarcomas, have been shown to facilitate an objective partial response in four of six patients with metastatic synovial cell sarcoma without significant toxicity [99]. The NY-ESO-1 antigen has been shown to be specific to a variety of tumors including osteosarcoma; however, the level of expression can vary between tumors. Interestingly, NY-ESO-1 expression can be upregulated in vitro in response to demethylating agent 5-aza-2′-deoxycytidine (decitabine), thus suggesting that the CAR-T cell therapy utilized against synovial cell sarcoma may be successful in osteosarcoma. Indeed, further in vivo studies showed that pretreatment with decitabine allowed for localization of the NY-ESO-1 CAR-T cells to the tumor site. The combination treatment of decitabine and CAR-T cells decreased tumor volume when compared to controls [100]. Due to its success with B cell malignancies, genetically modified T cell therapies represent a promising alternative for treatment of similarly complex malignancies such as osteosarcoma.

Macrophage Activation

Mifamurtide, a nonspecific immune modulator, is a synthetic Bacille Calmette-Guerin cell wall component that can be delivered to macrophages, particularly in the lung, in the liposomal form to stimulate an antitumor response [101–105]. Liposomal muramyl tripeptide phosphatidylethanolamine (L-MTP-PE) given to osteosarcoma patients increased circulating levels of TNFα and IL-6. Additionally, treatment with L-MTP-PE also increased circulating levels of neopterin, a marker of macrophage activation, independent of IFNγ secretion [103]. L-MTP-PE used in combination with a three or four drug chemotherapy regimen including doxorubicin, cisplatin, and high-dose methotrexate with or without ifosfamide (phase III trial INT-0133) showed improved overall 6-year survival and a trend toward improved event-free survival for patients with nonmetastatic osteosarcoma; no significant difference was noted between chemotherapy regimens [105]. While it was originally determined that there was no survival benefit to the addition of L-MTP-PE for patients with metastases, a long-term follow-up of those patients indicated a trend toward increased event-free survival and overall survival [106]. The toxicities associated with L-MTP-PE are limited to chills and fever, which are likely facilitated by the release of TNFα and IL-6 [93]. Due to the survival benefits and limited toxicity, MTP has been approved for treatment of patients with osteosarcoma by the European Medicine Agency [67, 102].

Peptide Vaccines

Peptide vaccines had previously gained popularity due to the potential for stimulating T cell response against tumor-associated antigens [107]. Vaccines involving tumor lysates, peptides, and proteins have been employed in sarcoma [67, 108–111]. The SART3 protein was found in osteosarcoma cell lines and thus may be a useful peptide vaccine to stimulate CTLs against osteosarcoma tumors [108]. Furthermore, papillomavirus binding factor protein was also found in osteosarcoma and may similarly serve as an antigenic peptide to stimulate an immunologic response against tumor tissue [109]. Indeed, in a small phase II trial in which patients with soft tissue sarcoma were treated with personalized peptide vaccines, 64.7% of patients developed an IgG response with 12 patients demonstrating epitope spreading. Furthermore, 70.6% of patients developed a CTL response following treatment. These patients with refractory sarcoma had a mean survival time of 9 months, while palliative care in similar instances facilitates an 8-month survival time [110]. Due to their positive antitumor effects, peptide vaccines potentially represent feasible alternative for treatment of osteosarcoma; however, additional studies are required to optimize outcomes [107].

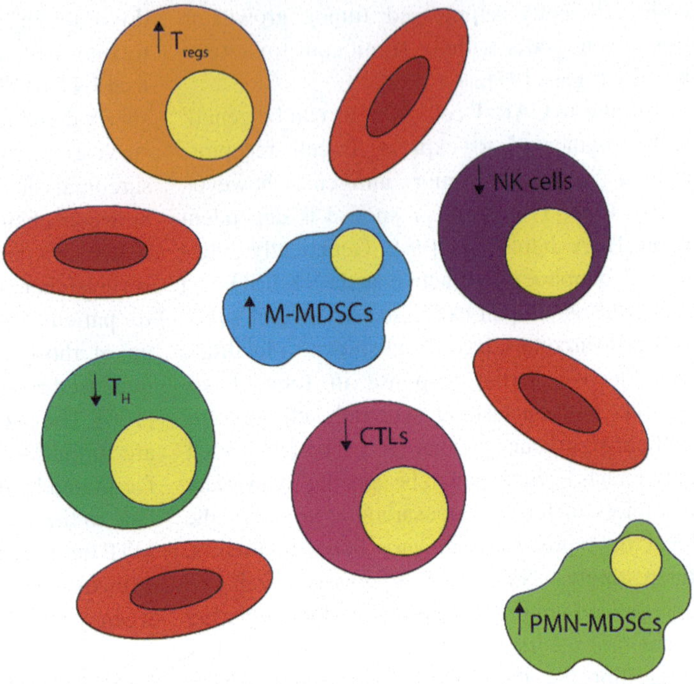

Fig. 6.7 Example osteosarcoma patient blood sample

Future Perspectives

As the field of flow cytometry evolves and we improve our ability to analyze single tissue samples with more and more antibodies and fluorochromes, the field of immunology and subsequently personalized medicine is also improving. However, these enhancements come with added responsibility as the proteins studied are highly labile ex vivo, which can result in poor data generation and thus false assumptions from the data. Therefore, a standardized process of tissue handling and preparation will become crucial to the process of data acquisition for this potential therapeutic tool. If done correctly, though, this tool has the potential to open many doors for the treatment of osteosarcoma using immunotherapeutics. Furthermore, previously popular treatments could be used much more efficiently and appropriately in dosing regiments as we can follow the patients' disease progression and recovery in real time as suggested in Fig. 6.7.

References

1. Wedekind MF, Wagner LM, Cripe TP (2018) Immunotherapy for osteosarcoma: where do we go from here? Pediatr Blood Cancer 65(9). https://doi.org/10.1002/pbc.27227
2. Chen YU, Xu SF, Xu M, Yu XC (2015) Post operative infection and survival in osteosarcoma patients: reconsideration of immunotherapy for osteosarcoma. Mol Clin Oncol 3(3):495–500
3. Liu B, Huang Y, Sun Y, Zhang J, Yao Y, Shen Z et al (2016) Prognostic value of inflammation-based scores in patients with osteosarcoma. Sci Rep 6:39862
4. Kushlinskii NE, Timofeev YS, Solov'ev YN, Gerstein ES, Lyubimova NV, Bulycheva IV (2014) Components of the RANK/RANKL/OPG system, IL-6, IL-8, IL-16, MMP-2, and calcitonin in the sera of patients with bone tumors. Bull Exp Biol Med 157(4):520–523
5. Chen Y, Yang Y, Liu S, Zhu S, Jiang H, Ding J (2016) Association between interleukin 8– 251 A/T and+ 781 C/T polymorphisms and osteosarcoma risk in Chinese population: a case–control study. Tumor Biol 37(5):6191–6196
6. Dang H, Wu W, Wang B, Cui C, Niu J, Chen J et al (2017) CXCL5 plays a promoting role in osteosarcoma cell migration and invasion in autocrine- and paracrine-dependent manners. Oncol Res 25(2):177–186

7. Holtermann OA, Djerassi I, Lisafeld BA, Elias EG, Papermaster BW, Klein E (1974) In vitro destruction of tumor cells by human monocyte. Proc Soc Exp Biol Med 147(2):456–459
8. King GW, File J, LoBuglio AF (1978) Normal human monocytes inhibit tumor cell growth in vitro. J Allergy Clin Immunol 62(5):283–288
9. Wang S, Zheng S, Hu K, Sun H, Zhang J, Rong G et al (2017) A predictive model to estimate the pre-test probability of metastasis in patients with osteosarcoma. Medicine 96(3):e5909
10. Murray PJ, Wynn TA (2017) Protective and pathogenic functions of macrophage subset. Nat Rev Immunol 11(11):723–737
11. Dumars C, Nguyen J, Gaultier A, Lanel R, Corradini N, Gouin F, Heymann D, Heymann M (2016) Dysregulation of macrophage polarization is associated with the metastatic process in osteosarcoma. Oncotarget 7(8):78343–78354
12. Buddingh EP, Kujjer ML, Duim R, Burger H, Agelopoulous K, Mykebost O et al (2011) Tumor-infiltrating macrophages are associated with metastasis suppression in high-grade osteosarcoma: a rationale for treatment with macrophage activating agents. Clin Cancer Res 17(8):2110–2119
13. Pahl JH, Kwappenberg KM, Varypataki EM, Satnos SJ, Kujjer ML, Mohamed S et al (2014) Macrophages inhibit human osteosarcoma cell growth after activation with the bacterial cell wall liposomal muramyl tripeptide combination with interferon-γ. J Exp Clin Cancer Res 33(1):27. https://doi.org/10.1186/1756-9966-33-27
14. Xiao Q, Zhang X, Wu T, Yang Y (2014) Inhibition of macrophage polarization prohibits growth of human osteosarcoma. Tumor Biol 35:7611–7616
15. Segaliny AI, Mohmadi A, Dizier B, Lokajczyk A, Brion R, Lanel R et al (2015) Interleukin-34 promotes tumor progression and metastatic process in osteosarcoma through induction of angiogenesis and macrophage recruitment. Int J Cancer 137:73–85
16. Gomez-Brouchet A, Illac C, Gilhodes J, Bouvier C, Aubert S, Guinebretiere J et al (2017) CD163-positive tumor-associated macrophages and CD8-positive cytotoxic lymphocytes are powerful diagnostic markers for the therapeutic stratification of osteosarcoma patients: an immunohistochemical analysis of the biopsies from a French OS2006 phase 3 trial. Onco Targets Ther 6(9):e1331193
17. Han Y, Guo W, Ren T, Huang Y, Wang S, Liu K (2019) Tumor-associated macrophages induce epithelial-mesenchymal transition in osteosarcoma by activating the COX-2/STAT3 axis. Cancer Lett 440–441:116–125
18. Heymann M-F, Lezot F, Heymann D (2017) The contribution of immune infiltrates and the local microenvironment in the pathogenesis of osteosarcoma. Cell Immunol 343:103711. https://doi.org/10.1016/j.cellimm.2017.10.011
19. Markel JE, Noore J, Emery EJ, Bobar HJ, Kleinerman ES, Lindsey BA (2018) Using the spleen as an in vivo systemic immune barometer alongside osteosarcoma disease progression and immunotherapy with α-PD-L1. Sarcoma 2018:8694397
20. Gardner A, Ruffell B (2016) Dendritic cells and cancer immunity. Trends Immunol 37(12):855–865
21. Chen DS, Mellman I (2013) Oncology meets immunology; the cancer-immunity cycle. Immunity 39:1–10
22. Zhang F, Zhu Y, Fan G, Hu S (2018) Photodynamic therapy reduces the inhibitory effect of osteosarcoma cells on dendritic cells by upregulating HSP70. Oncol Lett 16:5034–5040
23. Vakkila J, Jaffe R, Michelow M, Lotze M (2006) Pediatric cancers are infiltrated predominantly by macrophages and contain a paucicity of dendritic cells: a major nosologic difference with adult tumors. Clin Cancer Res 12(7):2049–2054
24. Cassetta L, Baekkevoid ES, Brandau S, Buljko A, Cassatella MA, Dorhoi A et al (2019) Deciphering myeloid-derived suppressor cells: isolation and markers in humans, mice, and non-human primates. Cancer Immunol Immunother 68:687–697
25. Gabrilovich DI (2017) Myeloid-derived suppressor cells. Cancer Immunol Res 5(1):3–8
26. Zhang S, Zhu C, Liu L, Wang G, Yuan X (2016) The role of myeloid-derived suppressor cells in patients with solid tumors: a meta-analysis. PLoS One 11(10):e0164514
27. Guan Y, Zhang R, Peng Z, Dong D, Wei G, Wang Y (2017) Inhibition of IL-18-mediated myeloid derived suppressor cell accumulation enhances anti-PD1 efficacy against osteosarcoma cancer. J Bone Oncol 9:59–64
28. Hingorani P, Maas ML, Gustafson MP, Dickman P, Adams RH, Watanabe M et al (2015) Increased CTLA-4+ T cells and an increased ratio of monocytes with loss of class II (CD14+ HLA-DRlo/neg) found in aggressive pediatric sarcoma patients. J Immunother Cancer 3(35):1–10
29. Bergenfelz C, Larsson A-M, von Stedingk K, Gruvberger-Saal S, Altonen K, Jansson S et al (2015) Systemic monocytic-MDSCs are generated from monocytes and correlate with disease progression in breast cancer patients. PLoS One 10(5):e0127028
30. Kim H-J (2014) CD4 T-cell subsets and tumor immunity: the helpful and the not-so-helpful. Cancer Immunol Res 2(2):91–98
31. Dummer W, Niethammer AG, Baccala R, Lawson BR, Wagner N, Reisfeld RA, Theofilopoulous AN (2002) T cell homeostatic proliferation elicits effective antitumor immunity. J Clin Investig 10(2):185–192
32. Borst J, Ahrends T, Babala N, Melief CJM, Kastenmüller W (2018) $CD4^+$ T cell help in cancer immunology and immunotherapy. Nat Rev 18:635–647

33. Tepper RI, Pattengale PK, Leder P (1989) Murine interleukin-4 displays potent anti-tumor activity in vivo. Cell 57:503–512
34. Zhou Q, Xian M, Xian S, Xiang D, Shao X, Wang J et al (2017) All-trans retinoic acid prevents osteosarcoma metastasis by inhibiting M2 polarization of tumor-associated macrophages. Cancer Immunol Res 5(7):547–559
35. Bailey SR, Nelson MH, Himes RA, Li Z, Mehrotra S, Paulos CM (2014) Th17 cells in cancer: the ultimate identity crisis. Front Immunol 5:1–13
36. Li P, Shi X, Xu Y, Zhong B, Lu Y, Sun Y (2018) Interleukin-22 promotes osteosarcoma cell proliferation and invasion via STAT3 activation. Med Sci Monit 24:7802–7808
37. Théoleyre S, Mori K, Cherrier B, Passuti N, Gouin F, Rédini F, Heymann D (2005) Phenotypic and functional analysis of lymphocytes infiltrating osteolytic tumors: use as a possible therapeutic approach of osteosarcoma. BMC Cancer 5:123. https://doi.org/10.1186/1471-2407-5-123
38. Zheng W, Xiao H, Liu H, Zhou Y (2014) Expression of programmed death 1 is correlated with progression of osteosarcoma. Acta Pathol Microbiol Immunol Scand 123:102–107
39. Ge W, Li J, Fan W, Xu D, Sun S (2017) Tim-3 as a diagnostic and prognostic biomarker of osteosarcoma. Tumor Biol 2017:1–8
40. Das M, Zhu C, Kuchroo VK (2017) Tim-3 and its role in regulating anti-tumor immunity. Immunol Rev 276(1):97–111
41. Liu H, Zhi L, Duan N, Su P (2016) Abnormal expression of Tim-3 antigen on peripheral blood T cells is associated with progressive disease in osteosarcoma patients. FEBS Open Bio 6:807–815
42. Xiao H, Luo G, Son H, Zhou Y, Zheng W (2014) Upregulation of peripheral CD4+CXCR5+ T cells in osteosarcoma. Tumor Biol 35:5273–5279
43. Crotty S (2014) T follicular helper cell differentiation, function, and roles in disease. Immunity 41(4):529–542
44. Galon J, Pagès F, Marincola FM, Angell HK, Thurin M, Lugli A et al (2012) Cancer classification using the Immunoscore: a worldwide task force. J Transl Med 10:205. https://doi.org/10.1186/1479-5876-10-205
45. Rivoltini L, Arienti F, Orazi A, Ceflo G, Gasparini M, Gambacorti-Passerini C, Fossati-Bellani F, Parmiani G (1992) Phenotypic and functional analysis of lymphocytes infiltrating paediatric tumours, with a characterization of the tumour phenotype. Cancer Immunol Immunother 34:241. https://doi.org/10.1007/BF01741792
46. Trieb K, Lechleitner T, Lang S, Windhager R, Kotz R, Dirnhofer S (1998) Evaluation of HLA-DR expression and T-lymphocyte infiltration in osteosarcoma. Pathol Res Pract 194:679–684
47. Fritzching B, Fellenberg J, Moskovszky L, Sàpi Z, Krenacs T, Machado I et al (2014) CD8+/FOXP3+-ratio in osteosarcoma microenvironment separates survivors from non- survivors: a multicenter validated retrospective study. Oncoimmunology 4(3):e990800
48. Spranger S, Gajewski TF (2016) Tumor-intrinsic oncogene pathways mediating immune avoidance. Oncoimmunology 5(3):e1086862
49. Sundara YT, Kostine M, Cleven AHG, Bovée JVMG, Schilman MW, Cleton-Jansen A-M (2017) Increased PD-L1 and T-cell infiltration in the presence of HLA class I expression in metastatic high-grade osteosarcoma: a rationale for T-cell-based immunotherapy. Cancer Immunol Immunother 66:119–128
50. Shen JK, Cote GM, Choy E, Yang P, Harmon D, Schwab J et al (2014) Programmed cell death 1 ligand 1 expression in osteosarcoma. Cancer Immunol Res 2(7):690–698
51. Antony PA, Piccirillo CA, Akpinaril A, Finkelstein SE, Speiss PJ, Surman DR et al (2005) CD8+ T cell immunity against a tumor/self-antigen is augmented by CD4 T helper cells and hindered by naturally occurring T regulatory cells. J Immunol 174(5):2591–2601
52. Piccirillo CA, Shevach EM (2001) Cutting edge: control of CD8+ T cell activation by CD4+CD25+ immunoregulatory cells. J Immunol 167(3):1137–1140
53. Nelson BH (2010) CD20+ B cells: the other tumor-infiltrating lymphocytes. J Immunol 185:4977–4982
54. Fonseca C, Dranoff G (2008) Capitalizing on the immunogenicity of dying tumor cells. Clin Cancer Res 14(6):1603–1608
55. Reuschenbach M, von Knebel Doeberitz M, Wentzensen N (2009) A systematic review of humoral immune responses against tumor antigens. Cancer Immunol Immunother 58:1535–1544
56. Al-Shibli KI, Donnem T, Al-Saad S, Persson M, Bremnes RM, Busund L-T (2008) Prognostic effect of epithelial and stromal lymphocyte infiltration in non small cell lung cancer. Clin Cancer Res 14(16):5220–5227
57. Milne K, Kobel M, Kalloger SE, Barnes RO, Gao D, Gilks CB, Watson PH, Nelson BH (2009) Systematic analysis of immune infiltrates in high-grade serous ovarian cancer reveals CD20, FoxP3 and TIA-1 as positive prognostic factors. PLoS One 4:e6412
58. Sorbye SW, Kilvaer T, Valkov A, Donnem T, Smeland E, Al-Shibli K, Bremnes RM, Busund L-T (2011) Prognostic impact of lymphocytes in soft tissue sarcomas. PLoS One 6:e14611
59. Gustafson MP, Lin Y, Maas ML, Van Keulen VP, Johnston PB, Peikert T, Gastineau DA, Dietz AB (2015) A method for identification and analysis of non-overlapping myeloid immunophenotypes in humans. PLoS One 10(3):e0121546
60. Souza-Fonseca-Guimaraes F, Cursons J, Huntington ND (2019) The emergence of natural killer cells as a major target in cancer immunotherapy. Trends Immunol 40(2):142–158

61. Tarek N, Lee DA (2014) Natural killer cells for osteosarcoma. In: Kleinerman MDE (ed) Current advances in osteosarcoma. Advances in experimental medicine and biology, vol 804. Springer, Cham
62. Tsukahara T, Kawaguchi S, Torigoe T, Asanuma H, Nakazawa E, Shimozawa K et al (2006) Prognostic significance of HLA class I expression in osteosarcoma defined by anti-pan HLA class I monoclonal antibody, EMR8-5. Cancer Sci 97(12):1374–1380
63. Delgado D, Webster DE, DeSantes KB, Durkin ET, Shaaban AF (2010) KIR receptor-ligand incompatibility predicts killing of osteosarcoma cell lines by allogeneic NK cells. Pediatr Blood Cancer 55(17):1300–1305
64. Buddingh EP, Schilham MW, Rusaln SEN, Berghuis D, Szuhai K, Suurmond J et al (2011) Chemotherapy-resistant osteosarcoma is highly susceptible to IL-15-activated allogeneic and autologous NK cells. Cancer Immunol Immunother 60:575–586
65. Markewicz K, Zeman K, Kozar A, Golebiowska-Wawrzyniak M, Wozniak W (2012) Evaluation of selected parameters of cellular immunity in children with osteosarcoma at diagnosis. Med Wieku Rozwoj 16:212–221
66. Wang Z, Wang Z, Li B, Wang S, Chen T, Ye Z (2019) Innate immune cells: a potential and promising cell population for treating osteosarcoma. Front Immunol 10:1114. https://doi.org/10.3389/fimmu.2019.01114
67. Miwa S, Shirai T, Yamamoto N, Hayashi K, Takuchi A, Igrashi K, Tsuchiya H (2019) Current and emerging targets in immunotherapy for osteosarcoma. J Oncol 2019. https://doi.org/10.1155/2019/7035045
68. Zheng B, Ren T, Huang Y, Sun K, Wang S, Bao X, Liu K, Guo W (2018) PD-1 axis expression in musculoskeletal tumors and antitumor effect of nivolumab in osteosarcoma model of humanized mouse. J Hematol Oncol 11(16). https://doi.org/10.1186/s13045-018-0560-1
69. Dhupkar P, Gordon N, Stewart J, Kleinerman ES (2018) Anti-PD-1 therapy redirects macrophages from an M2 to an M1 phenotype inducing regression of OS lung metastases. Cancer Med 7(6):2654–2664
70. Zhao Y, Yang W, Huang Y, Cui R, Li X, Li B (2018) Evolving roles for targeting CTLA-4 in cancer immunotherapy. Cell Physiol Biochem 47:721–734
71. Lussier DM, Johnson JL, Hingorani P, Blattman JN (2015) Combination immunotherapy with α-CTLA-4 and α-PD-L1 antibody blockade prevents immune escape and leads to complete control of metastatic osteosarcoma. J Immunother Cancer 3(21). https://doi.org/10.1186/s40425-015-0067-z
72. Brahmer JR, Tykodi SS, Chow LQM, Hwu W-J, Topalian SL, Hwu P (2012) Safety and activity of anti–PD-L1 antibody in patients with advanced cancer. N Engl J Med 366(26):2455 2465
73. Tawbi HA, Burgess M, Bolejack V, Van Tine BA, Scheutze SM, Hu J et al (2017) Pembrolizumab in advanced soft-tissue sarcoma and bone sarcoma (SARC028): a multicentre, two-cohort, single-arm, open-label, phase 2 trial. Lancent Oncol 18:1493–1501
74. Ratti C, Botti L, Cancila V, Galvan S, Torselli I, Garogalo C et al (2017) Trabectedin overrides osteosarcoma differentiative block and reprograms the tumor immune environment enabling effective combination with immune checkpoint inhibitors. Clin Cancer Res 23(17):5149–5161
75. Strander H, Einhorn S (1977) Effect of human leukocyte interferon on the growth of human osteosarcoma cells in tissue culture. Int J Cancer 19:468–473
76. Brosjö O, Bauer HC, Broström L-Å, Nilsonne U, Nilsson OS, Reinhold FP, Strander H, Tribukait B (1985) Influence of human α-interferon on four human osteosarcoma xenografts in nude mice. Cancer Res 45:5596–5602
77. Müller CR, Smeland S, Bauer HCF, Sæter G, Strander H (2005) Interferon-α as the only adjuvant treatment in high-grade osteosarcoma: long term results of the Karolinska Hospital series. Acta Oncol 44:475–480
78. Whelan J, Patterson D, Perisoglou M, Bielack S, Marina N, Smeland S, Bernstein P (2010) The role of interferons in treatment of osteosarcoma. Pediatr Blood Cancer 54(3):350–354
79. Brosjö O, Bauer HC, Nilsson OS, Halvorsen D, Reinholt FP, Tribukait B (1989) Effect of human interferon-alpha and interferon-gamma on growth, histology, and DNA content of human osteosarcomas in nude mice. J Interf Res 9(4):475–489
80. Strander H, Bauer HC, Brosjö O, Fernberg JO, Kreichbergs A, Nilsonne U et al (1995) Long-term adjuvant interferon treatment of human osteosarcoma: a pilot study. Acta Oncol 34(6):877–880
81. Bielack SS, Smeland S, Whelan JS, Marina N, Jovic G, Hook JM et al (2015) Methotrexate, doxorubicin, and cisplatin (MAP) plus maintenance pegylated interferon Alfa-2b versus MAP alone in patients with resectable high-grade osteosarcoma and good histologic response to preoperative MAP: first results of the EURAMOS-1 good response randomized controlled trial. J Clin Oncol 33(20):2279–2287
82. Rosenberg SA (2014) Progress in human tumour immunology and immunotherapy. Nature 411:380–384
83. Weigent DA, Stanton GJ, Johnson HM (1983) Interleukin 2 enhances natural killer cell activity through induction of gamma interferon. Infect Immun 41(3):992–997
84. Rosenberg SA, Lotze MT, Muul LM, Leitman S (1985) Observations on the systemic administration of autologous lymphokine-activated killer cells and recombinant interleukin-2 to patients with metastatic cancer. N Engl J Med 313:1485–1492

85. Meazza C, Cefalo G, Massinmino M, Daolio P, Pastorino U, Scanagatta P et al (2017) Primary metastatic osteosarcoma: results of a prospective study in children given chemotherapy and interleukin-2. Med Oncol 34(12). https://doi.org/10.1007/s12032-017-1052-9
86. Schwinger W, Klass V, Benesch M, Lackner H, Bornbusch HJ, Sovinz P, Moser A, Schwantzer G, Urban C (2005) Feasibility of high-dose interleukin-2 in heavily pretreated pediatric cancer patients. Ann Oncol 16:1199–1206
87. Saxena M, Bhardwaj N (2018) Re-emergence of dendritic cell vaccines for cancer treatment. Trends Cancer 4(2):119–137
88. He Y-T, Zhang Q-M, Kou Q-C, Tang B (2016) *In vitro* generation of cytotoxic T lymphocyte response using dendritic cell immunotherapy in osteosarcoma. Oncol Lett 12:1101–1106
89. Kawano M, Itonaga I, Iwasaki T, Tsumura H (2013) Enhancement of antitumor immunity by combining anti-cytotoxic T lymphocyte antigen-4 antibodies and cryotreated tumor lysate-pulsed dendritic cells in murine osteosarcoma. Oncol Rep 29:1001–1006
90. Kawano M, Tanaka K, Itonaga I, Iwasaki T, Miyazaki M, Ikeda S, Tsumura H (2015) Dendritic cells combined with anti-GITR antibody produce antitumor effects in osteosarcoma. Oncol Rep 34:1995–2001
91. Geiger JD, Hutchinson RJ, Hohenkirk LF, McKenna EA, Yanik GA, Levine JE et al (2001) Vaccination of pediatric solid tumor patients with tumor lysate-pulsed dendritic cells can expand specific T cells and mediate tumor regression. Cancer Res 61:8513–8519
92. Merchant MS, Bernstein D, Amoako M, Baird K, Fleisher TA, Morre M et al (2016) Adjuvant immunotherapy to improve outcome in high-risk pediatric sarcomas. Clin Cancer Res 22(13):3182–3191
93. Miwa S, Nishida H, Tanzawa Y, Takeuchi A, Hayashi K, Yamamoto N et al (2017) Phase 1/2 Study of immunotherapy with dendritic cells pulsed with autologous tumor lysate in patients with refractory bone and soft tissue sarcoma. Cancer 123(9):1576–1584
94. Köskal H, Müller E, Inderberg EM, Bruland Ø, Wälchi S (2018) Treating osteosarcoma with CAR T cells. Scand J Immunol 89:e12741
95. Ahmed N, Brawley VS, Hegde M, Robertson C, Ghazi A, Gerken C et al (2015) Human epidermal growth factor receptor 2 (HER2) –specific chimeric antigen receptor–modified T cells for the immunotherapy of HER2-positive sarcoma. J Clin Oncol 33(15):1688–1696
96. Long AH, Highfill SL, Cui Y, Smith JP, Walker AJ, Ramakrishna S et al (2016) Reduction of MDSCs with all-trans retinoic acid improves CAR therapy efficacy for sarcomas. Cancer Immunol Res 4(10):869–880
97. Wang Y, Yu W, Zhu J, Wang J, Xia K, Liang C, Tao H (2019) Anti-CD166/4-1BB chimeric antigen receptor T cell therapy for the treatment of osteosarcoma. J Exp Clin Cancer Res 38(168). https://doi.org/10.1186/s13046-019-1147-6
98. Wang Z, Li B, Ren Y, Ye Z (2016) T-cell-based immunotherapy for osteosarcoma: challenges and opportunities. Front Immunol 7(353). https://doi.org/10.3389/fimmu.2016.00353
99. Robbins PF, Morgan RA, Feldman SA, Yang JC, Sherry RM, Dudley ME et al (2011) Tumor regression in patients with metastatic synovial cell sarcoma and melanoma using genetically engineered lymphocytes reactive with NY-ESO-1. J Clin Oncol 29(7):917–924
100. Li B, Zhu X, Sun L, Yuan L, Zhang J, Hengyuan L, Ye Z (2014) Induction of a specific CD8+ T-cell response to cancer/testis antigens by demethylating pre-treatment against osteosarcoma. Oncotarget 5(21):10791–10802
101. Asano T, Kleinerman ES (1993) Liposome-encapsulated MTP-PE: a novel biologic agent for cancer therapy. J Immunother Emphasis Tumor Immunol 14(4):286–289. Meyers PA. (2015). Systemic therapy for osteosarcoma and ewing sarcoma. 2015 ASCO Educational Book: E644–7
102. Anderson PM, Meyers P, Kleinerman E, Venkatakrishnan K, Hughes DP, Herzog C (2014) Mifamurtide in metastatic and recurrent osteosarcoma: a patient access study with pharmacokinetic, pharmacodynamic, and safety assessments. Pediatr Blood Cancer 61(2):238–244
103. Meyers PA, Chou AJ (2014) Muramyl tripeptide-phosphatidyl ethanolamine encapsulated in liposomes (L-MTP-PE) in the treatment of osteosarcoma. In: Kleinerman MDE (ed) Current advances in osteosarcoma. Advances in experimental medicine and biology, vol 804. Springer, Cham
104. Kleinerman ES, Jia SF, Griffin J, Seibel NL, Benjamin RS, Jaffe N (1992) Phase II study of liposomal muramyl tripeptide in osteosarcoma: the cytokine cascade and monocyte activation following administration. J Clin Oncol 10(8):1310–1316
105. Meyers PA, Schwartz CL, Krailo MD, Healey JH, Bernstein ML, Betcher D et al (2008) Osteosarcoma: the addition of muramyl tripeptide to chemotherapy improves overall survival—a report from the Children's Oncology Group. J Clin Oncol 26(4):633–638
106. Chou AJ, Kleinerman ES, Krailo MD, Chen Z, Betcher DL, Healey JH et al (2009) Addition of muramyl tripeptide to chemotherapy for patients with newly diagnosed metastatic osteosarcoma: a report from the Children's Oncology Group. Cancer 1115(22):5339–5348
107. Kumai T, Kobayashi H, Harabuchi Y, Celis E (2017) Peptide vaccines in cancer – old concept revisited. Curr Opin Immunol 45:1–7

108. Tsuda N, Murayama K, Ishida H, Matsunaga K, Komiya S, Itoh K, Yamada A (2001) Expression of a newly defined tumor-rejection antigen SART3 in musculoskeletal tumors and induction of HLA class I-restricted cytotoxic T lymphocytes by SART3-derived peptides. J Orthop Res 19:346–351
109. Takahashi R, Ishibashi Y, Hiraoka K, Matsueda S, Kawano K, Kawahara A et al (2013) Phase II study of personalized peptide vaccination for refractory bone and soft tissue sarcoma patients. Cancer Sci 104(10):1285–1294
110. Kawaguchi S, Tsukahara T, Ida K, Kimura S, Murase M, Masanobu K (2012) SYT-SSX breakpoint peptide vaccines in patients with synovial sarcoma: a study from the Japanese Musculoskeletal Oncology Group. Cancer Sci 103(9):1625–1630
111. Tsukahara T, Nabeta Y, Kawaguchi S, Ikeda H, Sato Y, Shimozawa K et al (2004) Identification of human autologous cytotoxic T-lymphocyte-defined osteosarcoma gene that encodes a transcriptional regulator, papillomavirus binding factor. Cancer Res 64(15). https://doi.org/10.1158/0008-5472.CAN-04-0522

7 Targetable Intercellular Signaling Pathways Facilitate Lung Colonization in Osteosarcoma

James Brandon Reinecke and Ryan David Roberts

Abstract

Outcomes for young people diagnosed with osteosarcoma hinge almost exclusively on whether they develop lung metastasis. The striking predilection that osteosarcoma shows for metastatic spread to lung suggests properties and/or lung interactions that generate tissue-specific survival and proliferation advantages. While these mechanisms remain overall poorly defined, studies have begun to describe biological elements important to metastasis. Mechanisms described to date include both cell-autonomous adaptations that allow disseminated tumor cells to survive the stressors imposed by metastasis and intercellular signaling networks that tumor cells exploit to pirate needed signals from surrounding tissues or to recruit other cells that create a more favorable niche. Evidence suggests that cell-autonomous changes are largely driven by epigenetic reprogramming of disseminated tumor cells that facilitates resistance to late apoptosis, manages endoplasmic reticulum (ER) stressors, promotes translation of complex transcripts, and activates clotting pathways. Tumor-host signaling pathways important for lung colonization drive interactions with lung epithelium, mesenchymal stem cells, and mediators of innate and adaptive immunity. In this chapter, we highlight one particular pathway that integrates cell-autonomous adaptations with lung-specific tumor-host interactions. In this mechanism, aberrant ΔNp63 expression primes tumor cells to produce IL6 and CXCL8 upon interaction with lung epithelial cells. This tumor-derived IL6 and CXCL8 then initiates autocrine, osteosarcoma-lung paracrine, and osteosarcoma-immune paracrine interactions that facilitate metastasis. Importantly, many of these pathways appear targetable with clinically feasible therapeutics. Ongoing work to better understand metastasis is driving efforts to improve outcomes by targeting the most devastating complication of this disease.

J. B. Reinecke
Department of Pediatrics, Nationwide Children's Hospital and The Ohio State University, Columbus, OH, USA

R. D. Roberts (✉)
Division of Hematology, Oncology, and Blood and Marrow Transplant, Department of Pediatrics, Center for Childhood Cancer and Blood Diseases, Abigail Wexner Research Institute at Nationwide Children's Hospital and The James Comprehensive Cancer Center at The Ohio State University, Columbus, OH, USA
e-mail: ryan.roberts@nationwidechildrens.org

E. S. Kleinerman, R. Gorlick (eds.), *Current Advances in the Science of Osteosarcoma*, Advances in Experimental Medicine and Biology 1258, https://doi.org/10.1007/978-3-030-43085-6_7

Keywords

Osteosarcoma · Metastasis · Intercellular signaling · Tumor microenvironment · Cell-autonomous mechanism · Tumor-host interactions · Metastatic bottleneck · IL6 · CXCL8 · ΔNp63 · Lung colonization

Introduction

Mortality in osteosarcoma, the most common malignant bone tumor in children and adolescents, tracks strikingly well with metastasis, which exhibits an overwhelming tropism for lung tissues. Prior to the routine use of chemotherapy for the systemic treatment of osteosarcoma, 80–90% of patients would develop lung metastasis despite undergoing adequate surgical resection and having no signs of disseminated disease at diagnosis [1]. While the introduction of chemotherapy in the 1980s improved outcomes, 30–40% of osteosarcoma patients still experience relapse, with the vast majority experiencing metastatic relapse within the lung [2]. Overall survival after metastatic relapse is less than 20% [3–6]. The lack of improvement in outcomes for these patients underscores the desperate need for novel therapeutic approaches to combat this deadly disease [7–9].

One can clearly infer from the clinical data that, in most patients, disseminated osteosarcoma cells find refuge within the lung and evade chemotherapy. Furthermore, the strong proclivity that disseminated osteosarcoma has for lung tissue suggests that resident lung cells, those recruited to the metastatic niche, and the signaling milieu that they generate create an environment particularly favorable to metastatic colonization. If true, attempting to disrupt critical interactions between osteosarcoma cells and the cells that compose the metastatic niche within the lung could create novel therapeutic strategies for preventing metastatic disease or even for treating established metastatic lesions. However, a mechanistic understanding of the cellular and molecular features of the metastatic osteosarcoma niche that can facilitate such rational targeting is currently lacking but crucial to developing new approaches for patients with disseminated osteosarcoma.

In this chapter, we will outline the steps of the metastatic cascade that culminate in the formation of macro-metastatic disease in patients with osteosarcoma, as suggested by our best current knowledge. We will start by highlighting the stresses imposed on disseminated tumor cells, which must be overcome to colonize secondary organs. These stresses create profound inefficiencies that drive massive cellular attrition known as the metastatic bottleneck. We will provide several brief examples of cell-autonomous mechanisms that osteosarcoma cells utilize to overcome these stressors and traverse the metastatic bottleneck, then describe a series of recent observations that illustrate how reciprocal interactions between lung epithelial cells and osteosarcoma cells generate non-cell-autonomous pathways crucial for successful lung colonization.

While separated for the purpose of discussion, the thoughtful reader will recognize that cell-autonomous and non-cell-autonomous mechanisms of metastasis are not mutually exclusive, but inter-related and inter-dependent. We will conclude by illustrating how one particular mechanism links cell-autonomous and non-cell-autonomous pathways to effect osteosarcoma lung metastasis.

The Metastatic Bottleneck

Evidence from both clinical and preclinical models suggests that primary tumors release millions of cells into circulation [10]. Clinical specimens taken from patients with active disease show dissemination of cells from the primary tumor in the blood and bone marrow in most patients at time of diagnosis [11]. In orthotopic mouse models of osteosarcoma, cells can be found in circulation long before the primary tumor is detectable [12]. However, only 20–30% of patients with osteosarcoma initially present with overt evidence of metastasis [2, 7]. This discrepancy between the efficient hematogenous dissemination of cancer cells and the profoundly inefficient process of colonizing distant tissues illustrates a phenomenon known as the metastatic bottleneck [13].

Quantitative methods using mouse models have demonstrated that, while a large percentage of circulating cells will arrest in metastatic parenchyma, nearly all of these will perish. Classic experiments performed by Isaiah Fidler in the 1970s show that only 1.5% of highly selected, highly metastatic B16 melanoma cells injected intravenously will survive more than 24 hours in the lung [14]. By 2 weeks, less than 0.2% of the cells remain.

The stressors imposed by the transition from the favorable environment of the primary tumor to the relatively hostile environment of circulation and distant tissues prove insurmountable for most tumor cells, which undergo apoptosis within the first 48 hours post injection, regardless of their underlying metastatic potential [15, 16]. Indeed, simply overexpressing the anti-apoptosis gene Bcl-2 permitted survival of otherwise non-malignant rat embryonic fibroblasts in the lung, while all of these cells died in the absence of Bcl-2 overexpression [15].

In contrast, early metastatic events appear to be much more efficient. As noted above, large numbers of tumor cells acquire the ability to enter the bloodstream through intravasation, often facilitated by interactions with the surrounding stromal cells [17–19]. Transition back out of the bloodstream appears to be likewise a less intense barrier. Indeed, intravital microscopy of B16 melanoma cells injected into the portal vein found that 80% of cells can indeed extravasate [15].

Where such questions have been asked, osteosarcoma cells exhibit similar properties with respect to metastasis. Ex vivo monitoring of lung colonization by osteosarcoma cells strongly supports a model wherein cells with high and low metastatic capacity arrest in lung capillaries and extravasate efficiently [20]. Similar to findings in other tumor types, both high and low metastatic cells exhibit widespread early apoptosis. However, by 4–7 days, low metastatic cells cease to proliferate, while high metastatic cells potentially grow in number [21]. Similar findings with slightly longer kinetics were observed in vivo.

Currently, it remains unclear whether poorly metastatic osteosarcoma cells eventually undergo apoptosis in the lung or whether they remain dormant as solitary, non-proliferating cells. The molecular mechanisms that support growth through this tight metastatic bottleneck have been a topic of more intense investigation over the last decade, especially in osteosarcoma. It appears that diverse mechanisms contribute to this process. Several notable findings (Fig. 7.1) are highlighted below.

Cell-Autonomous Mechanisms that Facilitate Metastatic Lung Colonization in Osteosarcoma

The ability to survive the metastatic bottleneck depends on the tumor cell's response to stressors imposed upon arrival within a foreign tissue. That response can involve diverse molecular and cellular pathways. Studies identifying differences in the response of low and high metastatic cells as they arrive in metastatic target tissues have been a mainstay in the discovery of these mechanisms. Most studies published to date describe mechanisms that lie entirely within the tumor cells themselves—mechanisms that are cell-autonomous. In this section, we will discuss data from several studies that underscore the importance of these cell-autonomous mechanisms exhibited by highly metastatic osteosarcoma cells that allow them to survive the metastatic bottleneck.

Mechanisms that Resist Late Apoptosis Drive the Persistence of Disseminated Osteosarcoma Cells

Early insights into pathways required for osteosarcoma metastasis to the lung were obtained through differential gene expression analysis in mouse allograft models composed of the clonally related osteosarcoma cell lines K12 (low metastatic) and K7M2 (high metastatic) [22]. Expression arrays found that, while there was no difference in the proliferation and survival of the

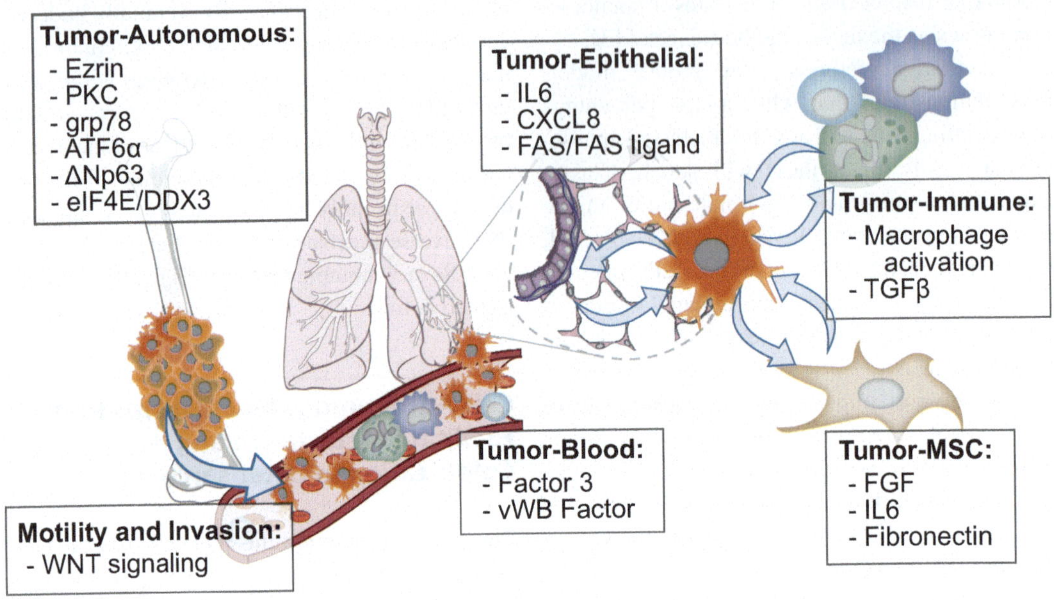

Fig. 7.1 Both cell-autonomous pathways and intercellular signaling networks have been implicated as mechanisms that drive osteosarcoma lung metastasis. Several pathways discussed are identified in the diagram

two cell lines in vitro, genes involved in cell survival, proliferation, adhesion/cytoskeleton, and angiogenesis were expressed at higher levels in the highly metastatic derivative cell line (K7M2) than in the poorly metastatic parent cells (K12).

Ultimately, these pioneering studies identified the cytoskeletal-associated protein ezrin as a key mediator of mechanisms important for osteosarcoma lung metastasis [23]. Studies showed that, when present, phosphorylation of ezrin could facilitate persistence of disseminated cells that survived the initial period of early attrition by preventing late apoptosis. Extensive work by Khanna and colleagues later elucidated mechanisms that linked phosphorylation of ezrin's c-terminus to activation of protein kinase C (PKC), which mediates ezrin's metastasis-promoting effector mechanisms [24]. Preclinical therapeutic studies performed by Khanna's group showed that inhibition of PKC, but not other cell survival signaling pathways (Akt, Src, mTOR, MAPK), decreased survival of highly metastatic cells within the metastatic bottleneck but did not affect cell survival or proliferation in vitro [25].

Increased Expression of HSAP5 Facilitates Management of ER Stress During Metastatic Progression

Several of the stressors encountered by disseminated tumor cells converge on the endoplasmic reticulum (ER) stress pathway, including hypoxia, nutrient deprivation, acidosis, and generation of reactive oxygen species—all of which disrupt protein folding within the ER [26]. When chaperone proteins such as the HSP70 protein glucose-response protein 78 (grp78/HSPA5) become overwhelmed by the amount of unfolded proteins, cells activate the unfolded response pathway (URP). This URP can trigger both pro- and anti-tumorigenic activities like angiogenesis and apoptosis, respectively [27]. In osteosarcoma, upregulation of HSPA5 and activation of the mTOR pathway in highly metastatic cells suggest that these cells actively inhibit URP during metastatic progression.

Khanna's group has demonstrated that MG63.3 cells (a highly metastatic subline derived from MG63 cells) exhibit marked upregulation of

HSPA5 upon arrest within lung tissues, a phenomenon not observed in the poorly metastatic parent cells [28]. This upregulation occurred after the early wave of cell attrition, suggesting a response that occurs to stressors encountered during lung colonization.

Investigation by other researchers has converged on this same mechanism. Work done by Sertil's group demonstrating a role for ATF6α in osteosarcoma metastasis supports the importance of suppressing the URP through upregulation of HSPA5, as HSPA5 is a direct transcriptional target of ATF6α and a key downstream effector of ATF6α activity [29]. The pro-metastatic activity of this transcription factor and chaperone protein appear to be closely linked.

Efficient Translation of Complex Transcripts Is Required for Osteosarcoma Metastatic Progression

While responses to ER stresses that ensure proper protein folding constitute important tumor cell reactions to stressors encountered during metastatic progression, increased and efficient translation of key mRNA transcripts is likewise important. Translation of complex 5'UTR-containing transcripts appears to be important for osteosarcoma cell survival and proliferation within the lung [30]. The increased expression of ezrin and activation of the mTOR pathway appear to converge on pathways that enhance the energetically costly translation of this important group of transcripts, which appears to be necessary for productive colonization of lung tissues [31, 32]. Evidence suggests that these mechanisms involve both the upregulation and modulation of DDX3 activity and the activation of eIF4E [32]. While it remains somewhat unclear how improved translation of this particular class of transcripts drives metastasis, panels of genes whose expression clearly benefits from this mechanism exist. Some of these, particularly DDIT4, have been used in high-throughput screens for inhibitors of ezrin activity.

Epigenetic Changes Influence Gene Expression Programs to Promote Metastasis

While investigators have uncovered several cell-autonomous mechanisms that facilitate osteosarcoma metastasis, almost none of these appear to result from existing or acquired metastasis-driving mutations. Indeed, apart from very frequent mutations in p53 and Rb, there are few recurrent mutations or structural rearrangements found in osteosarcoma tumor samples taken from patients [33, 34]. Indeed, osteosarcoma has become defined by frequent profound aneuploidy and large-scale disruption of chromosome structure [34, 35]. While it has become increasingly clear that specific patterns of copy number variation play an important role in osteosarcomagenesis [36], it remains unclear how these metastasis-driving genes become dysregulated. Many have suspected that the dysregulation of gene expression networks through epigenetic mechanisms might drive pro-metastatic gene programs.

Several studies have demonstrated changes that occur in histone occupancy on enhancers during tumorigenesis. These variant enhancer loci (VEL) invoke widespread changes in gene expression during osteosarcoma metastasis [37]. In one of the first studies of epigenetic dysregulation during malignant progression in osteosarcoma, Morrow et al. identified metastatic-VELs (Met-VELs) by comparing the histone landscape of patient tumor pairs (primary tumor and lung metastasis biopsies) as well as low metastatic potential cell lines with their highly metastatic clonal derivatives [37].

Beyond simply elucidating enhancer loci that become hyperactivated in the metastases, these analyses identified a small set of genes whose expression was recurrently increased by activation of these Met-VELs. These genes included, among others, Factor 3, a gene essential to effective blood clotting. Disruption of histone acetylation through bromodomain and extraterminal 4 (BRD4) inhibition, shRNA-knockdown of key met-VEL genes such as Factor 3, and genome-editing disruption of the F3 enhancer all significantly decreased osteosarcoma metastasis to lung without affecting primary tumor growth.

Illustrating another tumor-cell-autonomous epigenetic mechanism that promotes malignant progression, Hakan Cam's group has shown that a majority of osteosarcoma xenografts, cell lines, and human clinical samples aberrantly overexpress an alternate isoform of p63, called ΔNp63 [38]. While previous studies suggested that these ΔN isoforms of p53 family members (which retain DNA-binding structures but lack domains associated with transactivation) function primarily to counteract cell cycle arrest and apoptosis [39], these studies in osteosarcoma cells suggested distinct gain-of-function activity that drives features associated with malignant progression and metastasis. Similar findings in other tumor types now also support these findings, suggesting a key role for ΔN isoform expression in epithelial-to-mesenchymal transformations.

Among the roles described for ΔNp63 in osteosarcoma, Bid and colleagues showed that aberrant expression in these cells drives expression of several cytokines/chemokines directly and produces a gene expression pattern that promotes angiogenesis both in vitro and in vivo [38]. Two important mediators of this biology described in that work were IL6 and CXCL8, whose expression appears to result directly from binding of ΔNp63 to their respective promoters, driving transcriptional activation. Expression of these genes showed strong co-variability in human tumor samples and demonstrated intense upregulation in lung metastases relative to primary tumors collected from the same patients [38].

Interestingly, these same mechanisms appear to be recapitulated in canine osteosarcoma tumors, where aberrant ΔNp63 expression is observed in over 90% of primary tumors [40]. (Canine osteosarcoma has long been appreciated as an excellent model of spontaneous osteosarcoma with extremely strong clinical, histologic, and genetic similarity to the human disease [41, 42].) In addition to validating a role for ΔNp63 in cytokine/chemokine expression, these studies also showed how aberrant expression of ΔN can drive survival by blocking the p73-dependent expression of proapoptotic proteins PUMA and NOXA independent of p53 [40], which is almost universally lost in osteosarcoma [34].

Tumor-Host Interactions that Mediate Osteosarcoma Lung Metastasis

Considerable work by a few small groups of investigators has begun to reveal non-cell-autonomous mechanisms that drive osteosarcoma metastasis. The unusual predilection of osteosarcoma for colonizing lung tissue and bone (but not other organ systems) suggests that tumor-host interactions unique to those environments generate especially favorable growth conditions. Understanding the cell types and the mechanisms at play within the metastatic niche has potential to identify pathways critical to metastatic spread that may be vulnerable to therapeutic intervention.

When disseminated tumor cells arrest and establish within a metastatic niche, they become surrounded by the support cells and extracellular matrix that define the microenvironment of the metastatic target organ. These tissues supply cues that regulate survival, proliferation, migration, and angiogenesis as well as those that dictate immune reactivity and that identify abnormal cells for elimination [43]. Some elements unique to the lung metastatic niche have received attention from those studying other tumor types. These have identified cell-cell signaling pathways that mediate quiescence/dormancy [44], susceptibility to chemotherapy [45–48], metastatic outgrowth [49, 50], maintenance of stemness [50–52], and other metastasis-associated behaviors. In many of these cases, the cancer cells themselves manipulate the surrounding cells to elicit behaviors that promote survival and proliferation within the metastatic niche.

Elements of Both Host- and Tumor-Derived Extracellular Matrix Promote Osteosarcoma Survival and Stemness During Metastasis

Disseminated tumor cells maintain intimate association with the foreign extracellular matrix that surrounds them. Elements of that stroma may have importance for maintaining the survival of the stemlike metastasis initiating cells during the early stages of colonization within the micrometastatic niche. The Zhu group has shown that host-derived FGF signaling initiates a fibrogenic program in osteosarcoma cells that reprograms those cells in a way that maintains stemness and survival [53]. This signaling initially requires mTORC1 activation [52], but data suggests that the subsequent production of fibronectin by the tumor cells alters the surrounding matrix in a way that maintains this programming independent of those host signals [53]. This work provides an excellent illustration of how initially "normal" behaviors of the lung stromal cells can drive processes that, due to abnormalities in both the cellular context (osteoid-producing mesenchymal cells don't normally interact directly with pulmonary cells in a postnatal human) and in the dysregulated genetic programming of the tumor cell, drive a restructuring of the micrometastatic environment into something very abnormal that supports malignant progression.

Interactions with Resident Immune Surveillance Mechanisms and Infiltrating Cells Facilitate Immune Evasion

On establishing residence within the micrometastatic niche, tumor cells will interact with cells of the immune system, both those that reside within the lung and those that infiltrate the niche in response to signals from both host and malignant cells. Indeed, a key impediment to maintaining colonization is avoiding immune detection and/or suppressing the antitumor immune response. Extensive work led primarily by Eugenie Kleinerman's group has elucidated mechanisms important to this immune evasion, including some novel mechanisms that invoke barrier properties specific to the privileged sites of the lung [54–56]. Work in this area has led to several translational opportunities, including a large multinational trial investigating the effects of an immunomodulating drug—one of the few therapies that have demonstrated a positive effect on osteosarcoma metastasis since the advent of MAP (methotrexate, doxorubicin, cisplatin) therapy in the late 1970s [57–59]. These concepts are discussed in detail in Chapter ***.

Tumor-Educated Mesenchymal Stem Cells Facilitate Growth of Disseminated Osteosarcoma Cells Through IL6 Production

Nonimmune mesenchymal cells also infiltrate the metastatic niche and may play an important role in establishing a tumorigenic microenvironment. Such infiltration may recapitulate early embryonic developmental effects [60] in an inappropriate developmental context. Baglio et al. recently demonstrated that membrane-bound TGFβ on extracellular vesicles produced by osteosarcoma cells stimulates massive IL6 production by mesenchymal stem cells [61]. Activation of this multistep pathway, including the IL6-dependent activation of STAT3 within the tumor cells, directly promoted growth of disseminated osteosarcoma cells into metastatic lesions. Importantly, these pathways demonstrated targetability, as administration of anti-IL6 antibodies reversed the effect and reduced metastatic lesions in the mice [61].

Bidirectional Cytokine Signaling Between Osteosarcoma Cells and Lung Parenchyma Regulates Metastatic Progression

While significant attention has been given to cells that infiltrate the micrometastatic niche and the ways that these influence malignant progression, much less attention has gone to the ways that tumor cells might interact with the resident lung tissues that make up this niche. Defining and characterizing the signaling between lung cells and osteosarcoma may provide insight into potentially novel therapeutic approaches. It is likely that highly metastatic osteosarcoma cells elicit paracrine signaling pathways with lung parenchymal cells that promote metastatic colonization and progression. Identifying these mechanisms could not only point directly to targetable intercellular signaling pathways but may lead to identification of specific tumor-cell-intrinsic vulnerabilities. For instance, if a cytokine promotes survival within the metastatic niche, then the pathways activated within the tumor cell when that cytokine engages its receptor may be critical for survival of those cells elsewhere.

Work by the Roberts' group identified IL6 and CXCL8 as genes associated with osteosarcoma metastasis in a group of primary-metastatic tumor pairs from pediatric patients [62]. This work independently corroborated the findings of Bid et al. (who had identified both genes as important targets of ΔNp63 in osteosarcoma [38]), of Baglio et al. (who identified IL6 as a mediator of MSC-driven osteosarcoma cell survival within the metastatic niche [61]), and of Paoloni et al. (who identified CXCL8 as a poor prognostic marker in both canine and human osteosarcomas [42]).

These previous studies, however, presented important discrepancies, particularly with respect to the source of IL6 and CXCL8 and their effector mechanisms. Some studies suggested mesenchymal stem cells as the relevant source of IL6 [61], others that tumor cells produced their own IL6 [38, 40], while most evaluated only whole tumor extracts, leaving the source ambiguous. Using tumor-specific inducible knockdowns, Gross et al. showed that cytokine knockdown in the tumor cells was far more effective than receptor knockdown in the same cells, suggesting a critical role for tumor-derived IL6 and CXCL8 in metastatic lung colonization [62].

Further evidence for this hypothesis came from efforts to integrate the work previously done by Hakan Cam's group. Interestingly, while those previous studies demonstrated ΔNp63-dependent expression of IL6 and CXCL8 in osteosarcoma cells [38], Gross et al. showed that cells deprived of serum did not express these genes—that doing so required some second signal. ΔN-expressing cells responded nicely to serum stimulation, producing high levels of IL6 and CXCL8, and even higher levels were produced when the same osteosarcoma cells were exposed to supernatants from primary lung epithelial cells [62]. These results suggest that aberrant ΔNp63 expression primes osteosarcoma cells to produce IL6 and CXCL8 upon interaction with epithelial cells within the lung, an effect mediated by soluble lung-derived factors.

Mechanistic support for the importance of this ΔNp63/IL6/CXCL8 axis in osteosarcoma lung colonization comes from genetic manipulation experiments. Poorly metastatic OS-25 osteosarcoma cells, which do not express ΔNp63 at baseline, were endowed with IL6/CXCL8 expression and metastatic capacity when transduced with a virus that drove ΔNp63. Conversely, highly metastatic, ΔNp63-expressing OS-17 cells lost both IL6 and CXCL8 expression upon RNAi-mediated ΔNp63 knockdown, which abrogated their ability to colonize murine lungs. Virus-mediated IL6 and CXCL8 expression in the ΔN-knockdown cells restored metastatic capacity, suggesting that IL6 and CXCL8 function directly as mediators and effectors of ΔNp63-mediated metastatic lung colonization [62]. These studies provide a cogent illustration of pathways that can link cell-autonomous phenomena (like aberrant ΔNp63 expression) with non-cell-autonomous mechanisms (such as lung cell-derived factors that trigger IL6 and CXCL8 production in these cells). Fig. 7.2 illustrates how the end results of ΔNp63/IL6/CXCL8 pathway activation may link a number of the mechanisms previously described.

Importantly, these pathways have proved targetable. In the same work published by Gross et al., exploration of the therapeutic potential for IL6 and CXCL8 blockade yielded interesting results [62]. While inhibition of the receptors for IL6 or CXCL8 alone had very modest effects in their murine models, simultaneous targeting of both pathways demonstrated profoundly synergistic effects. Eighty-five percent of mice treated with simultaneous IL6 and CXCL8 pathway inhibition became long-term survivors, while 100% of mice receiving vehicle succumbed to metastatic disease. These treatments had no effect on proliferating cells in culture or on orthotopically growing tumors in mice, suggesting that the mechanism dealt with mechanisms derived from tumor-host interactions encountered in the

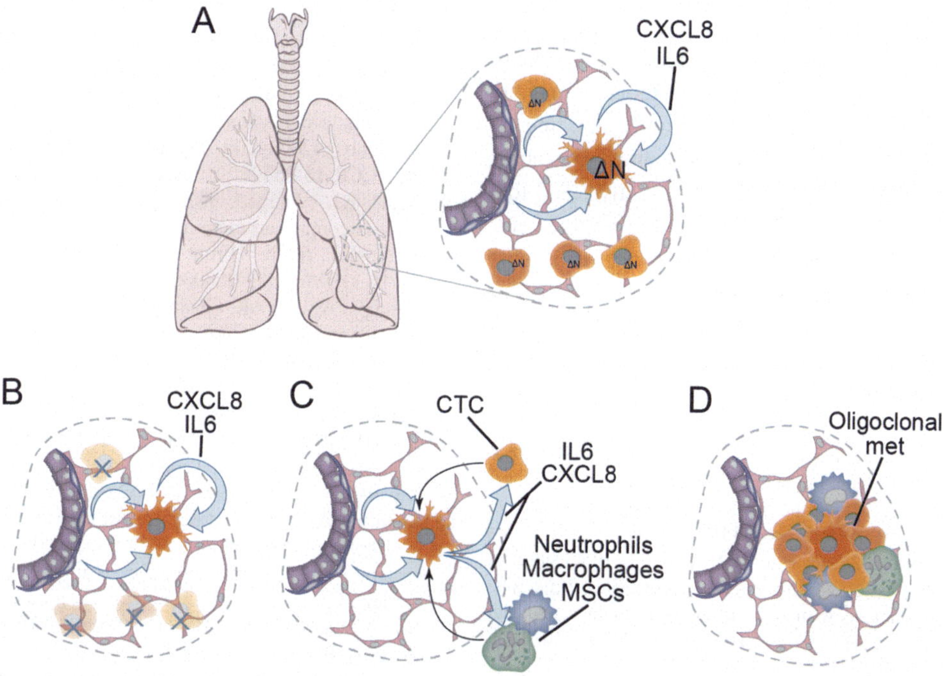

Fig. 7.2 The ΔNp63/IL6/CXCL8 axis drives lung colonization in some models of osteosarcoma metastasis. (**a**) A subset of tumor cells aberrantly expresses high levels of ΔNp63. This primes them to produce very high levels of IL6 and CXCL8 on interaction with pulmonary epithelium upon dissemination. IL6 and CXCL8 facilitate colonization of the lung by: (**b**) providing autocrine survival signals to a subset of metastasis initiating cells, allowing them to avoid apoptosis, and (**c**) recruiting circulating tumor cells and immune cells into the metastatic niche, activating them in ways that promote the formation of clinically relevant, mostly oligoclonal metastases (**d**)

lung microenvironment [62]. Similar reductions in tumor burden were seen in mice bearing osteosarcoma tumors representing several different cell lines, though it is unclear whether these results translate into the same profound effects seen with the OS-17 cells used in the long-term survival studies. Ongoing studies are evaluating similar therapeutic approaches in canine patients affected with osteosarcoma.

Unfortunately, the specific mechanism by which IL6 and CXCL8 affect metastasis remains elusive. While multiple investigators have independently identified these genes as important for the process by which osteosarcoma cells colonize lung tissues, none have demonstrated how they do so or why they are necessary. This remains an active area of research in multiple laboratories. Further investigation into the granular details of these mechanisms will enhance our understanding and may identify additional targetable vulnerabilities.

Conclusions

Mechanisms that allow disseminated osteosarcoma cells to persist and proliferate within the otherwise hostile microenvironment of the lung remain poorly understood. However, work completed to date would suggest the utility of such study. Genetic or epigenetic specialization of small subpopulations endow those cells with mechanisms for survival that allow this extremely small number of cells to persist when the vast majority of their counterparts cannot. Targeting osteosarcoma cells during this vulnerable transition may prove to be a unique therapeutic opportunity for preventing relapse and a unique scientific opportunity for discovering pathways most critical to the survival of all osteosarcoma cells. The integration of multiple emerging lines of research that results from rapidly improving technical capacity, from steadily increasing knowledge about the behavior of osteosarcoma cells within the lung environment, and from a recently precipitous improvement in our understanding of the mechanisms of osteosarcomagenesis promise to accelerate our understanding of these processes in the coming years. Translation of these findings into rational, effective, and tolerable treatments will be increasingly important. While the process of discovery can often surprise, optimism is warranted.

References

1. Marcove RC, Mike V, Hajek JV, Levin AG, Hutter RV (1970) Osteogenic sarcoma under the age of twenty-one. A review of one hunderd and forty-five operative cases. J Bone Joint Surg 52:411–423
2. Aljubran AH, Griffin A, Pintilie M, Blackstein M (2009) Osteosarcoma in adolescents and adults: survival analysis with and without lung metastases. Ann Oncol 20(6):1136–1141. https://doi.org/10.1093/annonc/mdn731
3. Crompton BD, Goldsby RE, Weinberg VK, Feren R, O'Donnell RJ, Ablin AR (2006) Survival after recurrence of osteosarcoma: a 20-year experience at a single institution. Pediatr Blood Cancer 47:255–259. https://doi.org/10.1002/pbc.20580
4. Gelderblom H, Jinks RC, Sydes M, Bramwell VHC, van Glabbeke M, Grimer RJ, Hogendoorn PCW, McTiernan A, Lewis IJ, Nooij MA, Taminiau AHM, Whelan J, Intergroup EO (2011) Survival after recurrent osteosarcoma: data from 3 European Osteosarcoma Intergroup (EOI) randomized controlled trials. Eur J Cancer 47:895–902. https://doi.org/10.1016/j.ejca.2010.11.036
5. Leary SES, Wozniak AW, Billups CA, Wu J, McPherson V, Neel MD, Rao BN, Daw NC (2013) Survival of pediatric patients after relapsed osteosarcoma: the St. Jude Children's Research Hospital experience. Cancer 119:2645–2653. https://doi.org/10.1002/cncr.28111
6. Perkins SM, Shinohara ET, DeWees T, Frangoul H (2014) Outcome for children with metastatic solid tumors over the last four decades. PLoS One 9:e100396. https://doi.org/10.1371/journal.pone.0100396
7. Allison DC, Carney SC, Ahlmann ER, Hendifar A, Chawla S, Fedenko A, Angeles C, Menendez LR (2012) A meta-analysis of osteosarcoma outcomes in the modern medical era. Sarcoma 2012:704872. https://doi.org/10.1155/2012/704872
8. Khanna C, Fan TM, Gorlick R, Helman LJ, Kleinerman ES, Adamson PC, Houghton PJ, Tap WD, Welch DR, Steeg PS, Merlino G, Sorensen PH, Meltzer P, Kirsch DG, Janeway KA, Weigel B, Randall L, Withrow SJ, Paoloni M, Kaplan R, Teicher BA, Seibel NL, Smith M, Uren A, Patel SR, Trent J, Savage SA, Mirabello L, Reinke D, Barkaukas DA, Krailo M, Bernstein M (2014) Toward a drug development path that targets metastatic progression in osteosarcoma. Clin Cancer Res 20(16):4200–4209. https://doi.org/10.1158/1078-0432.CCR-13-2574
9. Saraf AJ, Fenger JM, Roberts RD (2018) Osteosarcoma: accelerating progress makes for a hopeful future. Front Oncol 8:4. https://doi.org/10.3389/fonc.2018.00004
10. Dasgupta A, Lim AR, Ghajar CM (2017) Circulating and disseminated tumor cells: harbingers or initiators of metastasis? Mol Oncol 11:40–61. https://doi.org/10.1002/1878-0261.12022
11. Bruland ØS, Høifødt H, Sæter G, Smeland S, Fodstad Ø (2005) Hematogenous micrometastases in osteosarcoma patients. Clin Cancer Res 11:4666–4673. https://doi.org/10.1158/1078-0432.CCR-05-0165
12. Chalopin A, Tellez-Gabriel M, Brown HK, Vallette F, Heymann MF, Gouin F, Heymann D (2018) Isolation of circulating tumor cells in a preclinical model of osteosarcoma: effect of chemotherapy. J Bone Oncol 12:83–90. https://doi.org/10.1016/j.jbo.2018.07.002
13. Turajlic S, Swanton C (2016) Metastasis as an evolutionary process. Science (New York, NY) 352:169–175. https://doi.org/10.1126/science.aaf2784
14. Fidler IJ (1970) Metastasis: quantitative analysis of distribution and fate of tumor emboli labeled with 125 I-5-iodo-2'-deoxyuridine. J Natl Cancer Inst 45:773–782
15. Wong CW, Lee A, Shientag L, Yu J, Dong Y, Kao G, Al-Mehdi AB, Bernhard EJ, Muschel RJ (2001) Apoptosis: an early event in metastatic inefficiency. Cancer Res 61:333–338
16. Cameron MD, Schmidt EE, Kerkvliet N, Nadkarni KV, Morris VL, Groom AC, Chambers AF, MacDonald IC (2000) Temporal progression of metastasis in lung: cell survival, dormancy, and location dependence of metastatic inefficiency. Cancer Res 60:2541–2546
17. Goldstein SD, Trucco M, Bautista Guzman W, Hayashi M, Loeb DM (2016) A monoclonal antibody against the Wnt signaling inhibitor dickkopf-1 inhibits osteosarcoma metastasis in a preclinical model. Oncotarget 7:21114–21123. https://doi.org/10.18632/oncotarget.8522
18. Harper KL, Sosa MS, Entenberg D, Hosseini H, Cheung JF, Nobre R, Avivar-Valderas A, Nagi C, Girnius N, Davis RJ, Farias EF, Condeelis J, Klein CA, Aguirre-Ghiso JA (2016) Mechanism of early dissemination and metastasis in Her2(+) mammary cancer. Nature 540(7634):588–592. https://doi.org/10.1038/nature20609
19. Sahai E (2007) Illuminating the metastatic process. Nat Rev Cancer 7:737–749. https://doi.org/10.1038/nrc2229

20. Mendoza A, Hong SH, Osborne T, Khan MA, Campbell K, Briggs J, Eleswarapu A, Buquo L, Ren L, Hewitt SM, Dakir e-H, Garfield S, Walker R, Merlino G, Green JE, Hunter KW, Wakefield LM, Khanna C (2010) Modeling metastasis biology and therapy in real time in the mouse lung. J Clin Invest 120(8):2979–2988. https://doi.org/10.1172/JCI40252
21. Mendoza A, Hong S-H, Osborne T, Khan MA, Campbell K, Briggs J, Eleswarapu A, Buquo L, Ren L, Hewitt SM, Dakir E-H, Garfield S, Walker R, Merlino G, Green JE, Hunter KW, Wakefield LM, Khanna C (2010) Modeling metastasis biology and therapy in real time in the mouse lung. J Clin Invest 120:2979–2988. https://doi.org/10.1172/JCI40252
22. Khanna C, Prehn J, Yeung C, Caylor J, Tsokos M, Helman L (2000) An orthotopic model of murine osteosarcoma with clonally related variants differing in pulmonary metastatic potential. Clin Exp Metastasis 18:261–271. https://doi.org/10.1023/A:1006767007547
23. Khanna C, Wan X, Bose S, Cassaday R, Olomu O, Mendoza A, Yeung C, Gorlick R, Hewitt SM, Helman LJ (2004) The membrane-cytoskeleton linker ezrin is necessary for osteosarcoma metastasis. Nat Med 10(2):182–186. https://doi.org/10.1038/nm982
24. Ren L, Mendoza A, Zhu J, Briggs JW, Halsey C, Hong ES, Burkett SS, Morrow J, Lizardo MM, Osborne T, Li SQ, Luu HH, Meltzer P, Khanna C (2015) Characterization of the metastatic phenotype of a panel of established osteosarcoma cells. Oncotarget 6:29469–29481. https://doi.org/10.18632/oncotarget.5177
25. Hong S-H, Ren L, Mendoza A, Eleswarapu A, Khanna C (2012) Apoptosis resistance and PKC signaling: distinguishing features of high and low metastatic cells. Neoplasia (New York, NY) 14:249–258
26. Cubillos-Ruiz JR, Bettigole SE, Glimcher LH (2017) Tumorigenic and immunosuppressive effects of endoplasmic reticulum stress in cancer. Cell 168(4):692–706. https://doi.org/10.1016/j.cell.2016.12.004
27. Vandewynckel YP, Laukens D, Geerts A, Bogaerts E, Paridaens A, Verhelst X, Janssens S, Heindryckx F, Van Vlierberghe H (2013) The paradox of the unfolded protein response in cancer. Anticancer Res 33(11):4683–4694
28. Lizardo MM, Morrow JJ, Miller TE, Hong ES, Ren L, Mendoza A, Halsey CH, Scacheri PC, Helman LJ, Khanna C (2016) Upregulation of glucose-regulated protein 78 in metastatic cancer cells is necessary for lung metastasis progression. Neoplasia 18(11):699–710. https://doi.org/10.1016/j.neo.2016.09.001
29. Yarapureddy S, Abril J, Foote J, Kumar S, Asad O, Sharath V, Faraj J, Daniel D, Dickman P, White-Collins A, Hingorani P, Sertil AR (2019) ATF6alpha activation enhances survival against chemotherapy and serves as a prognostic indicator in osteosarcoma. Neoplasia 21(6):516–532. https://doi.org/10.1016/j.neo.2019.02.004
30. Briggs JW, Ren L, Nguyen R, Chakrabarti K, Cassavaugh J, Rahim S, Bulut G, Zhou M, Veenstra TD, Chen Q, Wei JS, Khan J, Uren A, Khanna C (2012) The Ezrin metastatic phenotype is associated with the initiation of protein translation. Neoplasia (New York, NY) 14:297–310
31. Celik H, Bulut G, Han J, Graham GT, Minas TZ, Conn EJ, Hong SH, Pauly GT, Hayran M, Li X, Ozdemirli M, Ayhan A, Rudek MA, Toretsky JA, Uren A (2016) Ezrin inhibition up-regulates stress response gene expression. J Biol Chem 291(25):13257–13270. https://doi.org/10.1074/jbc.M116.718189
32. Morrow JJ, Mendoza A, Koyen A, Lizardo MM, Ren L, Waybright TJ, Hansen RJ, Gustafson DL, Zhou M, Fan TM, Scacheri PC, Khanna C (2016) mTOR inhibition mitigates enhanced mRNA translation associated with the metastatic phenotype of osteosarcoma cells in vivo. Clin Cancer Res 22:6129–6141. https://doi.org/10.1158/1078-0432.CCR-16-0326
33. Perry JA, Kiezun A, Tonzi P, Van Allen EM, Carter SL, Baca SC, Cowley GS, Bhatt AS, Rheinbay E, Pedamallu CS, Helman E, Taylor-Weiner A, McKenna A, DeLuca DS, Lawrence MS, Ambrogio L, Sougnez C, Sivachenko A, Walensky LD, Wagle N, Mora J, de Torres C, Lavarino C, Dos Santos AS, Yunes JA, Brandalise SR, Mercado-Celis GE, Melendez-Zajgla J, Cardenas-Cardos R, Velasco-Hidalgo L, Roberts CW, Garraway LA, Rodriguez-Galindo C, Gabriel SB, Lander ES, Golub TR, Orkin SH, Getz G, Janeway KA (2014) Complementary genomic approaches highlight the PI3K/mTOR pathway as a common vulnerability in osteosarcoma. Proc Natl Acad Sci U S A 111(51):E5564–E5573. https://doi.org/10.1073/pnas.1419260111
34. Chen X, Bahrami A, Pappo A, Easton J, Dalton J, Hedlund E, Ellison D, Shurtleff S, Wu G, Wei L, Parker M, Rusch M, Nagahawatte P, Wu J, Mao S, Boggs K, Mulder H, Yergeau D, Lu C, Ding L, Edmonson M, Qu C, Wang J, Li Y, Navid F, Daw NC, Mardis ER, Wilson RK, Downing JR, Zhang J, Dyer MA, St. Jude Children's Research Hospital-Washington University Pediatric Cancer Genome P (2014) Recurrent somatic structural variations contribute to tumorigenesis in pediatric osteosarcoma. Cell Rep 7(1):104–112. https://doi.org/10.1016/j.celrep.2014.03.003
35. Behjati S, Tarpey PS, Haase K, Ye H, Young MD, Alexandrov LB, Farndon SJ, Collord G, Wedge DC, Martincorena I, Cooke SL, Davies H, Mifsud W, Lidgren M, Martin S, Latimer C, Maddison M, Butler AP, Teague JW, Pillay N, Shlien A, McDermott U, Futreal PA, Baumhoer D, Zaikova O, Bjerkehagen B, Myklebost O, Amary MF, Tirabosco R, Van Loo P, Stratton MR, Flanagan AM, Campbell PJ (2017) Recurrent mutation of IGF signalling genes and distinct patterns of genomic rearrangement in osteosarcoma. Nat Commun 8:15936. https://doi.org/10.1038/ncomms15936

36. Sayles LC, Breese MR, Koehne AL, Leung SG, Lee AG, Liu HY, Spillinger A, Shah AT, Tanasa B, Straessler K, Hazard FK, Spunt SL, Marina N, Kim GE, Cho SJ, Avedian RS, Mohler DG, Kim MO, DuBois SG, Hawkins DS, Sweet-Cordero EA (2019) Genome-informed targeted therapy for osteosarcoma. Cancer Discov 9(1):46–63. https://doi.org/10.1158/2159-8290.CD-17-1152
37. Morrow JJ, Bayles I, Funnell APW, Miller TE, Saiakhova A, Lizardo MM, Bartels CF, Kapteijn MY, Hung S, Mendoza A, Dhillon G, Chee DR, Myers JT, Allen F, Gambarotti M, Righi A, DiFeo A, Rubin BP, Huang AY, Meltzer PS, Helman LJ, Picci P, Versteeg HH, Stamatoyannopoulos JA, Khanna C, Scacheri PC (2018) Positively selected enhancer elements endow osteosarcoma cells with metastatic competence. Nat Med 24(2):176–185. https://doi.org/10.1038/nm.4475
38. Bid HK, Roberts RD, Cam M, Audino A, Kurmasheva RT, Lin J, Houghton PJ, Cam H (2014) DeltaNp63 promotes pediatric neuroblastoma and osteosarcoma by regulating tumor angiogenesis. Cancer Res 74(1):320–329. https://doi.org/10.1158/0008-5472.CAN-13-0894
39. Engelmann D, Putzer BM (2014) Emerging from the shade of p53 mutants: N-terminally truncated variants of the p53 family in EMT signaling and cancer progression. Sci Signal 7(345):re9. https://doi.org/10.1126/scisignal.2005699
40. Cam M, Gardner HL, Roberts RD, Fenger JM, Guttridge DC, London CA, Cam H (2016) DeltaNp63 mediates cellular survival and metastasis in canine osteosarcoma. Oncotarget 7(30):48533–48546. https://doi.org/10.18632/oncotarget.10406
41. Fenger JM, London CA, Kisseberth WC (2014) Canine osteosarcoma: a naturally occurring disease to inform pediatric oncology. ILAR J 55(1):69–85. https://doi.org/10.1093/ilar/ilu009
42. Paoloni M, Davis S, Lana S, Withrow S, Sangiorgi L, Picci P, Hewitt S, Triche T, Meltzer P, Khanna C (2009) Canine tumor cross-species genomics uncovers targets linked to osteosarcoma progression. BMC Genomics 10:625. https://doi.org/10.1186/1471-2164-10-625
43. Liu Y, Cao X (2016) Characteristics and significance of the pre-metastatic niche. Cancer Cell 30(5):668–681. https://doi.org/10.1016/j.ccell.2016.09.011
44. Ghajar CM, Peinado H, Mori H, Matei IR, Evason KJ, Brazier H, Almeida D, Koller A, Hajjar KA, Stainier DY, Chen EI, Lyden D, Bissell MJ (2013) The perivascular niche regulates breast tumour dormancy. Nat Cell Biol 15(7):807–817. https://doi.org/10.1038/ncb2767
45. Carlson P, Dasgupta A, Grzelak CA, Kim J, Barrett A, Coleman IM, Shor RE, Goddard ET, Dai J, Schweitzer EM, Lim AR, Crist SB, Cheresh DA, Nelson PS, Hansen KC, Ghajar CM (2019) Targeting the perivascular niche sensitizes disseminated tumour cells to chemotherapy. Nat Cell Biol 21(2):238–250. https://doi.org/10.1038/s41556-018-0267-0
46. Weekes CD, Kuszynski CA, Sharp JG (2001) VLA-4 mediated adhesion to bone marrow stromal cells confers chemoresistance to adherent lymphoma cells. Leuk Lymphoma 40(5–6):631–645. https://doi.org/10.3109/10428190109097661
47. Lane SW, Scadden DT, Gilliland DG (2009) The leukemic stem cell niche: current concepts and therapeutic opportunities. Blood 114(6):1150–1157. https://doi.org/10.1182/blood-2009-01-202606
48. Berg SL, Winick N, Ingle AM, Adamson PC, Blaney SM (2010) Reasons for participation in optional pharmacokinetic studies in children with cancer: a Children's Oncology Group Phase 1 Consortium study. Pediatr Blood Cancer 55:119–122. https://doi.org/10.1002/pbc.22529
49. Minn AJ, Gupta GP, Siegel PM, Bos PD, Shu W, Giri DD, Viale A, Olshen AB, Gerald WL, Massagué J (2005) Genes that mediate breast cancer metastasis to lung. Nature 436:518–524. https://doi.org/10.1038/nature03799
50. Oskarsson T, Acharyya S, Zhang XH-F, Vanharanta S, Tavazoie SF, Morris PG, Downey RJ, Manova-Todorova K, Brogi E, Massagué J (2011) Breast cancer cells produce tenascin C as a metastatic niche component to colonize the lungs. Nat Med 17:867–874. https://doi.org/10.1038/nm.2379
51. Chiquet-Ehrismann R, Orend G, Chiquet M, Tucker RP, Midwood KS (2014) Tenascins in stem cell niches. Matrix Biol 37:112–123. https://doi.org/10.1016/j.matbio.2014.01.007
52. Zhang W, Ding M-L, Zhang J-N, Qiu J-R, Shen Y-H, Ding X-Y, Deng L-F, Zhang W-B, Zhu J (2015) mTORC1 maintains the tumorigenicity of SSEA-4+ high-grade osteosarcoma. Sci Rep 5(1):9604. https://doi.org/10.1038/srep09604
53. Zhang W, Zhao J-M, Lin J, Hu C-Z, Zhang W-B, Yang W-L, Zhang J, Zhang J-W, Zhu J (2018) Adaptive fibrogenic reprogramming of osteosarcoma stem cells promotes metastatic growth. Cell Rep 24(5):1266–1277.e1265. https://doi.org/10.1016/j.celrep.2018.06.103
54. Koshkina NV, Khanna C, Mendoza A, Guan H, DeLauter L, Kleinerman ES (2007) Fas-negative osteosarcoma tumor cells are selected during metastasis to the lungs: the role of the Fas pathway in the metastatic process of osteosarcoma. Mol Cancer Res 5:991–999. https://doi.org/10.1158/1541-7786.MCR-07-0007
55. Lafleur EA (2004) Increased Fas expression reduces the metastatic potential of human osteosarcoma cells. Clin Cancer Res 10(23):8114–8119. https://doi.org/10.1158/1078-0432.ccr-04-0353
56. Gordon N, Koshkina NV, Jia SF, Khanna C, Mendoza A, Worth LL, Kleinerman ES (2007) Corruption of the Fas pathway delays the pulmonary clearance of murine osteosarcoma cells, enhances their metastatic potential, and reduces the effect of aerosol gemcitabine. Clin Cancer Res 13(15):4503–4510. https://doi.org/10.1158/1078-0432.ccr-07-0313

57. Chou AJ, Kleinerman ES, Krailo MD, Chen Z, Betcher DL, Healey JH, Conrad EU 3rd, Nieder ML, Weiner MA, Wells RJ, Womer RB, Meyers PA, Children's Oncology Group (2009) Addition of muramyl tripeptide to chemotherapy for patients with newly diagnosed metastatic osteosarcoma: a report from the Children's Oncology Group. Cancer 115(22):5339–5348. https://doi.org/10.1002/cncr.24566
58. Meyers PA, Schwartz CL, Krailo M, Kleinerman ES, Betcher D, Bernstein ML, Conrad E, Ferguson W, Gebhardt M, Goorin AM, Harris MB, Healey J, Huvos A, Link M, Montebello J, Nadel H, Nieder M, Sato J, Siegal G, Weiner M, Wells R, Wold L, Womer R, Grier H (2005) Osteosarcoma: a randomized, prospective trial of the addition of ifosfamide and/or muramyl tripeptide to cisplatin, doxorubicin, and high-dose methotrexate. J Clin Oncol 23:2004–2011. https://doi.org/10.1200/JCO.2005.06.031
59. Meyers PA, Schwartz CL, Krailo MD, Healey JH, Bernstein ML, Betcher D, Ferguson WS, Gebhardt MC, Goorin AM, Harris M, Kleinerman E, Link MP, Nadel H, Nieder M, Siegal GP, Weiner MA, Wells RJ, Womer RB, Grier HE (2008) Osteosarcoma: the addition of muramyl tripeptide to chemotherapy improves overall survival–a report from the Children's Oncology Group. J Clin Oncol 26:633–638. https://doi.org/10.1200/JCO.2008.14.0095
60. Rackley CR, Stripp BR (2012) Building and maintaining the epithelium of the lung. J Clin Investig 122(8):2724–2730. https://doi.org/10.1172/jci60519
61. Baglio SR, Lagerweij T, Perez-Lanzon M, Ho XD, Leveille N, Melo SA, Cleton-Jansen AM, Jordanova ES, Roncuzzi L, Greco M, van Eijndhoven MAJ, Grisendi G, Dominici M, Bonafede R, Lougheed SM, de Gruijl TD, Zini N, Cervo S, Steffan A, Canzonieri V, Martson A, Maasalu K, Koks S, Wurdinger T, Baldini N, Pegtel DM (2017) Blocking tumor-educated MSC paracrine activity halts osteosarcoma progression. Clin Cancer Res 23(14):3721–3733. https://doi.org/10.1158/1078-0432.CCR-16-2726
62. Gross AC, Cam H, Phelps DA, Saraf AJ, Bid HK, Cam M, London CA, Winget SA, Arnold MA, Brandolini L, Mo X, Hinckley JM, Houghton PJ, Roberts RD (2018) IL-6 and CXCL8 mediate osteosarcoma-lung interactions critical to metastasis. JCI Insight 3(16). https://doi.org/10.1172/jci.insight.99791

Wnt Signaling in Osteosarcoma

8

Amit Singla, Jichuan Wang, Rui Yang,
David S. Geller, David M. Loeb,
and Bang H. Hoang

Abstract

Wnt molecules are a class of cysteine-rich secreted glycoproteins that participate in various developmental events during embryogenesis and adult tissue homeostasis. Since its discovery in 1982, the roles of Wnt signaling have been established in various key regulatory systems in biology. Wnt signals exert pleiotropic effects, including mitogenic stimulation, cell fate specification, and differentiation. The Wnt signaling pathway in humans has been shown to be involved in a wide variety of disorders including colon cancer, sarcoma, coronary artery disease, tetra-amelia, Mullerian duct regression, eye vascular defects, and abnormal bone mass. The canonical Wnt pathway functions by regulating the function of the transcriptional coactivator β-catenin, whereas noncanonical pathways function independent of β-catenin. Although the role of Wnt signaling is well established in epithelial malignancies, its role in mesenchymal tumors is more controversial. Some studies have suggested that Wnt signaling plays a pro-oncogenic role in various sarcomas by driving cell proliferation and motility; however, others have reported that Wnt signaling acts as a tumor suppressor by committing tumor cells to differentiate into a mature lineage. Wnt signaling pathway also plays an important role in regulating cancer stem cell function. In this review, we will discuss Wnt signaling pathway and its role in osteosarcoma.

Keywords

Osteosarcoma · Wnt · β-Catenin · Dickkopf · Cancer stem cell · Wnt inhibitory protein · Frizzled- related protein

A. Singla · R. Yang · D. S. Geller · B. H. Hoang (✉)
Department of Orthopedic Surgery, Montefiore Medical Center, Albert Einstein College of Medicine, Bronx, NY, USA
e-mail: bahoang@montefiore.org

J. Wang
Department of Orthopedic Surgery, Montefiore Medical Center, Albert Einstein College of Medicine, Bronx, NY, USA

Musculoskeletal Tumor Center, Beijing Key Laboratory for Musculoskeletal Tumors, Peking University People's Hospital, Beijing, China

D. M. Loeb
Departments of Pediatrics and Developmental and Molecular Biology, Children's Hospital at Montefiore, Albert Einstein College of Medicine, Bronx, NY, USA

E. S. Kleinerman, R. Gorlick (eds.), *Current Advances in the Science of Osteosarcoma*, Advances in Experimental Medicine and Biology 1258, https://doi.org/10.1007/978-3-030-43085-6_8

Introduction

Wnt molecules are a class of cysteine-rich secreted glycoproteins that participate in various developmental events during embryogenesis and adult tissue homeostasis. Since its discovery in 1982, the roles of Wnt signaling have been established in various key regulatory systems in biology. The term Wnt combines the *Drosophila* segment polarity gene *Wingless* and the mouse proto-oncogene *Int-1*. Currently, 19 different Wnt proteins have been identified in the human genome [1, 2]. Wnt signals exert pleiotropic effects, including mitogenic stimulation, cell fate specification, and differentiation. It is a complex, tightly regulated pathway with many functions, whose involvement in cancer reinforces the notion that oncogenesis is a form of development gone awry. The Wnt signaling pathway in humans has been shown to be involved in a wide variety of disorders including colon cancer, coronary artery disease, tetra-amelia, Mullerian duct regression, eye vascular defects, and abnormal bone mass [2, 3].

The canonical Wnt pathway functions by regulating the function of the transcriptional coactivator β-catenin, whereas noncanonical pathways function independent of β-catenin. In the development of the bone, the canonical Wnt/β-catenin pathway is required for osteoblast differentiation, enhanced ossification, and suppression of chondrocyte formation. Recent reports of conditional inactivation of β-catenin in skeletal progenitors using Cre lines have revealed that β-catenin is essential for differentiation of mature osteoblasts and consequently for bone formation. Lack of β-catenin leads to a failure of perichondral and periosteal cells to express the osteoblast commitment factor Osterix, resulting in a chondrogenic fate [4, 5]. Loss-of-function and gain-of-function mutations in low-density lipoprotein receptor-related protein 5 (LRP5), a Wnt receptor, are associated with osteoporosis-pseudoglioma syndrome and high bone mass phenotypes, respectively [6, 7]. LRP5 expression has also been shown to be a marker for disease progression in high-grade osteosarcoma (OS), and its suppression may lead to reduction in local or systemic disease burden [8, 9].

The association between Wnt/β-catenin signaling and colon cancer is well recognized. Blocking β-catenin signaling has generated a significant interest in colon cancer treatment [10]. Although the role of Wnt signaling is well established in epithelial malignancies, its role in mesenchymal tumors is more controversial. Some studies have suggested that Wnt signaling plays a pro-oncogenic role in various sarcomas by driving cell proliferation and motility [11–14]; however, others have reported that Wnt signaling acts as a tumor suppressor by committing tumor cells to differentiate into a mature lineage [15–18]. In addition, the Wnt signaling pathway plays an important role in regulating cancer stem cell (CSC) function [19]. In osteosarcoma, stem cells have activated Wnt/β-catenin signaling, and Wnt inhibition can thus reduce drug resistance [20, 21].

Osteosarcoma (OS) is the most common primary bone malignancy in children and young adults. With the current multidisciplinary treatments, 60–70% of patients with localized disease survive [22]. Unfortunately, the long-term survival of patients with relapsed disease is only about 20% [23]. Despite aggressive efforts to strengthen and modify chemotherapy, the outcome of patients with OS has not significantly improved over the past few decades [24]. In this review, we will discuss Wnt signaling pathway and its role in osteosarcoma.

Overview of Wnt/β-Catenin Signaling Pathway

The Canonical Wnt Pathway

In the inactive state, there is an absence or inhibition of Wnt, which enables cytoplasmic β-catenin to form a complex with multiple proteins, including Axin, adenomatous polyposis coli gene product (APC), casein kinase 1 (CK1), and glycogen synthase kinase 3β (GSK3β) [2, 25–27] (see Fig. 8.1). Within this complex, CK1

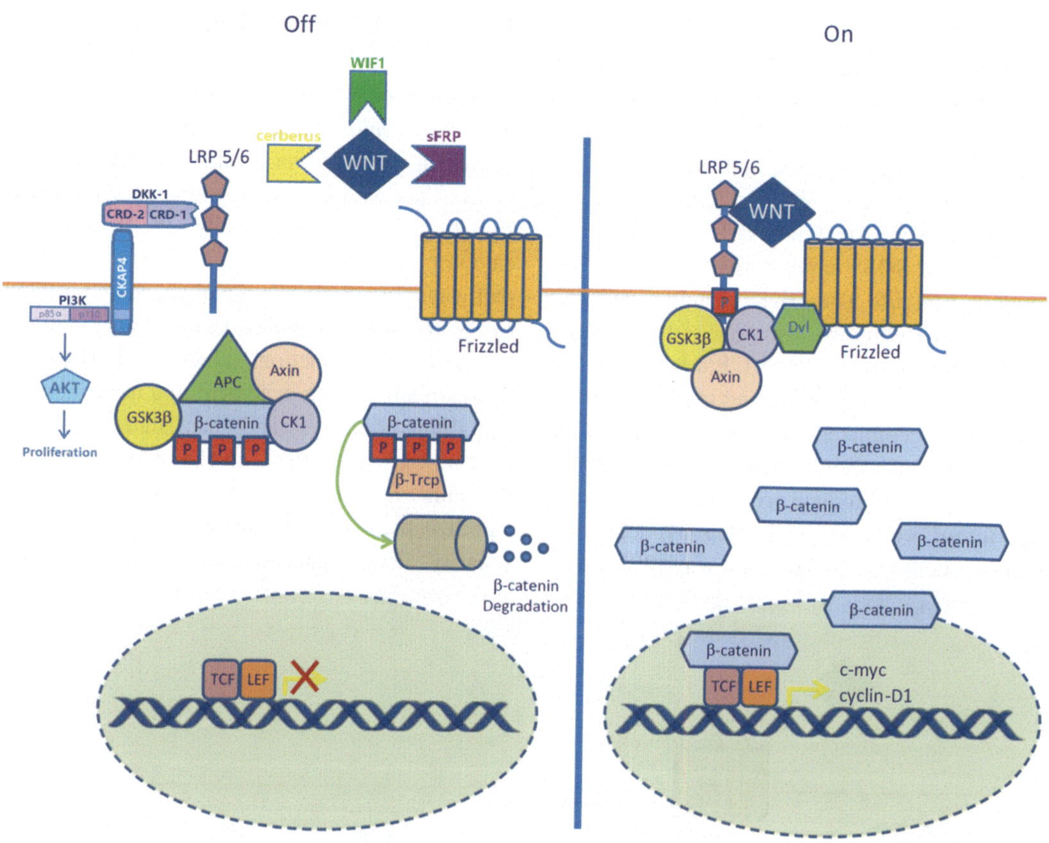

Fig. 8.1 Overview of Wnt/β-catenin signaling
In the absence or inhibition of Wnt, the cytoplasmic β-catenin forms a complex with Axin, adenomatous polyposis coli (APC), casein kinase 1 (CK1), and glycogen synthase kinase 3β (GSK3β). CK1 and GSK3β phosphorylate β-catenin. β-Trcp (E3 ubiquitin ligase subunit) recognizes this complex and targets β-catenin for proteasomal degradation. The Wnt antagonists WIF1, sFRP, and cerberus bind directly to Wnt ligands. The DKK family proteins competitively bind to Wnt receptor LRP5/6. The DKK proteins have two cysteine-rich domains, cysteine-rich domain 1 (CRD1) and CRD2. DKK-1 binds to LRP5/6 through the CRD1 domain, and DKK-1 binds to cytoskeleton-associated protein (CKAP) 4 through CRD2 to induce proliferation in normal and tumor cells in a β-catenin-independent manner via the PI3K/AKT pathway
In the presence of Wnt binding to targeted receptors frizzleds, low-density lipoprotein receptor-related protein 5 and 6 (LPR 5/6), and disheveled (Dvl), the complex becomes phosphorylated, leading to the inhibition of GSK3β. Cytoplasmic non-phosphorylated β-catenin accumulates, inhibiting its degradation and promoting translocation to the nucleus. A complex with transcription factors, including T-cell transcription factor (TCF), lymphoid enhancer-binding factor (LEF), and transcriptional coactivators, leads to transcriptional activity of multiple downstream target oncogenes

and GSK3β act in concert to phosphorylate β-catenin, which is then targeted by the E3 ubiquitin ligase β-Trcp for proteasomal degradation. When Wnt ligands bind to their target membrane receptors frizzled and LRP5/6, cytoplasmic disheveled (Dvl) causes phosphorylation of the complex, leading to an inhibition of GSK3β. The resulting cytoplasmic accumulation of non-phosphorylated β-catenin promotes its translocation into the nucleus. Within the nucleus, β-catenin forms a complex with T-cell transcription factor (TCF), lymphoid enhancer-binding factor (LEF), and other transcriptional coactivators to induce the transcription of multiple downstream target genes (e.g., c-Myc, cyclin D1) that promote cellular proliferation [2, 28].

In addition to Wnt ligands and receptors, five families of Wnt antagonists have been identified: Wnt inhibitory factor 1 (WIF1), secreted frizzled-related proteins (sFRP1–5), Dickkopf (DKK) proteins, cerberus, and the Wise/SOST family. Among these, WIF1, sFRP, and cerberus bind directly to Wnt ligands. The DKK family proteins and SOST competitively bind to Wnt receptor LRP5/6. The DKK proteins have two cysteine-rich domains, cysteine-rich domain 1 (CRD1) and CRD2. DKK-1 binds to LRP5/6 through the CRD1 domain, and more recently, it has been recognized that DKK-1 binds to cytoskeleton-associated protein (CKAP) 4 through CRD2 to induce proliferation in normal and tumor cells in a β-catenin-independent manner via the PI3K/AKT pathway [29].

The Noncanonical Wnt Pathways

Noncanonical Wnt pathways play important roles in embryonic and tissue development, homeostasis, and bone formation. Wnt5a is one of the major ligands for noncanonical Wnt signaling. Planar cell polarity (PCP) and Wnt/calcium are the major noncanonical Wnt pathways (Fig. 8.2). These pathways are initiated by Wnt/frizzled signaling rather than β-catenin transcriptional function. Disheveled (Dvl) is downstream of both canonical and noncanonical signaling pathways and has three different domains, i.e., DIX, PDZ, and DEP. DIX and PDZ domains function in canonical pathway to stabilize β-catenin. In noncanonical planar cell polarity (PCP) pathway, activation of small

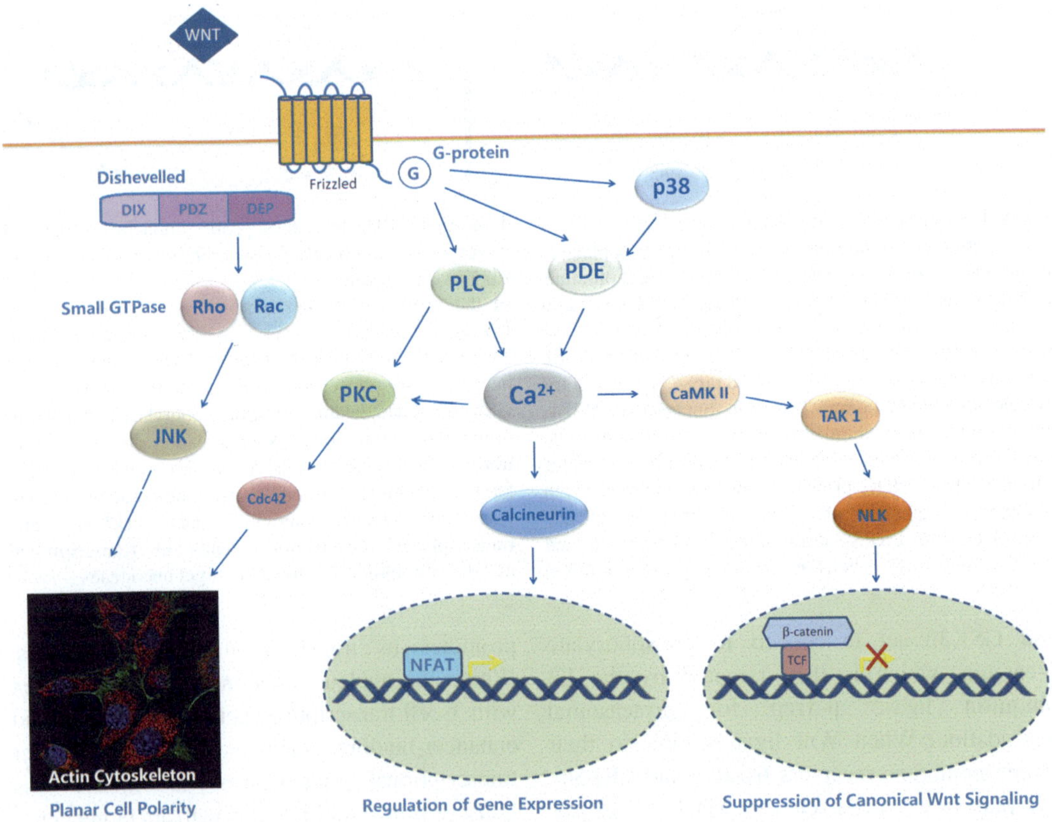

Fig. 8.2 Noncanonical Wnt signaling. The activation of frizzled by Wnt is mediated by disheveled or heterotrimeric G-proteins. The planar cell polarity pathway is mediated by small GTPase (Rho and Rac), JNK, and Cdc42, which is activated by PKC. Ca^{2+}-calcineurin pathway activates NFAT to regulate the gene expression. Ca^{2+}-induced CaMKII-TAK1-NLK pathway suppresses canonical Wnt signaling by inhibiting β-catenin-dependent transcription

GTPases, Rho and Rac, occurs downstream of DEP domain. The PCP pathway regulates the cytoskeletal architecture to affect cell migration via the activation of c-Jun NH2-terminal kinase (JNK).

In the Wnt/Ca^{2+} pathway, frizzled activation by Wnt leads to activation of heterotrimeric G-proteins. Activated G-protein regulates phospholipase C (PLC), phosphodiesterase (PDE), and p38, which on activation release intracellular Ca^{2+} to activate calcium-sensitive enzymes such as protein kinase C (PKC), calcineurin (CaCN), and Ca^{2+}/calmodulin-dependent kinase II (CaMKII). PKC activates the small GTPase Cdc42, which is a key regulator of PCP to remodel the actin cytoskeleton and control the polarity of cells. Nuclear factor of activated T-cells (NFAT) is the downstream target of CaCN. NFAT enters the nucleus to regulate target gene expression and regulates cell proliferation and differentiation. TGF-β-activated kinase-1 (TAK1)/Nemo-like kinase (NLK) are downstream targets of CaMKII that antagonize β-catenin-mediated canonical signaling [30, 31].

Table 8.1 Recent studies suggesting Wnt is pro-tumorigenic

Author/year	Protein of interest	Clinical relevance
Fang et al. 2018 [11]	Small molecular inhibitor of Wnt, PRI-724	Treatment with it leads to decreased cell proliferation, migration, invasion, colony formation
Neves et al. 2018 [21]	IWR-1 tankyrase inhibitor	IWR-1 inhibits translocation of β-catenin to the nucleus. It impairs self-renewal capacity of stem cells and hampers the activity and expression of stemness-related markers
Zhao et al. 2015 [113]	Naked cuticle homolog 2 (NKD2) gene, a down regulator of Wnt signaling	Downregulation of NKD2 expression found in metastatic and recurrent OS. Overexpression of NKD2 decreases cell proliferation and metastasis ability in vivo and in vitro by inhibiting Wnt signaling
Neves et al. 2015 [34]	Nuclear β-catenin	Exposure to conventional chemotherapy induces transition to stem-like phenotype associated with activation of Wnt/β-catenin signaling
Neves et al. 2015 [20]	Nuclear β-catenin	Cancer stem cells show activation of Wnt/β-catenin signaling as evident by increased nuclear β-catenin, TCF/LEF activity, and Axin2 expression

Wnt Signaling in Osteosarcoma

The role of Wnt signaling in osteosarcoma is controversial. The majority of studies report that Wnt signaling is pro-tumorigenic (Table 8.1); however, other studies also suggest that Wnt is tumor suppressive (Table 8.2). Tissue samples from osteosarcoma patients have been used to correlate various components of the Wnt pathway with clinical outcomes. In our study, RNA isolated from fresh-frozen osteosarcoma tissue was used to examine the expression of the Wnt receptor LRP5 by RT-PCR. LRP5 RNA expression statistically correlated with worse event-free survival in patients [8], and a dominant-negative LRP5 decreased tumorigenicity and metastasis of OS in vivo in nude mouse experiments [9]. Furthermore, it appears that blocking Wnt/LRP5 signaling can reduce tumor invasiveness by reversing the epithelial to mesenchymal transition [32]. Some recent studies also report Wnt/β-catenin signaling activation in osteosarcoma [13, 33]; however, others have shown that the Wnt/β-catenin signaling activation occurs only in the cancer stem cell (CSC) subpopulation of osteosarcoma cells and not in parental cells. It is the CSCs that are thought to be responsible for relapse, metastasis, and resistance to chemotherapy, so if targeting Wnt signaling can eliminate these cells, this may offer a new therapeutic approach that may improve patient outcomes [20, 21, 34]. There are some reports of DKK-1, a Wnt inhibitor, possibly playing a pro-tumorigenic role [15–17, 35, 36].

By using siRNA to suppress Wnt5a, Enomoto et al. demonstrated reduced invasiveness and

Table 8.2 Recent studies suggesting Wnt is anti-tumorigenic in osteosarcoma

Author/year	Protein of interest	Clinical relevance
Goldstein et al. 2016 [16]	Wnt signaling inhibitor DKK-1	Anti-DKK-1 antibody slowed the growth of tumor and inhibits metastasis. This effect was correlated with increased nuclear beta-catenin and increased expression of bone differentiation marker osteopontin
Cai et al. 2010 [18]	GIN (GSK3β inhibitor)	Absence of nuclear staining of β-catenin is found in 90% of OS cell lines, whereas all osteoblastomas demonstrated strong nuclear β-catenin staining. GIN activates Wnt/β-- catenin pathway as shown by translocation of β-catenin into the nucleus. GIN significantly reduces cell proliferation and enhances differentiation in OS cell lines
Lee et al. 2007 [35]	DKK-1	DKK-1 is highly expressed by OS tumors, and the level in blood is proportional to number of surviving OS cells in the tumor. DKK-1 is maximally expressed by OS in tumor periphery suggesting that it may have a role to prevent the repair of surrounding osteoid as the tumor expands
Krause et al. 2014 [36]	DKK-1	DKK-1 decreases cell differentiation potential, increases proliferation, and enhances osteolytic capacity. DKK-1 acts by shifting the balance of Wnt signaling in favor of Jun-mediated noncanonical Wnt pathways. This results in activation of RhoA and JNK and transcriptional activation of ALDH1 through Jun-responsive promoter elements
Gregory et al. 2003 [17]	Antibody to DKK-1	Anti-DKK-1 antibody increases the lag phase of OS cells

invadopodia formation in OS cells. These results suggest the role of noncanonical Wnt in conferring the invasive properties of osteosarcoma [37].

Wnt Antagonists in Osteosarcoma

WIF1

The antagonist Wnt inhibitor factor 1 (WIF1) is frequently downregulated in cancer cells, including prostate, breast, lung, and bladder cancer and in osteosarcoma [38, 39]. In these cancers, silencing of the WIF1 promoter by hypermethylation is associated with Wnt signaling activation [40–43]. In human osteosarcoma, silencing of WIF1 by promotor hypermethylation was shown to be associated with loss of differentiation, increased β-catenin levels, and increased proliferation, and in mice experiments, disruption of WIF1 accelerated osteosarcomagenesis [44]. Recently, we demonstrated that re-expressing WIF1 in OS cell lines inhibits anchorage-independent growth and cellular motility and decreases proteolytic enzyme matrix metalloproteinases (MMP-9 and MMP-14). In vivo, injecting WIF1-transfected OS cells into nude mice showed reduced tumorigenesis and pulmonary metastasis [39].

sFRP

Frzb, a member of the secreted frizzled-related protein (sFRP) family, is another Wnt antagonist that has been associated with cancer. It has an amino-terminal cysteine-rich domain (CRD) that is homologous to the ligand-binding domain of frizzled [45]. Frzb prevents receptor signaling primarily by binding to extracellular Wnt ligands, preventing the ligand-receptor interaction [46]. Frzb re-expression has been shown to inhibit tumorigenesis and invasiveness in both prostate and fibrosarcoma cancer cells. Recently, systemic and local levels of sFRP3 were found to be decreased in osteosarcoma [47].

DKK-1: Tumor Suppressor or Pro-tumorigenic Factor?

The role of DKK-1 in Wnt signaling pathways is complex (Fig. 8.3). Human DKK-1 inhibits the canonical Wnt signaling pathway by binding to the transmembrane receptor LRP5/6, preventing interaction with Wnt ligands [48]. More recently, it has been recognized that DKK-1 also binds to cytoskeleton-associated protein (CKAP) 4 to induce cell proliferation in normal cells and in tumor cells in a β-catenin-independent manner by activating the PI3K/AKT pathway [29]. DKK-1 has immunomodulatory role by attenuating the canonical Wnt signaling pathway, thereby facilitating cell-mediated immune evasion by natural killer cells [49].

Inhibition of Wnt by recombinant DKK-1 decreases both nuclear β-catenin and cytoplasmic β-catenin. Cytoplasmic β-catenin helps in formation of adherens junctions. DKK-1 decreases cell-cell adherence which is required for differentiation of the cells. Inhibiting Wnt signaling by DKK-1 in human mesenchymal stem cells can transform them to form high-grade undifferentiated sarcoma-like tumors in mice, and conversely, re-establishing Wnt signaling in these tumors can differentiate them along mature connective tissue lineage [15]. DKK-1 has contrasting effects on tumors and surrounding stroma. They may not only slow down tumor cell proliferation but also exert potent osteo-inhibitory effects on the stroma and maintain the tumor niche. DKK-1 was shown to inhibit osteogenesis in osteosarcoma cells and the surrounding tissue when implanted in vivo. DKK-1 also had the unexpected effect of increasing proliferation and resistance to metabolic stress in vitro. This effect was attributed to the upregulation of the stress response enzyme and cancer stem cell marker aldehyde dehydrogenase-1 (ALDH-1) via noncanonical planar cell polarity Wnt signaling [36].

Inhibiting DKK-1 leads to positive signaling through canonical Wnt/β-catenin/LEF pathway which drove cells out of cell cycle toward differentiation and postmitotic state. The link between DKK-1 and noncanonical Wnt pathway was also suggested as DKK-1 downregulates Ap-1/JNK pathway and thereby decreases the expression of cell adhesion protein VCAM-1 [17].

Whether DKK-1 is pro- or anti-tumorigenic may depend on cellular context also. A tumor-suppressive role has been demonstrated in renal cell carcinoma and in colon cancer [50–53], but there are reports suggesting that DKK-1 is pro-tumorigenic in esophageal, pancreatic, hepatocellular, gastric, and prostate cancers and in multiple myeloma [54–63].

In prostate cancer, DKK-1 is expressed at high levels in early-stage disease and decreases once the primary tumor progresses to metastasize, which can unmask the Wnt-mediated osteoblastic activity and promote the development of osteoblastic osseous metastases [56]. DKK-1 has a complex role in organotropic metastasis in breast cancer, suppressing lung metastasis by suppressing noncanonical Wnt-JNK and Wnt/Ca^{2+} signaling and promoting bone metastasis through canonical Wnt signaling [64].

In osteosarcoma, DKK-1 has been shown to have an unexpected role in cancer survival and resistance to stress via tipping the balance of Wnt signaling in favor of the noncanonical planar cell polarity pathway [36]. Another recent report has shown increased nuclear β-catenin and expression of the bone differentiation marker osteopontin on

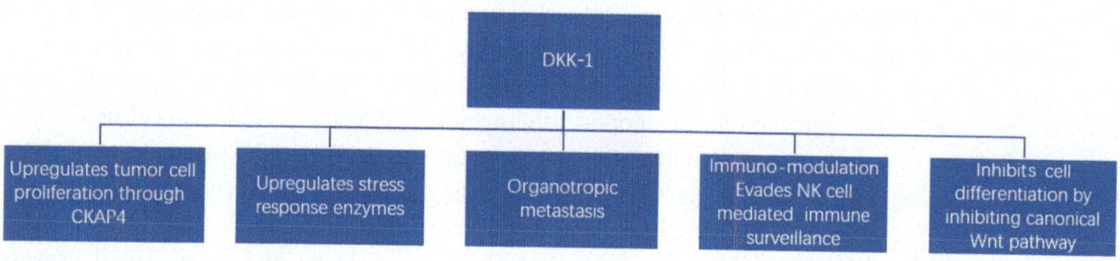

Fig. 8.3 Functions of DKK-1

treating the osteosarcoma with a neutralizing antibody against -DKK-1, resulting in decreased tumor growth and metastasis, suggesting that DKK-1 also regulates the canonical pathway [16].

Taken together, these studies have called into question the role of DKK-1 as a tumor suppressor and suggest that DKK-1 may be pro-tumorigenic in certain contexts [15–17, 35, 36].

DKK-3

Dickkopf-3 (DKK-3), also known as reduced expression in immortalized cells (REIC), is downregulated in multiple cancer cell lines, although its exact oncogenic suppressive mechanism is still unknown [65].

In osteosarcoma cells, DKK-3 has been shown to block β-catenin nuclear translocation, leading to inhibition of downstream LEF/TCF activation. The ectopic expression of DKK-3 and dominant-negative LRP5 mutant in osteosarcoma cell lines substantially decreases cell invasion and motility [66]. Furthermore, DKK-3 suppresses tumorigenesis and pulmonary metastasis in nude mice when transfected into osteosarcoma cells [67].

SOST

Sclerostin is another glycoprotein known to antagonize Wnt/β-catenin signaling in osteoblasts by binding to LRP5/6 and inhibiting osteoblast differentiation, activity, and survival [68, 69]. The SOST gene encodes for sclerostin, and its inhibition has been an area of interest for the treatment of osteoporosis [70, 71]. The FDA recently approved romosozumab, a monoclonal antibody against sclerostin, in postmenopausal women with osteoporosis at high risk of fracture. This agent has shown promising results in recent clinical trials for fracture prevention in osteoporosis [72, 73]. A recent study has shown that after SOST gene silencing, mRNA and protein expression of Wnt-1, β-catenin, c-Myc, cyclin D1, MMP-7 and more, which promotes proliferation, invasion, and migration, inhibits apoptosis of osteosarcoma cells [74].

Naturally Occurring Small Molecules

The small-molecule compound curcumin is a natural ingredient in turmeric, which shows an inhibitory effect against β-catenin/TCF signaling among several cancer cell lines [75]. Hallett et al. found that PKIF118-310, a natural compound of microbial origin and a small-molecule inhibitor of Wnt signaling (β-catenin/TCF inhibitor II), given to breast-tumor-bearing syngeneic mice arrested tumor growth in vivo [76]. In osteosarcoma, Leow et al. demonstrated that both curcumin and PKIF118-310 suppressed both intrinsic and activated β-catenin/TCF transcriptional activities using luciferase reporter assays. These compounds also reduce nuclear β-catenin and inhibit osteosarcoma cell migration and invasion in a dose-dependent manner. These anticancer effects are associated with decreased expression of several oncogenes, such as cyclin D1, c-Myc, and survivin [77]. Resveratrol, a natural grape product, has shown to inhibit proliferation of osteosarcoma cells by downregulating the expression of β-catenin [78].

Other Small Molecules

Besides naturally occurring antagonists, two new classes of small molecules that perturb the Wnt pathway have been reported. The first class of compound inhibits the membrane-bound acyltransferase Porcupine, which is involved in Wnt protein posttranslational modification. The second class nullifies the destruction of Axin, which is known to suppress the Wnt/β-catenin signaling [79]. Huang et al. [80] have described tankyrase inhibition by stabilizing Axin. Tankyrase interacts with Axin and stimulates its degradation through the ubiquitin-proteasome pathway. Tankyrase inhibitors, by attenuating Wnt/β-catenin signaling, have shown potential therapeutic effects in hepatocellular and colorectal cancers [81, 82]. Stratford et al. [83] demonstrated the efficacy of tankyrase inhibitor in three osteosarcoma cell lines. They reported stabilization of Axin2 and reduced cell growth due to delay in cell cycle progression and induction of caspase-3-mediated apoptosis. They also noticed that the miRNA of let-7 family was upregulated following treatment. Small-molecule inhibition of Wnt signaling by inhibiting tankyrase 1/2 enzymes was found to be cytotoxic to multiple osteosarcoma cell lines [84]. IWR-1, another tankyrase inhibitor, was shown to be specifically cytotoxic to osteosarcoma cancer stem cells [21] (Fig. 8.4).

A study by Grandy et al. identified another small-molecule inhibitor of Wnt which interacts with the PDZ domain of dishevelled [85]. Dishevelled (Dvl) is an essential component of the Wnt signaling pathway, which transduces Wnt signals from the frizzled receptor to downstream targeted components. Through structure-based ligand screening and NMR spectroscopy, these investigators were able to discover a small-molecule inhibitor (3289-8625) with an affinity to the PDZ domain of Dvl. It was shown to suppress the tumorigenesis of prostate cancer PC-3 cells and decrease Wnt signaling in the hyaloid vessel system and may prove to have similar affects in osteosarcoma cells.

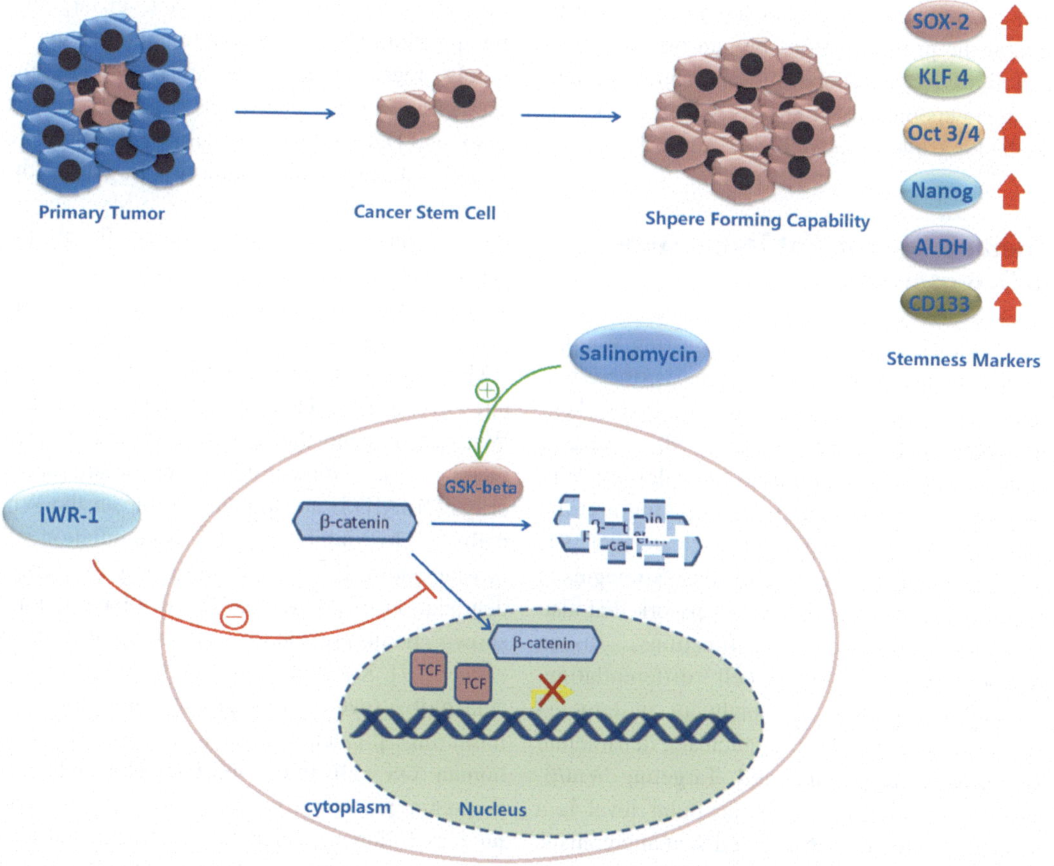

Fig. 8.4 Primary tumor contains few cancer stem cells which have the ability to self-renew. These stem cells have sphere-forming capability and have upregulated stem cell markers such as SOX-2, KLF4, Oct 3/4, Nanog, ALDH, and CD133 IWR-1, tankyrase inhibitor, inhibits the translocation of β-catenin from cytoplasm to nucleus, and salinomycin is a small-molecule inhibitor of LRP6 and it activates GSK3β to degrade the β-catenin in cytoplasm

Other Drugs Recently Shown to Inhibit OS

Tegavivint, a novel β-catenin/transducing β-like protein 1 (TBL1) inhibitor, has been shown to have antitumor activity in acute myeloid leukemia and multiple myeloma in preclinical trials and in a clinical trial for desmoid tumor. In osteosarcoma, this agent exhibits antiproliferative activity in vitro and reduces micro- and macrometastatic disease development in vivo. Metastatic osteosarcoma cell lines exhibited increased ALDH1 and β-catenin expression which was suppressed by tegavivint [86].

Niclosamide is a drug that inhibits Wnt/β-- catenin signaling by suppressing LRP6 expression. This compound has been shown to inhibit the proliferation of human osteosarcoma cell lines by targeting multiple pathways and inducing apoptosis [87, 88]. Its role is also being studied in triple-negative breast cancer [89].

Therapy Against Wnt Target Genes in Osteosarcoma

Given the abundance of data suggesting Wnt/β-- catenin involvement in tumorigenesis, there is a need to discover ways to mitigate Wnt transcriptional activation [27, 90]. Several strategies have been proposed to exploit the Wnt pathway for cancer therapy by targeting it at the extracellular, cytoplasmic, or nuclear level [91, 92]. The challenge to some of these strategies is that the Wnt pathway is a vast network that also regulates normal cell functions, tissue regeneration, and stem cell differentiation. Depending on how this pathway is targeted (extracellular, cytoplasmic, nuclear), detrimental side effects may be incurred. Targeting Wnt/β-catenin signaling at the extracellular level is a strategy that focuses on secreted Wnt antagonists, including WIF1, DKK-1, and sFRPs. Restoring the expression of these antagonists in antagonist-deficient tumors may prove to be helpful in reducing the proliferation of OS cells. Another strategy that simulates the mechanisms of Wnt antagonists is to create anti-Wnt monoclonal antibodies that can induce apoptosis of OS cells by blocking the Wnt-frizzled interaction. Therapeutic monoclonal antibodies against Wnt-1 and Wnt-2 have demonstrated inhibition of Wnt signaling and suppression of tumor growth in hepatocellular carcinoma and melanoma, respectively [93, 94]. This model can also be explored and potentially replicated for OS. Besides the extracellular level, we can aim to target the cytoplasmic components, such as β-catenin-binding domain of APC, for tumor suppression. Shih et al. showed that in colon cancer cells, re-expression of a recombinant adenovirus with APC (with known β-catenin-binding repeats) can inhibit nuclear translocation of β-catenin as well as β-catenin/TCF-mediated transactivation [95]. At the nuclear level, targeting the β-catenin/TCF transcriptional activity is widely regarded as impossible because the interacting surface between the transcription factor and DNA is huge and subject to significant changes during DNA binding. Targeting the downstream mediators, such as c-Myc, cyclin D1, and survivin, is being explored. In OS, the hepatocyte growth factor receptor c-Met is another Wnt target gene that is frequently overexpressed. Recent evidence suggests that c-Met can transform normal human osteoblasts into OS cells [96]. Therefore, c-Met is a candidate Wnt-related gene that can be explored for OS therapeutics. Nonsteroidal anti-inflammatory drugs (NSAIDS) are thought to impact the Wnt pathway by inhibiting the accumulation of prostaglandin E2, which ultimately decreases degradation of β-catenin. NSAIDs have mainly shown chemo-preventative effects against colon cancer [97]. Xia et al. demonstrated the effects of celecoxib (cyclo-oxygenase-2 inhibitor) on inhibiting β-catenin-dependent survival of a human OS cell line (MG-63). Not only did β-catenin protein decrease in the cytosol and nucleus following celecoxib treatment, but also there was a significant reduction of the Wnt target gene c-Myc and CCND1 [98]. As mentioned previously, using small-molecule inhibitors identified by high-throughput screens can be helpful to make an impact on OS therapy. These inhibitors are known to target β-catenin/TCF antagonists and transcriptional coactivator modulators along with the PDZ domain of Dvl [85]. Juan et al. have shown antitumor effect of blockade of Porcupine, an acyl-transferase essential for Wnt ligand secretion and activity, that diminished WNT → β-catenin → c-MYC signaling [99].

Wnt/B-catenin Signaling and Stem Cells

The Wnt/β-catenin pathway not only has a role in tumorigenesis but also has been suggested to exert diverse regulatory effects on cancer stem cells (CSC) [100]. Stem cells in general are defined as having the ability to self-renew along with creating specialized cells. Several groups of investigators have examined the Wnt pathway and its effects on specific stem cell functions [101]. As early as the 1990s, Korinek et al. demonstrated the association between mutated TCF4 and subsequent depletion of intestinal stem cells. Studies on the role of stem cells in hair follicle formation have suggested that Wnt inhibitors play a prominent role in regulating the stem cell microenvironments [102]. Gibbs et al. were first to identify a subpopulation of cells in osteosarcoma which were able to grow in spheres in serum-free conditions [103, 104]. In OS cell lines, Tirino et al. identified a subpopulation of CD133+ cells with self-renewal properties, higher proliferation, spherical formation, and expression of the stem cell-associated gene OCT3/4 [105]. In addition, elevated aldehyde dehydrogenase (ALDH) activity in normal stem cells and solid tumor CSC has led to the use of ALDH as a means of identifying CSC in sarcomas. Wang et al. found that OS cell lines containing a subpopulation of cells with high ALDH activity possess increased tumorigenic capacity, proliferative capacities, self-renewal, and expression of stem cell markers [106]. Neves et al. isolated the cancer stem cells from MNNG/HOS cell lines using the sphere formation assay. These cells possessed self-renewal and multipotential differentiation capabilities [107]. They expressed several markers of pluripotent embryonic stem cells such as Oct4, Nanog, ABC transporter P-glycoprotein, and BRCP. Compared to parental cells, CSC exhibits low metabolic activity and is more resistant to chemotherapy and irradiation. In subsequent studies, these investigators found that CSC has active Wnt/β-catenin signaling and overexpress SOX2 and KLF4, which are stemness-related genes. In osteosarcoma, chemotherapeutic drugs promote a stem-like phenotype through Wnt/β-catenin pathway, and hence targeting this pathway might be effective in overcoming the stemness that non-stem cells might acquire after treatment [20, 21, 34].

IWR-1, tankyrase inhibitor, was shown to be specifically cytotoxic to osteosarcoma cancer stem cells by inhibiting the translocation of β-catenin from cytoplasm to nucleus [21]. Salinomycin is a novel small-molecule inhibitor of LRP6, and it activates GSK3β in cancer cells [108]. Salinomycin can also block β-catenin/TCF4 complex formation and has demonstrated to selectively inhibit stem cells in breast cancer, colorectal cancer, and leukemia [109]. Tang et al. found that salinomycin selectively targets osteosarcoma stem cells both in vitro and in vivo, potentially through Wnt/β-catenin signaling pathway. They demonstrated that tumor samples treated with salinomycin have decreased expression of both β-catenin and cyclin D1 by immunohistochemistry confirmed with western blotting [110]. To overcome the poor solubility of salinomycin, Ni et al. developed salinomycin-entrapped nanoparticles labeled with CD133 aptamers which could target and kill CD133+ osteosarcoma CSCs [111]. Chen et al. recently constructed salinomycin-entrapped nanoparticles labeled with EGFR and CD133+ aptamers to simultaneously target both osteosarcoma cells and CSCs [112].

Summary

Wnt signaling plays an important role in osteosarcoma proliferation, metastasis, and cancer stem cells. The Wnt ligands can play their role through canonical and noncanonical signaling pathways which are tightly regulated and demonstrate cross talk with each other. Wnt signaling also plays a role in the tumor microenvironment and immunomodulation. Most of the studies which suggest that Wnt signaling is tumor suppressive studied the Wnt antagonist DKK-1, which affects both canonical and noncanonical Wnt signaling. There is probably a fine balance between these different pathways, and it is tipping the balance in one way or the other that can affect the response. This is a promising area for the development of targeted therapies, though with concern for toxicities given the key role Wnt signaling plays in normal stem cell function. Future studies are needed to study this balance more closely and create therapeutic interventions to help patients with osteosarcoma and other cancers related to this important signaling pathway.

References

1. Nusse R, Varmus HE (1982) Many tumors induced by the mouse mammary tumor virus contain a provirus integrated in the same region of the host genome. Cell 31(1):99–109
2. MacDonald BT, Tamai K, He X (2009) Wnt/beta-catenin signaling: components, mechanisms, and diseases. Dev Cell 17(1):9–26
3. Kubota T, Michigami T, Ozono K (2010) Wnt signaling in bone. Clin Pediatr Endocrinol 19(3):49–56
4. Day TF et al (2005) Wnt/beta-catenin signaling in mesenchymal progenitors controls osteoblast and chondrocyte differentiation during vertebrate skeletogenesis. Dev Cell 8(5):739–750
5. Hill TP et al (2005) Canonical Wnt/beta-catenin signaling prevents osteoblasts from differentiating into chondrocytes. Dev Cell 8(5):727–738
6. Gong Y et al (2001) LDL receptor-related protein 5 (LRP5) affects bone accrual and eye development. Cell 107(4):513–523
7. Boyden LM et al (2002) High bone density due to a mutation in LDL-receptor-related protein 5. N Engl J Med 346(20):1513–1521
8. Hoang BH et al (2004) Expression of LDL receptor-related protein 5 (LRP5) as a novel marker for disease progression in high-grade osteosarcoma. Int J Cancer 109(1):106–111
9. Guo Y et al (2008) Dominant negative LRP5 decreases tumorigenicity and metastasis of osteosarcoma in an animal model. Clin Orthop Relat Res 466(9):2039–2045
10. Polakis P (2007) The many ways of Wnt in cancer. Curr Opin Genet Dev 17(1):45–51
11. Fang F et al (2018) Targeting the Wnt/beta-catenin pathway in human osteosarcoma cells. Oncotarget 9(95):36780–36792
12. Trautmann M et al (2014) SS18-SSX fusion protein-induced Wnt/beta-catenin signaling is a therapeutic target in synovial sarcoma. Oncogene 33(42):5006–5016
13. Vijayakumar S et al (2011) High-frequency canonical Wnt activation in multiple sarcoma subtypes drives proliferation through a TCF/beta-catenin target gene, CDC25A. Cancer Cell 19(5):601–612
14. Uren A et al (2004) Wnt/Frizzled signaling in Ewing sarcoma. Pediatr Blood Cancer 43(3):243–249
15. Matushansky I et al (2007) Derivation of sarcomas from mesenchymal stem cells via inactivation of the Wnt pathway. J Clin Invest 117(11):3248–3257
16. Goldstein SD et al (2016) A monoclonal antibody against the Wnt signaling inhibitor dickkopf-1 inhibits osteosarcoma metastasis in a preclinical model. Oncotarget 7(16):21114–21123
17. Gregory CA et al (2003) The Wnt signaling inhibitor dickkopf-1 is required for reentry into the cell cycle of human adult stem cells from bone marrow. J Biol Chem 278(30):28067–28078
18. Cai Y et al (2010) Inactive Wnt/beta-catenin pathway in conventional high-grade osteosarcoma. J Pathol 220(1):24–33
19. Yu Z et al (2012) Cancer stem cells. Int J Biochem Cell Biol 44(12):2144–2151
20. Martins-Neves SR et al (2016) Osteosarcoma stem cells have active Wnt/beta-catenin and Overexpress SOX2 and KLF4. J Cell Physiol 231(4):876–886
21. Martins-Neves SR et al (2018) IWR-1, a tankyrase inhibitor, attenuates Wnt/beta-catenin signaling in cancer stem-like cells and inhibits in vivo the growth of a subcutaneous human osteosarcoma xenograft. Cancer Lett 414:1–15
22. Bacci G et al (2000) Long-term outcome for patients with nonmetastatic osteosarcoma of the extremity treated at the istituto ortopedico rizzoli according to the istituto ortopedico rizzoli/osteosarcoma-2 protocol: an updated report. J Clin Oncol 18(24):4016–4027
23. Mialou V et al (2005) Metastatic osteosarcoma at diagnosis: prognostic factors and long-term outcome--the French pediatric experience. Cancer 104(5):1100–1109
24. Lewis IJ et al (2007) Improvement in histologic response but not survival in osteosarcoma patients treated with intensified chemotherapy: a randomized phase III trial of the European Osteosarcoma Intergroup. J Natl Cancer Inst 99(2):112–128
25. Nusse R (2012) Wnt signaling. Cold Spring Harb Perspect Biol 4(5):a011163
26. Jamieson C, Sharma M, Henderson BR (2012) Wnt signaling from membrane to nucleus: beta-catenin caught in a loop. Int J Biochem Cell Biol 44(6):847–850

27. Takahashi-Yanaga F, Sasaguri T (2007) The Wnt/beta-catenin signaling pathway as a target in drug discovery. J Pharmacol Sci 104(4):293–302
28. Lustig B, Behrens J (2003) The Wnt signaling pathway and its role in tumor development. J Cancer Res Clin Oncol 129(4):199–221
29. Kimura H et al (2016) CKAP4 is a Dickkopf1 receptor and is involved in tumor progression. J Clin Invest 126(7):2689–2705
30. Sugimura R, Li L (2010) Noncanonical Wnt signaling in vertebrate development, stem cells, and diseases. Birth Defects Res C Embryo Today 90(4):243–256
31. Piters E, Boudin E, Van Hul W (2008) Wnt signaling: a win for bone. Arch Biochem Biophys 473(2):112–116
32. Guo Y et al (2007) Blocking Wnt/LRP5 signaling by a soluble receptor modulates the epithelial to mesenchymal transition and suppresses met and metalloproteinases in osteosarcoma Saos-2 cells. J Orthop Res 25(7):964–971
33. Chen C et al (2015) Aberrant activation of Wnt/beta-catenin signaling drives proliferation of bone sarcoma cells. Oncotarget 6(19):17570–17583
34. Martins-Neves SR et al (2016) Chemotherapy induces stemness in osteosarcoma cells through activation of Wnt/beta-catenin signaling. Cancer Lett 370(2):286–295
35. Lee N et al (2007) A potential role for Dkk-1 in the pathogenesis of osteosarcoma predicts novel diagnostic and treatment strategies. Br J Cancer 97(11):1552–1559
36. Krause U et al (2014) An unexpected role for a Wnt-inhibitor: Dickkopf-1 triggers a novel cancer survival mechanism through modulation of aldehyde-dehydrogenase-1 activity. Cell Death Dis 5:e1093
37. Enomoto M et al (2009) Autonomous regulation of osteosarcoma cell invasiveness by Wnt5a/Ror2 signaling. Oncogene 28(36):3197–3208
38. Wissmann C et al (2003) WIF1, a component of the Wnt pathway, is down-regulated in prostate, breast, lung, and bladder cancer. J Pathol 201(2):204–212
39. Rubin EM et al (2010) Wnt inhibitory factor 1 decreases tumorigenesis and metastasis in osteosarcoma. Mol Cancer Ther 9(3):731–741
40. Ai L et al (2006) Inactivation of Wnt inhibitory factor-1 (WIF1) expression by epigenetic silencing is a common event in breast cancer. Carcinogenesis 27(7):1341–1348
41. Mazieres J et al (2004) Wnt inhibitory factor-1 is silenced by promoter hypermethylation in human lung cancer. Cancer Res 64(14):4717–4720
42. Taniguchi H et al (2005) Frequent epigenetic inactivation of Wnt inhibitory factor-1 in human gastrointestinal cancers. Oncogene 24(53):7946–7952
43. Lee BB et al (2009) Aberrant methylation of APC, MGMT, RASSF2A, and Wif-1 genes in plasma as a biomarker for early detection of colorectal cancer. Clin Cancer Res 15(19):6185–6191
44. Kansara M et al (2009) Wnt inhibitory factor 1 is epigenetically silenced in human osteosarcoma, and targeted disruption accelerates osteosarcomagenesis in mice. J Clin Invest 119(4):837–851
45. Hoang B et al (1996) Primary structure and tissue distribution of FRZB, a novel protein related to Drosophila frizzled, suggest a role in skeletal morphogenesis. J Biol Chem 271(42):26131–26137
46. Leyns L et al (1997) Frzb-1 is a secreted antagonist of Wnt signaling expressed in the Spemann organizer. Cell 88(6):747–756
47. Bravo D et al (2018) Decreased local and systemic levels of sFRP3 protein in osteosarcoma patients. Gene 674:1–7
48. Ahn VE et al (2011) Structural basis of Wnt signaling inhibition by Dickkopf binding to LRP5/6. Dev Cell 21(5):862–873
49. Malladi S et al (2016) Metastatic latency and immune evasion through autocrine inhibition of WNT. Cell 165(1):45–60
50. Aguilera O et al (2006) Epigenetic inactivation of the Wnt antagonist DICKKOPF-1 (DKK-1) gene in human colorectal cancer. Oncogene 25(29):4116–4121
51. Hirata H et al (2011) Wnt antagonist DKK1 acts as a tumor suppressor gene that induces apoptosis and inhibits proliferation in human renal cell carcinoma. Int J Cancer 128(8):1793–1803
52. Mikheev AM et al (2004) A functional genomics approach for the identification of putative tumor suppressor genes: Dickkopf-1 as suppressor of HeLa cell transformation. Carcinogenesis 25(1):47–59
53. Qi L et al (2012) Dickkopf-1 inhibits epithelial-mesenchymal transition of colon cancer cells and contributes to colon cancer suppression. Cancer Sci 103(4):828–835
54. Begenik H et al (2014) The association between serum Dickkopf-1 levels and esophageal squamous cell carcinoma. Hum Exp Toxicol 33(8):785–788
55. Gomceli I et al (2012) A novel screening biomarker in gastric cancer: serum Dickkopf-1. Hepato-Gastroenterology 59(117):1661–1664
56. Hall CL et al (2008) Dickkopf-1 expression increases early in prostate cancer development and decreases during progression from primary tumor to metastasis. Prostate 68(13):1396–1404
57. Liu DJ et al (2017) The role of Dickkopf-1 as a potential prognostic marker in pancreatic ductal adenocarcinoma. Cell Cycle 16(17):1622–1629
58. Peng YH et al (2016) Combined detection of serum Dickkopf-1 and its autoantibodies to diagnose esophageal squamous cell carcinoma. Cancer Med 5(7):1388–1396
59. Rachner TD et al (2014) High serum levels of Dickkopf-1 are associated with a poor prognosis in prostate cancer patients. BMC Cancer 14:649

60. Watany M et al (2017) Study of Dickkopf-1 (DKK-1) gene expression in hepatocellular carcinoma patients. J Clin Diagn Res 11(2):OC32–OC34
61. Chen L et al (2013) DKK1 promotes hepatocellular carcinoma cell migration and invasion through beta-catenin/MMP7 signaling pathway. Mol Cancer 12:157
62. Fulciniti M et al (2009) Anti-DKK1 mAb (BHQ880) as a potential therapeutic agent for multiple myeloma. Blood 114(2):371–379
63. Makino T et al (2009) Dickkopf-1 expression as a marker for predicting clinical outcome in esophageal squamous cell carcinoma. Ann Surg Oncol 16(7):2058–2064
64. Zhuang X et al (2017) Differential effects on lung and bone metastasis of breast cancer by Wnt signalling inhibitor DKK1. Nat Cell Biol 19(10):1274–1285
65. Veeck J, Dahl E (2012) Targeting the Wnt pathway in cancer: the emerging role of Dickkopf-3. Biochim Biophys Acta 1825(1):18–28
66. Hoang BH et al (2004) Dickkopf 3 inhibits invasion and motility of Saos-2 osteosarcoma cells by modulating the Wnt-beta-catenin pathway. Cancer Res 64(8):2734–2739
67. Lin CH et al (2013) Dkk-3, a secreted wnt antagonist, suppresses tumorigenic potential and pulmonary metastasis in osteosarcoma. Sarcoma 2013:147541
68. Li X et al (2005) Sclerostin binds to LRP5/6 and antagonizes canonical Wnt signaling. J Biol Chem 280(20):19883–19887
69. Holdsworth G et al (2012) Characterization of the interaction of sclerostin with the low density lipoprotein receptor-related protein (LRP) family of Wnt co-receptors. J Biol Chem 287(32):26464–26477
70. Clarke BL (2014) Anti-sclerostin antibodies: utility in treatment of osteoporosis. Maturitas 78(3):199–204
71. Lewiecki EM (2011) Sclerostin monoclonal antibody therapy with AMG 785: a potential treatment for osteoporosis. Expert Opin Biol Ther 11(1):117–127
72. Solling ASK, Harslof T, Langdahl B (2018) The clinical potential of romosozumab for the prevention of fractures in postmenopausal women with osteoporosis. Ther Adv Musculoskelet Dis 10(5–6):105–115
73. Saag KG et al (2017) Romosozumab or alendronate for fracture prevention in women with osteoporosis. N Engl J Med 377(15):1417–1427
74. Zou J, Zhang W, Li XL (2017) Effects of SOST gene silencing on proliferation, apoptosis, invasion, and migration of human osteosarcoma cells through the Wnt/beta-catenin signaling pathway. Calcif Tissue Int 100(6):551–564
75. Park CH et al (2005) The inhibitory mechanism of curcumin and its derivative against beta-catenin/Tcf signaling. FEBS Lett 579(13):2965–2971
76. Hallett RM et al (2012) Small molecule antagonists of the Wnt/beta-catenin signaling pathway target breast tumor-initiating cells in a Her2/Neu mouse model of breast cancer. PLoS One 7(3):e33976
77. Leow PC et al (2010) Antitumor activity of natural compounds, curcumin and PKF118-310, as Wnt/beta-catenin antagonists against human osteosarcoma cells. Investig New Drugs 28(6):766–782
78. Zou Y, Yang J, Jiang D (2015) Resveratrol inhibits canonical Wnt signaling in human MG-63 osteosarcoma cells. Mol Med Rep 12(5):7221–7226
79. Chen B et al (2009) Small molecule-mediated disruption of Wnt-dependent signaling in tissue regeneration and cancer. Nat Chem Biol 5(2):100–107
80. Huang H et al (2013) Structure-based design of 2-aminopyridine oxazolidinones as potent and selective tankyrase inhibitors. ACS Med Chem Lett 4(12):1218–1223
81. Ma L et al (2015) Tankyrase inhibitors attenuate WNT/beta-catenin signaling and inhibit growth of hepatocellular carcinoma cells. Oncotarget 6(28):25390–25401
82. Mizutani A et al (2018) RK-287107, a potent and specific tankyrase inhibitor, blocks colorectal cancer cell growth in a preclinical model. Cancer Sci 109(12):4003–4014
83. Stratford EW et al (2014) The tankyrase-specific inhibitor JW74 affects cell cycle progression and induces apoptosis and differentiation in osteosarcoma cell lines. Cancer Med 3(1):36–46
84. Gustafson CT et al (2017) FH535 suppresses osteosarcoma growth in vitro and inhibits Wnt signaling through tankyrases. Front Pharmacol 8:285
85. Grandy D et al (2009) Discovery and characterization of a small molecule inhibitor of the PDZ domain of dishevelled. J Biol Chem 284(24):16256–16263
86. Nomura M et al (2019) Tegavivint and the beta-catenin/ALDH Axis in chemotherapy-resistant and metastatic osteosarcoma. J Natl Cancer Inst 111(11):1216–1227
87. Li Z et al (2015) Niclosamide inhibits the proliferation of human osteosarcoma cell lines by inducing apoptosis and cell cycle arrest. Oncol Rep 33(4):1763–1768
88. Liao Z et al (2015) The anthelmintic drug niclosamide inhibits the proliferative activity of human osteosarcoma cells by targeting multiple signal pathways. Curr Cancer Drug Targets 15(8):726–738
89. Yin L et al (2016) Niclosamide sensitizes triple-negative breast cancer cells to ionizing radiation in association with the inhibition of Wnt/beta-catenin signaling. Oncotarget 7(27):42126–42138
90. Barker N, Clevers H (2006) Mining the Wnt pathway for cancer therapeutics. Nat Rev Drug Discov 5(12):997–1014
91. Moon RT et al (2004) WNT and beta-catenin signalling: diseases and therapies. Nat Rev Genet 5(9):691–701

92. Luo J et al (2007) Wnt signaling and human diseases: what are the therapeutic implications? Lab Investig 87(2):97–103
93. Wei W et al (2009) Blockade of Wnt-1 signaling leads to anti-tumor effects in hepatocellular carcinoma cells. Mol Cancer 8:76
94. You L et al (2004) An anti-Wnt-2 monoclonal antibody induces apoptosis in malignant melanoma cells and inhibits tumor growth. Cancer Res 64(15):5385–5389
95. Shih IM et al (2000) The beta-catenin binding domain of adenomatous polyposis coli is sufficient for tumor suppression. Cancer Res 60(6):1671–1676
96. Patane S et al (2006) MET overexpression turns human primary osteoblasts into osteosarcomas. Cancer Res 66(9):4750–4757
97. Thun MJ, Henley SJ, Patrono C (2002) Nonsteroidal anti-inflammatory drugs as anticancer agents: mechanistic, pharmacologic, and clinical issues. J Natl Cancer Inst 94(4):252–266
98. Xia JJ et al (2010) Celecoxib inhibits beta-catenin-dependent survival of the human osteosarcoma MG-63 cell line. J Int Med Res 38(4):1294–1304
99. Juan J et al (2014) Diminished WNT -> beta-catenin -> c-MYC signaling is a barrier for malignant progression of BRAFV600E-induced lung tumors. Genes Dev 28(6):561–575
100. Dravid G et al (2005) Defining the role of Wnt/beta-catenin signaling in the survival, proliferation, and self-renewal of human embryonic stem cells. Stem Cells 23(10):1489–1501
101. Clevers H, Nusse R (2012) Wnt/beta-catenin signaling and disease. Cell 149(6):1192–1205
102. Fuchs E, Tumbar T, Guasch G (2004) Socializing with the neighbors: stem cells and their niche. Cell 116(6):769–778
103. Gibbs CP Jr, Levings PP, Ghivizzani SC (2011) Evidence for the osteosarcoma stem cell. Curr Orthop Pract 22(4):322–326
104. Gibbs CP et al (2005) Stem-like cells in bone sarcomas: implications for tumorigenesis. Neoplasia 7(11):967–976
105. Tirino V et al (2008) Detection and characterization of CD133+ cancer stem cells in human solid tumours. PLoS One 3(10):e3469
106. Wang L et al (2011) Prospective identification of tumorigenic osteosarcoma cancer stem cells in OS99-1 cells based on high aldehyde dehydrogenase activity. Int J Cancer 128(2):294–303
107. Martins-Neves SR et al (2012) Therapeutic implications of an enriched cancer stem-like cell population in a human osteosarcoma cell line. BMC Cancer 12:139
108. Lu W, Li Y (2014) Salinomycin suppresses LRP6 expression and inhibits both Wnt/beta-catenin and mTORC1 signaling in breast and prostate cancer cells. J Cell Biochem 115(10):1799–1807
109. Wang Z et al (2019) Salinomycin exerts anti-colorectal cancer activity by targeting the beta-catenin/TCF complex. Br J Pharmacol 176(17):3390–3406
110. Tang QL et al (2011) Salinomycin inhibits osteosarcoma by targeting its tumor stem cells. Cancer Lett 311(1):113–121
111. Ni M et al (2015) Poly(lactic-co-glycolic acid) nanoparticles conjugated with CD133 aptamers for targeted salinomycin delivery to CD133+ osteosarcoma cancer stem cells. Int J Nanomedicine 10:2537–2554
112. Chen F et al (2018) Targeted salinomycin delivery with EGFR and CD133 aptamers based dual-ligand lipid-polymer nanoparticles to both osteosarcoma cells and cancer stem cells. Nanomedicine 14(7):2115–2127
113. Zhao S et al (2015) NKD2, a negative regulator of Wnt signaling, suppresses tumor growth and metastasis in osteosarcoma. Oncogene 34(39):5069–5079

Receptor Tyrosine Kinases in Osteosarcoma: 2019 Update

9

Edward M. Greenfield, Christopher D. Collier, and Patrick J. Getty

Abstract

The primary conclusions of our 2014 contribution to this series were as follows:

- Multiple receptor tyrosine kinases (RTKs) likely contribute to aggressive phenotypes in osteosarcoma and, therefore, inhibition of multiple RTKs is likely necessary for successful clinical outcomes.
- Inhibition of multiple RTKs may also be useful to overcome resistance to inhibitors of individual RTKs as well as resistance to conventional chemotherapies.
- Different combinations of RTKs are likely important in individual patients.
- AXL, EPHB2, FGFR2, IGF1R, and RET were identified as promising therapeutic targets by our in vitro phosphoproteomic/siRNA screen of 42 RTKs in the highly metastatic LM7 and 143B human osteosarcoma cell lines.

This chapter is intended to provide an update on these topics as well as the large number of osteosarcoma clinical studies of inhibitors of multiple tyrosine kinases (multi-TKIs) that were recently published.

Keywords

Osteosarcoma · Receptor tyrosine kinases · Multi-TKIs · AXL · EPHB2 · FGFR2 · IGF1R · RET

E. M. Greenfield (✉)
Department of Orthopaedic Surgery, Indiana University School of Medicine and Indiana University Melvin and Bren Simon Cancer Center, Indianapolis, IN, USA
e-mail: egreenf@iu.edu

C. D. Collier
Department of Orthopaedics, University of Chicago Pritzker School of Medicine, Chicago, IL, USA

P. J. Getty
Department of Orthopaedics, Case Western Reserve University, Cleveland, OH, USA

Seidman Cancer Center, University Hospitals – Cleveland Medical Center, Cleveland, OH, USA
e-mail: Patrick.Getty@UHhospitals.org

AXL

AXL, from *a*ne*x*e*l*ekto, the Greek word for uncontrolled, was originally identified as a transforming gene in chronic myelogenous leukemia. It is the primary member in the mesenchymal lineage of the TAM family of RTKs that also includes TYRO3 and MER. GAS6 is the primary ligand

E. S. Kleinerman, R. Gorlick (eds.), *Current Advances in the Science of Osteosarcoma*, Advances in Experimental Medicine and Biology 1258, https://doi.org/10.1007/978-3-030-43085-6_9

for the TAM RTKs. The initial evidence suggesting that AXL might be important in osteosarcoma was that *AXL* is the most highly upregulated (~40-fold) of the 637 measured cancer-related mRNAs in highly metastatic subclones of the HuO9 human osteosarcoma cell line [5]. Osteosarcoma cell lines also had the second highest level of *AXL* mRNA of the 37 types of cancer cell lines included in the Broad Institute Cancer Cell Line Encyclopedia [1]. A phosphoproteomics study found abundant AXL phosphorylation in all four human osteosarcoma cell lines that were studied [6]. *AXL* expression may be higher in tumors than in those cell lines as its transcription is induced by hypoxia, at least in epithelial cancers [7]. In that regard, AXL was detected by immunohistochemistry in 30 out of 40 human osteosarcomas but in only 8 out of the 40 adjacent noncancerous tissues [8]. Most importantly, high levels of *AXL* mRNA correlated with poor clinical outcomes in a study of 68 osteosarcoma patients [9]. Osteosarcoma cell lines also had the seventh highest level of *GAS6* mRNA of the human cancer cell lines included in the Broad Institute Cancer Cell Line Encyclopedia [1]. In contrast, *GAS6* mRNA is downregulated in primary osteosarcoma biopsies and human osteosarcoma cell lines compared with both bone marrow-derived stromal cells and osteoblasts [10]. Moreover, low levels correlated with poor clinical outcomes in that study of 83 osteosarcoma patients [10]. A high level of immunostaining for active phosphorylated AXL was also reported to correlate with poor clinical outcomes in osteosarcoma patients [11]. However, we (unpublished data) found that the anti-phospho-AXL antibody used in that study is not specific when used for immunohistochemistry.

Our in vitro phosphoproteomic/siRNA screen identified AXL as contributing to migration, invasion, and non-adherent colony formation, but not to cell growth, by the highly metastatic 143B human osteosarcoma cell line [4]. More recently, we found that *AXL* shRNA also inhibits migration, non-adherent colony formation, and growth of sarcospheres generated from highly metastatic human osteosarcoma cell lines [12]. Other investigators reported that *AXL* shRNA inhibits proliferation and induces apoptosis of the MG63 human osteosarcoma cell line [8] and GAS6 inhibits apoptosis and increases migration by the MG63 and U2OS human osteosarcoma cell lines [11]. All of those in vitro results are consistent with our finding that stable transfection of two different *AXL* shRNA constructs reduced tumor growth by ~70% and the number of metastases by ~90% by the 143B cell line in orthotopic murine xenografts [12]. A miR-199a-3p mimic downregulates *AXL* mRNA and inhibits in vitro migration by the MG63 and U2OS human osteosarcoma cell lines [13]. Moreover, high levels of that miR correlated with better clinical outcomes in a study of 30 osteosarcoma patients [13]. The same group of investigators went on to show that overexpression of the lncRNA *DANCR* upregulates *AXL*; increases proliferation, migration, invasion, and expression of stemness genes by the HOS and 143B human osteosarcoma cell lines in vitro; and increases tumor growth and the number of metastases formed by the 143B cell line in subcutaneous murine xenografts [9]. Moreover, high levels of *DANCR* correlated with poor clinical outcomes in osteosarcoma patients [9].

Multiple small-molecule inhibitors that target the ATP-binding domain of AXL are in development [14, 15]. Most, if not all, of them target multiple RTKs [14, 15]. More specific inhibition can be achieved by targeting the extracellular domain of AXL and the other TAM family RTKs with small molecules [16], neutralizing antibodies [17], decoy receptors [18], or nucleic acid aptamers [19]. However, the polypharmacology of the more common inhibitors that target the intracellular ATP-binding domain may contribute to their potential clinical efficacy [2, 3]. For example, BGB324 (previously known as R428), which is often considered to be specific for AXL, also potently inhibits a number of other RTKs, including RET [16, 20]. Indeed, BGB324 inhibits growth in our in vitro 3D sarcosphere platform [21] by both AXL-dependent and AXL-independent mechanisms [12].

AXL and the other TAM RTKs can cause resistance to conventional chemotherapeutics and kinase inhibitors in many other cancers [15, 22, 23]. Molecular mechanisms responsible for that resistance include feedback loops that increase expression of the TAM RTKs or their ligand, GAS6, cross talk with other kinases or other oncogenes, and induction of dormancy [15, 22–28]. AXL and the other TAM RTKs also repress innate immunity [29], and targeting their activity might therefore be especially useful in combination therapy with liposomal muramyl tripeptide, a macrophage activator approved for osteosarcoma therapy in Europe [30]. Activation of innate immunity by targeting AXL or the other TAM RTKs may also increase the efficacy of T cell-mediated immune checkpoint therapy [31, 32]. The discovery of T cell-mediated cancer immunotherapy received the 2018 Nobel Prize in Physiology or Medicine [33] and has also received considerable attention as a potential therapy for osteosarcoma [34, 35].

EPHB2

*EPH*s were originally discovered in an *e*rythropoietin-*p*roducing *h*epatocellular carcinoma cell line as a homologue of the viral oncogene v-*fps*. The 14 mammalian EPHs comprise the largest RTK family [36]. *EPHA3*, *EPHB2*, and *EPHB3* mRNAs were highly expressed in human osteosarcoma tissue samples when compared to primary human osteoblasts [37]. Proteomic studies showed that cell surface levels of EPHA2, EPHB2, and EPHB4 are, respectively, 12-, 43-, and 20-fold more abundant on five human osteosarcoma cell lines than on primary human osteoblasts [38] and found abundant EPHB2 phosphorylation in one of the four tested human osteosarcoma cell lines [6]. Our in vitro phosphoproteomic/siRNA screen detected higher levels of EPHA2, EPHA4, and EPHB2 in the highly metastatic LM7 human osteosarcoma cell line than in its nonmetastatic parental SAOS-2 cell line and identified EPHB2 as contributing to migration and non-adherent colony formation, but not to cell growth or invasion, by the LM7 cell line [4]. We confirmed the siRNA results with *EPHB2* antisense experiments [4]. Other investigators showed that mRNAs encoding EFNA5 and EFNB1, two of the ligands that activate EPHB2 as well as a number of other EPH RTKs, are upregulated in human osteosarcomas and *EFNB1* mRNA level was prominent in samples from patients with poor clinical outcomes [39]. *EPHB2* is also highly expressed in *SYT-SSX2*-positive synovial sarcoma tissues, and SYT-SSX2-induced stabilization of the microtubule network is blocked by soluble forms of the extracellular domain of EPHB2 that bind and inactivate its ligands [40]. Given that osteosarcomas arise from relatively immature members of the osteoblast lineage [41], it is intriguing that EPHB2 and the other EPH RTKs modulate differentiation at multiple steps in that lineage [36, 42, 43].

FGFR2

FGFRs were originally identified biochemically on fibroblasts and muscle cells as membrane receptors that bind fibroblast growth factors. All four of the FGFRs are amplified in human osteosarcomas [44–47]. Those amplifications can predict responsiveness to NVP-BGJ398, a fairly specific inhibitor of FGFR1-3, and are associated with a poor response to conventional osteosarcoma chemotherapy [45, 46]. A phosphoproteomics study found abundant FGFR1 phosphorylation in all four human osteosarcoma cell lines that were studied and abundant phosphorylation of FGFR2 and FGFR4 in two of them [6]. A separate study found abundant FGFR1 phosphorylation in ~70% of human osteosarcomas but did not examine the other FGFRs [48]. Moreover, the intensity of total FGFR immunostaining in primary osteosarcomas correlated with metastasis and reduced survival [49]. Both FGFR1 and FGFR2 were identified as contributing to viability of human osteosarcoma cell lines in a kinome-wide siRNA screen [50]. Our in vitro phosphoproteomic/siRNA screen detected higher levels of FGFR2 and FGFR3 in the highly metastatic LM7 human osteosarcoma cell line than in its nonmetastatic parental SAOS-2 cell line and identified FGFR2 as contributing to migration and non-adherent colony formation, but not to cell growth or invasion, by the LM7 cell line [4]. We confirmed the siRNA results with FGFR2 antisense experiments [4].

Signalling by FGFR2 can support stemness in many cancers, including osteosarcoma [51, 52]. An elegant study recently showed that FGFR2 signalling induces fibrogenic reprogramming in human osteosarcoma cell line-derived stem cells, which, in turn, induces growth of metastases in the lung microenvironment without affecting growth of the primary tumor [49]. Those results led to experiments in which nintedanib, an inhibitor of FGFR1–3, reduced stemness and the fibrogenic reprogramming and increased apoptosis in the stem cells derived from human osteosarcoma cell lines as well as in stem cells derived from all six of the primary human osteosarcomas that were studied [49]. Moreover, a preventive regimen of nintedanib blocked lung metastasis following tibial or tail vein injection of the Well5 human osteosarcoma cell line, and even more impressively, a therapeutic regimen of nintedanib caused regression of lung metastases [49]. A preventive regimen of another FGFR inhibitor, AZD4547, reduced metastasis from an orthotopic human osteosarcoma xenograft model [53]. PD173074, in combination with doxorubicin, inhibited growth and stemness of the primary tumors in a murine syngeneic subcutaneous model, while neither agent had detectable effects as monotherapies [52]. It should however be noted that nintedanib, AZD4547, and PD173074 inhibit multiple tyrosine kinases with similar or greater potency than the FGFRs [54, 55].

IGF1R

IGF1R was originally identified biochemically as the type *1* membrane *r*eceptor that binds *i*nsulin-like *g*rowth *f*actors I and II. Amplification of *IGF1R* occurs in 14–31% of osteosarcomas, depending on the threshold used to define amplification [56, 57]. Those studies also found other genetic events predicted to activate IGF1R (amplifications of *IGF1* or *IGF2* and deletions of either *IGF2R, IGFBP3,* or *IGFBP5*) in an additional 4.5–19% of the osteosarcomas. *IGF1R* mRNA and IGF1R protein levels are substantially increased in human osteosarcomas compared with adjacent noncancerous tissues [58], and a phosphoproteomics study found abundant IGF1R phosphorylation in three of the four human osteosarcoma cell lines that were studied [6]. *IGF1R* mRNA and IGF1R protein levels are substantially increased in human osteosarcomas compared with adjacent noncancerous tissues [58]. Moreover, higher IGFIR protein levels in the tumors are associated with poor clinical outcomes in both human [58, 59] and canine osteosarcomas [60]. At least eight miRs have been reported to inhibit proliferation and other in vitro measures of osteosarcoma aggressiveness in part by targeting IGF1R [61–68]. *IGF2* siRNA substantially reduced growth of the MG63 human osteosarcoma cell line in low-serum cultures [69], and exogenous IGF2 can induce dormancy in both human and murine osteosarcoma cell lines and thereby induce resistance to methotrexate, doxorubicin, and cisplatin [70]. Consistent with those in vitro findings, elevated IGF2 serum levels are associated with decreased event-free survival in osteosarcoma patients [69], and *IGF2* mRNA tumor levels were reduced post-chemotherapy in the five osteosarcoma patients who responded well to chemotherapy but were either unchanged or increased 13-fold in the two osteosarcoma patients who did not respond [70].

IGF1R is one of the most studied RTKs in osteosarcoma [71]. We therefore consider the identification of IGF1R as contributing to cell growth, migration, invasion, and non-adherent colony formation by the highly metastatic LM7 human osteosarcoma cell line as validation of our in vitro phosphoproteomic/siRNA screen [4]. We confirmed the siRNA results with an IGF1R-neutralizing antibody [4]. Other investigators found that stable transfection of *IGF1R* shRNA reduced adhesion, migration, and invasion in vitro as well as the number of metastases and increased survival of mice following tail vein injection of the U2OS human osteosarcoma cell line [58]. A recent study showed that IGF1R upregulation is responsible for the increased in vitro measures of osteosarcoma aggressiveness that are induced by overexpression of CYR61/CCN1 [72]. We [73] and other investigators [74] found that picropodophyllin, which was originally described as an IGF1R inhibitor [75], reduced growth, migration, and non-adherent colony formation and induced apoptosis by multiple human osteosarcoma cell lines. However, subsequent studies showed that the effects of picropodophyllin are primarily due to microtubule destabilization, rather than inhibition of IGF1R [76, 77].

IGF-binding proteins (IGFBPs) can inhibit IGF1R activity by sequestering IGFs [78]. In that regard, *IGFBP3*, *IGFBP4*, *IGFBP6*, and *IGFBP7* mRNA levels were downregulated in primary osteosarcomas and in two osteosarcoma-patient-derived xenografts compared with mesenchymal stem cells before and after osteogenic differentiation [37, 79]. Similarly, *IGFBP5* mRNA and IGFBP5 protein levels were substantially reduced in highly metastatic human osteosarcoma cell lines compared with isogenic, but weakly metastatic, cell lines, and immunostaining for IGFBP5 was reduced in metastases compared with matched primary osteosarcomas from the same patients [80]. Low levels of *IGFBP4* mRNA correlated with poor clinical outcomes in the study of 83 osteosarcoma patients described above in the section on AXL [10]. Moreover, IGFBP5 overexpression induced apoptosis and inhibited primary tumor growth and metastasis by the highly metastatic cell lines in orthotopic murine xenografts, and *IGFBP5* siRNA had the opposite effects [80].

An IGF1R-neutralizing antibody inhibited primary tumor growth in subcutaneous xenografts of multiple human osteosarcoma cell lines [81, 82]. In a similar xenograft model, the combination of two neutralizing antibodies that bind to different epitopes on IGF1R inhibited primary tumor growth more effectively than either agent as monotherapy [83]. Three different IGF1R-neutralizing antibodies in combination with an mTOR inhibitor reduced primary tumor growth more effectively than either agent as monotherapy in multiple subcutaneous xenograft osteosarcoma models [84–86]. Nonetheless, multiple IGF1R-neutralizing antibodies showed little clinical efficacy against osteosarcoma in Phase II studies, either alone [87, 88] or in combination with an mTOR inhibitor [89, 90]. Targeting IGF1R along with other RTKs might be more effective as dual IGF1R/IR inhibitors resensitized doxorubicin-resistant and cisplatin-resistant subclones of human osteosarcoma cell lines in vitro [91, 92]. Moreover, the combinations of *IGF1R* siRNA and insulin receptor siRNA or neutralizing antibodies against IGF1R and HER2 were more effective in combination than alone at reducing in vitro growth of human osteosarcoma cell lines [69, 93]. A bispecific IGF1R/EGFR-neutralizing antibody inhibited both tumor growth and the number of metastases from the 143B human osteosarcoma cell line in an orthotopic murine xenograft model [94]. Antibodies against either of those RTKs had less effect, either alone or in combination, and the authors suggest that the recruitment of natural killer (NK) cells by the bispecific antibody may account for its increased efficacy [94]. The EGFR-neutralizing antibody used in that study stimulates NK cell-mediated cytotoxicity against the SJSA-1 human osteosarcoma cell line in vitro [95], but we are unaware of similar studies with the bispecific IGF1R/EGFR-neutralizing antibody.

Identification of biomarkers that predict which osteosarcoma patients will respond robustly is another approach that could increase the clinical efficacy of IGF1R inhibitors [56, 96]. In the osteosarcoma clinical studies, however, responses to IGF1R-neutralizing antibodies, either alone or in combination with the mTOR inhibitor, did not correlate with *IGF1R* mutations or amplifications or with levels of *IGF1R* mRNA or IGF1R protein [89, 97, 98]. However, nuclear immunostaining for IGF1R in the absence of cytoplasmic staining is associated with sixfold longer progression-free survival and fourfold higher overall survival in a study of soft tissue sarcoma ($n = 9$), Ewing sarcoma (n = 3), and osteosarcoma (n = 4) patients treated with IGF1R-neutralizing antibodies [97]. In that regard, a number of recent studies found that nuclear IGF1R can contribute to in vitro measures of aggressiveness in epithelial cancers [99–101].

RET

RET (*re*arranged during *t*ransfection) was originally identified as a transforming gene in lymphoma. Translocation-induced *RET* fusion genes are well-known oncogenes in epithelial cancers such as thyroid and non-small-cell lung cancer [102, 103]. Although *RET* fusion proteins have not been identified in osteosarcoma [56], *RET* point mutations or overexpression can also be oncogenic in the absence of translocations [103, 104]. Our in vitro phosphoproteomic/siRNA screen detected higher levels of RET in the highly metastatic LM7 and 143B human osteosarcoma cell lines than in their non-metastatic parental SAOS-2 and HOS-TE85 cell lines and identified RET as contributing to migration and, to a lesser extent, non-adherent colony formation, but not to cell growth or invasion, by the LM7 cell line [4]. We confirmed the siRNA results with *RET* antisense experiments [4]. Chen and colleagues reported that *RET* siRNA can also decrease migration, invasion, and colony formation by other human osteosarcoma cell lines [105]. Most importantly, high levels of *RET* mRNA are associated with poor clinical outcomes in studies of 68 and 19 osteosarcoma patients [105, 106].

Overexpression of the lncRNA MALAT1 upregulates *RET* in human osteosarcoma cell lines in vitro, at least in part, by inhibiting miR-129-5p [105]. MALAT1 overexpression increases and MALAT1 knockdown decreases proliferation, invasion, and colony formation by multiple human osteosarcoma cell lines in vitro as well as tumor growth in subcutaneous or peritoneal murine xenografts [105, 106]. Moreover, MALAT1 expression correlated with *RET* expression and negatively correlated with expression of miR-129-5p and survival in the study of 68 osteosarcoma patients [105].

Multi-TKIs

This section will focus on the multi-TKIs evaluated in clinical studies that included patients with osteosarcoma (Table 9.1). All eleven of those multi-TKIs can inhibit at least one of the RTKs identified in our original phosphoproteomic/siRNA screen [4]. For example, AXL and IGF1R were among the eight RTKs inhibited by imatinib in the HOS human osteosarcoma cell line, as assessed by phospho-RTK arrays [107]. Moreover, live cell, biochemical, and proteomic profiling as well as X-ray crystallography revealed that, among many other RTK targets, sunitinib can potently inhibit AXL, EPHB2, FGFR2, IGF1R, and RET; dasatinib can potently inhibit AXL, EPHB2, FGFR2, and RET; cabozantinib can potently inhibit AXL, EPHB2, and RET; sorafenib can potently inhibit AXL, FGFR2, and RET; pazopanib can potently inhibit FGFR2, IGF1R, and RET; cediranib can potently inhibit AXL and RET; axitinib and regorafenib can potently inhibit FGFR2 and RET; crizotinib can potently inhibit AXL; and apatinib can potently inhibit RET [2, 54, 103, 108–112]. The polypharmacology of the multi-TKIs likely contributes to their potential clinical efficacy [2, 3] but also can contribute to serious "off-target" toxicities [103, 113].

Cediranib, dasatinib, and sunitinib were among the most effective drugs in a screen that measured viability of monolayer cultures obtained from four primary canine osteosarcomas [114]. Sorafenib, the only other multi-TKI in Table 9.1 included in that screen, had no detectable effects on viability of cultures from any of the canine osteosarcomas. Those results led to dasatinib treatment of four canines with osteosarcoma following limb amputation and carboplatin chemotherapy, which is a standard-of-care chemotherapy for canine osteosarcoma [115]. In two of the four canines, initial results suggest that dasatinib led to stable disease or partial remission [115]. Many multi-TKIs are more effective against epithelial cancers in hypoxic conditions [116]. Similarly, gefitinib is substantially more potent against human osteosarcoma cell lines in low-serum cultures and in

Table 9.1 Clinical studies of multi-TKIs in osteosarcoma

Multi-TKI	Study type	No. of evaluable osteosarcoma patients/disease status	Outcomes (no. of patients, %)	References
Apatinib	Case report	1/Metastatic	Partial response (1, 100%)	[145]
	Retrospective	2/Metastatic or recurrent	No objective response	[146]
	Retrospective	4/Refractory and progressive	Partial response (2, 50%) Stable disease (2, 50%)	[147]
	Observational	10/Refractory and metastatic	Partial response (2, 20%) Stable disease (5, 50%)	[148]
	Retrospective	22/Refractory and either local unresectable or metastatic	Partial response (9, 41%)	[149]
	Retrospective	27/Refractory and metastatic	Partial response (7, 26%) Stable disease (11, 41%)	[150]
	Phase II	11/Refractory and metastatic	Stable disease (10, 91%)	[134]
	Phase II	37/Refractory and either locally advanced, unresectable, or metastatic	Partial response (16, 43%) Stable disease (8, 22%)	[135]
Axitinib	Phase I	2/Refractory	Stable disease (2, 100%)	[151]
Cabozantinib	Phase I	2/Relapsed or refractory	No objective response	[152]
Cediranib	Phase I	4/Refractory	34% reduction in size of lung metastases (1, 25%)	[153]
Crizotinib	Phase I	7/Elapsed or refractory	Stable disease (3, 43%)	[154]
Dasatinib	Phase I	5/Refractory	No objective response	[155]
	Phase II	46/Unresectable, recurrent, or metastatic	Clinical benefit response (CBR)[a] (6, 13%)	[156]
Imatinib	Phase II	10/Refractory or recurrent	No objective response	[157]
	Phase II	27/Metastatic or locally advanced	Clinical benefit response (CBR)[b] (5, 19%)	[158]
Pazopanib	Case report	1/Refractory and relapsed	No objective response	[159]
	Case report	2/Recurrent and metastatic	Partial response (1, 50%) Stable disease (1, 50%)	[160]
	Case report	3/Second recurrence	Stabilization of serum alkaline phosphatase level (1, 33%)	[161]
	Case report	3/Refractory and metastatic	Stable disease (2, 67%)	[162]
	Retrospective	6/Advanced, after 1–4 lines of therapy	Stable disease (2, 33%)	[163]
	Case report	15/Refractory and metastatic	Partial response (1, 7%) Stable disease (8, 53%)	[164]
	Phase I	4/Recurrent or refractory	Stable disease (1, 25%)	[165]
Regorafenib	Phase I	Not stated/refractory	Partial response (1)	[166]
	Randomized Phase II	22 + 10 in placebo group who crossed over after progression/progressive and either advanced or metastatic, after ≥1 lines of therapy	Improved mean progression-free survival (3.6 months vs 1.7 months with placebo group)	[136]
	Randomized phase II	26/Progressive and metastatic, after 1–2 lines of therapy	Increased stable disease (7, 27% vs 0% with placebo)	[137]

(continued)

Table 9.1 (continued)

Multi-TKI	Study type	No. of evaluable osteosarcoma patients/disease status	Outcomes (no. of patients, %)	References
Sorafenib	Case report	1/Refractory, progressive, and metastatic	Partial response (1, 100%)	[167]
	Case report	4/Refractory and relapsed	Stable disease (3, 75%)	[159]
	Case report	8/Metastatic (six patients) or local (two patients)	Partial response (6, 75%)	[168]
	Case report, combo with denosumab	1/Relapsed and unresectable	Stable disease (1, 100%)	[169]
	Phase I	10/Refractory	No objective response	[170]
	Phase I, combo with bevacizumab and cyclophosphamide	2/Recurrent or refractory	Stable disease (2, 100%)	[171]
	Phase II	35/Metastatic, relapsed, unresectable, and progressive	Progression-free survival at 6 months (10, 29%)	[138]
	Phase II, combo with everolimus	38/Progressive and either locally advanced, unresectable, or metastatic	Progression-free survival at 6 months (17, 45%)	[139]
Sunitinib	Case report	5/Refractory and relapsed	Partial response (1, 20%) Stable disease (1, 20%)	[159]
	Phase I	2/Refractory	Stable disease (1, 50%)	[172]

[a]CBR: Dasatinib – Objective response within 6 months or stable disease for ≥6 months
[b]CBR: Imatinib – Complete or partial response at 2 or 4 months or stable disease at 2 and 4 months

the presence of doxorubicin or methotrexate (but not cisplatin), compared with cultures containing 10% serum without chemotherapeutics [117]. Since 3D cultures mimic the oxygen, nutrient, and drug gradients found in sarcomas and other solid tumors [41], it is therefore not surprising that multi-TKIs were one of the most effective drug classes in our screen of FDA-approved oncology drugs that measured effects on the in vitro growth of 3D sarcospheres in both the absence and presence of MAP (methotrexate, doxorubicin, and cisplatin) standard-of-care chemotherapeutics [118]. Moreover, six (cabozantinib, crizotinib, dasatinib, pazopanib, regorafenib, and sunitinib) of the nine multi-TKIs in Table 9.1 that were included in our screen were among the top hits in at least one of the three tested highly metastatic human osteosarcoma cell lines [118]. The three other multi-TKIs in Table 9.1 that were included in our screen (axitinib, imatinib, sorafenib) had modest effects. Regorafenib was also the fourth most effective drug in a screen that measured viability of monolayer cultures of five human osteosarcoma cell lines [119].

To evaluate the potential clinical relevance of the in vitro screening results described in the previous paragraph, it is important to determine whether the drugs are effective in vivo. Imatinib reduced growth of primary osteosarcomas in a syngeneic murine model [107]. Moreover, preventive regimens of cediranib, dasatinib, sorafenib, and sunitinib each had intermediate to high activity in multiple subcutaneous xenograft primary osteosarcoma models evaluated by the Pediatric Preclinical Testing Program [120], and crizotinib, pazopanib, and regorafenib reduced tumor growth in similar xenograft models [121–123]. However, none of those studies [107, 120–123] determined whether the multi-TKIs also block growth of osteosarcoma metastases – the life-threatening process in osteosarcoma. In contrast, a therapeutic regimen of sorafenib caused regression in a subcutaneous xenograft primary tumor model and reduced the number and size of lung metastases in mice after tail vein injections of the SJSA-1 and MMNG human osteosarcoma cell lines [124, 125], and a therapeutic regimen of pazopanib reduced the number of lung metastases in mice after subcutaneous injection of the

LM8 murine osteosarcoma cell line and resection of the resultant primary tumor [126]. Similarly, a therapeutic regimen of sunitinib reduced primary tumor growth and the number of detectable metastases derived from intratibial injection of the 143B human osteosarcoma cell line in mice [127], but no effect was seen in response to dasatinib [128], imatinib [129], or sorafenib [130] as monotherapies in similar models. In the later studies, however, combinations of doxorubicin with either sorafenib or imatinib were more effective than the monotherapies [129, 130]. Given the potential translational relevance [131], it is surprising that none of the multi-TKIs have been tested in animal models in combination with all three components of MAP chemotherapy. In other combinations, sorafenib either with the mTOR inhibitor everolimus or with the CDK inhibitor palbociclib blocked growth in an MNNG human osteosarcoma cell line subcutaneous xenograft primary tumor model and in a patient-derived osteosarcoma orthotopic xenograft model [125, 132, 133]. More importantly, the therapeutic regimen of sorafenib with everolimus inhibited the number and size of lung metastases more effectively than either agent as monotherapy following tail vein injection of the MNNG human osteosarcoma cell line [125]. To maximize clinical relevance, it will be important for future murine studies to focus on therapeutic rather than preventive regimens.

Although the available clinical trials are limited in size, some of multi-TKIs appear promising as monotherapies (Table 9.1). The most encouraging are the Phase II studies of apatinib [134, 135], regorafenib [136, 137], and sorafenib, both alone [138] and in combination with everolimus [139]. Those studies recently led to designation of regorafenib as a category 1 recommendation by the National Comprehensive Cancer Network for second-line therapy of osteosarcoma patients with relapsed/refractory or metastatic disease (NCCN Guidelines Version 1.2020, Bone Cancer). Sorafenib alone and in combination with everolimus are included, respectively, as category 2A and 2B recommendations. Multi-TKIs in ongoing clinical trials listed in ClinicalTrials.gov for osteosarcoma patients include apatinib plus gemcitabine and docetaxel (Phase II, NCT03742193); apatinib plus anti-PD1 (Phase II, NCT03359018); cabozantinib (Phase II, NCT02243605 and NCT02867592); dasatinib plus ifosfamide, carboplatin, and etoposide (Phase II, NCT00788125); famitinib plus anti-PD1 (Phase I/II, NCT04044378); lenvatinib plus ifosfamide and etoposide (Phase I/II, NCT02432274); pazopanib plus topotecan (Phase II, NCT02357810); regorafenib (Phase II, NCT02048371 and NCT03277924); sunitinib plus anti-PD1 (Phase I/II, NCT03277924); and sunitinib plus losartan (Phase I, NCT03900793). In addition, the Pediatric Molecular Analysis for Therapy Choice (MATCH) screening trial (NCT03155620) includes osteosarcoma patients in sub-studies of ensartinib, erdafitinib, larotrectinib, ulixertinib, and vemurafenib. Future studies will be needed to determine whether the multi-TKIs are more effective in combination with other agents and whether a subset of osteosarcoma patients can be identified that will respond to individual multi-TKIs. For example, levels of RTKs or their ligands might serve as biomarkers to predict responsiveness to appropriate multi-TKIs [45–47, 56, 96].

Systemic toxicities are a major limitation regarding multi-TKI therapies. Strategies are therefore being developed to target multi-TKIs and other drugs to the involved tissue. For example, intranasal administration can directly target multi-TKIs to osteosarcoma metastases in the lung [140, 141]. Another potential approach is to target the multi-TKIs to the tumor and/or metastases following systemic administration. For example, a liposomal formulation of ponatinib inhibited primary tumor growth by the K7M2 murine osteosarcoma cell line in a subcutaneous syngeneic model more effectively than a tenfold higher dose of free ponatinib without inducing the systemic toxicity caused by the free drug [142]. A high dose but pulsatile (once every 2 weeks) regimen has also shown promise to increase efficacy and decrease toxicity of multi-TKIs in epithelial cancers [143, 144].

Much work, both preclinical and clinical, remains to be done to identify optimal multi-TKIs, optimal regimens, and the most responsive patients for each multi-TKI. We are nonetheless cautiously optimistic that multi-TKIs will ultimately improve survival for osteosarcoma patients and/or will allow use of lower doses of conventional chemotherapeutics and thereby reduce their systemic toxicity.

References

1. Rettew AN, Getty PJ, Greenfield EM (2014) Receptor tyrosine kinases in osteosarcoma: not just the usual suspects. Adv Exp Med Biol 804:47–66
2. Klaeger S et al (2017) The target landscape of clinical kinase drugs. Science 358(6367):eaan4368
3. Sumi NJ et al (2019) Divergent polypharmacology-driven cellular activity of structurally similar multi-kinase inhibitors through cumulative effects on individual targets. Cell Chem Biol 26(9):1240–1252.e11
4. Rettew A et al (2012) Multiple receptor tyrosine kinases promote the in vitro phenotype of metastatic human osteosarcoma cell lines. Oncogenesis 1:e34:1–9
5. Nakano T et al (2003) Biological properties and gene expression associated with metastatic potential of human osteosarcoma. Clin Exp Metastasis 20:665–674
6. Bai Y et al (2012) Phosphoproteomics identifies driver tyrosine kinases in sarcoma cell lines and tumors. Cancer Res 72(10):2501–2511
7. Rankin EB et al (2014) Direct regulation of GAS6/AXL signaling by HIF promotes renal metastasis through SRC and MET. Proc Natl Acad Sci U S A 111(37):13373–13378
8. Zhang Y et al (2013) Knockdown of AXL receptor tyrosine kinase in osteosarcoma cells leads to decreased proliferation and increased apoptosis. Int J Immunopathol Pharmacol 26(1):179–188
9. Jiang N et al (2017) lncRNA DANCR promotes tumor progression and cancer stemness features in osteosarcoma by upregulating AXL via miR-33a-5p inhibition. Cancer Lett 405:46–55
10. Kuijjer ML et al (2013) IR/IGF1R signaling as potential target for treatment of high-grade osteosarcoma. BMC Cancer 13:245
11. Han J et al (2013) Gas6/Axl mediates tumor cell apoptosis, migration and invasion and predicts the clinical outcome of osteosarcoma patients. Biochem Biophys Res Commun 435(3):493–500
12. Rettew A et al (2020) The receptor tyrosine kinase Axl promotes tumor growth and metastasis and is a novel therapeutic target for osteosarcoma. in preparation
13. Tian R et al (2014) miR-199a-3p negatively regulates the progression of osteosarcoma through targeting AXL. Am J Cancer Res 4(6):738–750
14. Giaccia A, Tabibiazar R, Miao R (2017) Targeting GAS6 in cancer, fibrosis, and viral infection. Drug Target Rev 4(1):33–36
15. Schoumacher M, Burbridge M (2017) Key roles of AXL and MER receptor tyrosine kinases in resistance to multiple anticancer therapies. Curr Oncol Rep 19(3):19
16. Kimani SG et al (2017) Small molecule inhibitors block Gas6-inducible TAM activation and tumorigenicity. Sci Rep 7:43908
17. Ye X et al (2010) An anti-Axl monoclonal antibody attenuates xenograft tumor growth and enhances the effect of multiple anticancer therapies. Oncogene 29(38):5254–5264
18. Kariolis MS et al (2014) An engineered Axl 'decoy receptor' effectively silences the Gas6-Axl signaling axis. Nat Chem Biol 10(11):977–983
19. Kanlikilicer P et al (2017) Therapeutic targeting of AXL receptor tyrosine kinase inhibits tumor growth and intraperitoneal metastasis in ovarian cancer models. Mol Ther Nucleic Acids 9:251–262
20. Holland SJ et al (2010) R428, a selective small molecule inhibitor of Axl kinase, blocks tumor spread and prolongs survival in models of metastatic breast cancer. Cancer Res 70(4):1544–1554
21. Collier C et al (2018) Micrometastatic drug-screening platform for osteosarcoma demonstrates heterogeneous response to MAP chemotherapy in osteosarcoma cell lines. Clin Orthop Relat Res 476:1400–1411
22. Mahadevan D et al (2007) A novel tyrosine kinase switch is a mechanism of imatinib resistance in gastrointestinal stromal tumors. Oncogene 26(27):3909–3919
23. Croucher PI, McDonald MM, Martin TJ (2016) Bone metastasis: the importance of the neighbourhood. Nat Rev Cancer 16:373
24. Brand TM et al (2017) The receptor tyrosine kinase AXL mediates nuclear translocation of the epidermal growth factor receptor. Sci Signal 10(460):eaag1064
25. Kariolis MS et al (2017) Inhibition of the GAS6/AXL pathway augments the efficacy of chemotherapies. J Clin Invest 127(1):183–198
26. Hong J, Maacha S, Belkhiri A (2018) Transcriptional upregulation of c-MYC by AXL confers epirubicin resistance in esophageal adenocarcinoma. Mol Oncol 12(12):2191–2208
27. McDaniel NK et al (2018) MERTK mediates intrinsic and adaptive resistance to AXL-targeting agents. Mol Cancer Ther 17(11):2297–2308
28. Yang H et al (2019) HSP90/AXL/eIF4E-regulated unfolded protein response as an acquired vulnerability in drug-resistant KRAS-mutant lung cancer. Oncogenesis 8(9):45
29. Myers KV, Amend SR, Pienta KJ (2019) Targeting Tyro3, Axl and MerTK (TAM receptors):

implications for macrophages in the tumor microenvironment. Mol Cancer 18(1):94
30. Meyers PA, Chou AJ (2014) Muramyl tripeptide-phosphatidyl ethanolamine encapsulated in liposomes (L-MTP-PE) in the treatment of osteosarcoma. Adv Exp Med Biol 804:307–321
31. Akalu YT, Rothlin CV, Ghosh S (2017) TAM receptor tyrosine kinases as emerging targets of innate immune checkpoint blockade for cancer therapy. Immunol Rev 276(1):165–177
32. Aguilera TA, Giaccia AJ (2017) Molecular pathways: oncologic pathways and their role in T-cell exclusion and immune evasion-a new role for the AXL receptor tyrosine kinase. Clin Cancer Res 23(12):2928–2933
33. Ljunggren HG, Jonsson R, Hoglund P (2018) Seminal immunologic discoveries with direct clinical implications: the 2018 Nobel Prize in Physiology or Medicine honours discoveries in cancer immunotherapy. Scand J Immunol 88(6):e12731
34. DeRenzo C, Gottschalk S (2014) Genetically modified T-cell therapy for osteosarcoma. Adv Exp Med Biol 804:323–340
35. Roberts RD et al (2019) Provocative questions in osteosarcoma basic and translational biology: a report from the Children's Oncology Group. Cancer 125(20):3514–3525
36. Pasquale EB (2008) Eph-ephrin bidirectional signaling in physiology and disease. Cell 133(1):38–52
37. Fritsche-Guenther R et al (2010) De novo expression of EphA2 in osteosarcoma modulates activation of the mitogenic signalling pathway. Histopathology 57(6):836–850
38. Posthumadeboer J et al (2013) Surface proteomic analysis of osteosarcoma identifies EPHA2 as receptor for targeted drug delivery. Br J Cancer 109(8):2142–2154
39. Varelias A et al (2002) Human osteosarcoma expresses specific ephrin profiles: implications for tumorigenicity and prognosis. Cancer 95(4):862–869
40. Barco R et al (2007) The synovial sarcoma SYT-SSX2 oncogene remodels the cytoskeleton through activation of the ephrin pathway. Mol Biol Cell 18(10):4003–4012
41. Gaebler M et al (2017) Three-dimensional patient-derived in vitro sarcoma models: promising tools for improving clinical tumor management. Front Oncol 7:203
42. Tonna S, Sims NA (2014) Talking among ourselves: paracrine control of bone formation within the osteoblast lineage. Calcif Tissue Int 94(1):35–45
43. Arthur A et al (2018) The osteoprogenitor-specific loss of ephrinB1 results in an osteoporotic phenotype affecting the balance between bone formation and resorption. Sci Rep 8(1):12756
44. Baird K et al (2005) Gene expression profiling of human sarcomas: insights into sarcoma biology. Cancer Res 65(20):9226–9235
45. Guagnano V et al (2012) FGFR genetic alterations predict for sensitivity to NVP-BGJ398, a selective pan-FGFR inhibitor. Cancer Discov 2(12):1118–1133
46. Fernanda Amary M et al (2014) Fibroblastic growth factor receptor 1 amplification in osteosarcoma is associated with poor response to neo-adjuvant chemotherapy. Cancer Med 3(4):980–987
47. Baroy T et al (2016) Genome analysis of osteosarcoma progression samples identifies FGFR1 overexpression as a potential treatment target and CHM as a candidate tumor suppressor gene. PLoS One 11(9):e0163859
48. Chaiyawat P et al (2017) Activation status of receptor tyrosine kinases as an early predictive marker of response to chemotherapy in osteosarcoma. Transl Oncol 10(5):846–853
49. Zhang W et al (2018) Adaptive fibrogenic reprogramming of osteosarcoma stem cells promotes metastatic growth. Cell Rep 24(5):1266–1277.e5
50. Campbell J et al (2016) Large-scale profiling of kinase dependencies in cancer cell lines. Cell Rep 14(10):2490–2501
51. Basu-Roy U et al (2012) Sox2 maintains self renewal of tumor-initiating cells in osteosarcomas. Oncogene 31(18):2270–2282
52. Shimizu T et al (2012) Fibroblast growth factor-2 is an important factor that maintains cellular immaturity and contributes to aggressiveness of osteosarcoma. Mol Cancer Res 10(3):454–468
53. Weekes D et al (2016) Regulation of osteosarcoma cell lung metastasis by the c-Fos/AP-1 target FGFR1. Oncogene 35(22):2852–2861
54. Davis MI et al (2011) Comprehensive analysis of kinase inhibitor selectivity. Nat Biotechnol 29(11):1046–1051
55. Gudernova I et al (2016) Multikinase activity of fibroblast growth factor receptor (FGFR) inhibitors SU5402, PD173074, AZD1480, AZD4547 and BGJ398 compromises the use of small chemicals targeting FGFR catalytic activity for therapy of short-stature syndromes. Hum Mol Genet 25(1):9–23
56. Behjati S et al (2017) Recurrent mutation of IGF signalling genes and distinct patterns of genomic rearrangement in osteosarcoma. Nat Commun 8:15936
57. Cheng L et al (2019) Integration of genomic copy number variations and chemotherapy-response biomarkers in pediatric sarcoma. BMC Med Genet 12(Suppl 1):23
58. Wang YH et al (2012) Increased expression of insulin-like growth factor-1 receptor is correlated with tumor metastasis and prognosis in patients with osteosarcoma. J Surg Oncol 105(3):235–243
59. Jentzsch T et al (2014) Worse prognosis of osteosarcoma patients expressing IGF-1 on a tissue microarray. Anticancer Res 34(8):3881–3889
60. Maniscalco L et al (2015) Increased expression of insulin-like growth factor-1 receptor is correlated with worse survival in canine appendicular osteosarcoma. Vet J 205(2):272–280
61. Chen L et al (2013) miR-16 inhibits cell proliferation by targeting IGF1R and the Raf1-MEK1/2-ERK1/2

pathway in osteosarcoma. FEBS Lett 587(9):1366–1372
62. Zhao H et al (2013) MiR-133b is down-regulated in human osteosarcoma and inhibits osteosarcoma cells proliferation, migration and invasion, and promotes apoptosis. PLoS One 8(12):e83571
63. Tan X et al (2015) MicroRNA-26a inhibits osteosarcoma cell proliferation by targeting IGF-1. Bone Res 3:15033
64. Chen G et al (2016) MicroRNA-133a inhibits osteosarcoma cells proliferation and invasion via targeting IGF-1R. Cell Physiol Biochem 38(2):598–608
65. Liu Y et al (2016) MiR-100 inhibits osteosarcoma cell proliferation, migration, and invasion and enhances chemosensitivity by targeting IGFIR. Technol Cancer Res Treat 15(5):NP40–NP48
66. Wang Z et al (2017) MicroRNA-503 suppresses cell proliferation and invasion in osteosarcoma via targeting insulin-like growth factor 1 receptor. Exp Ther Med 14(2):1547–1553
67. Zhang K et al (2019) Let-7b acts as a tumor suppressor in osteosarcoma via targeting IGF1R. Oncol Lett 17(2):1646–1654
68. Zhao X, Li J, Yu D (2019) MicroRNA-939-5p directly targets IGF-1R to inhibit the aggressive phenotypes of osteosarcoma through deactivating the PI3K/Akt pathway. Int J Mol Med 44(5):1833–1843
69. Avnet S et al (2009) Insulin receptor isoform A and insulin-like growth factor II as additional treatment targets in human osteosarcoma. Cancer Res 69(6):2443–2452
70. Shimizu T et al (2014) IGF2 preserves osteosarcoma cell survival by creating an autophagic state of dormancy that protects cells against chemotherapeutic stress. Cancer Res 74(22):6531–6541
71. Mancarella C, Scotlandi K (2018) IGF system in sarcomas: a crucial pathway with many unknowns to exploit for therapy. J Mol Endocrinol 61(1):T45–T60
72. Habel N et al (2019) CYR61 triggers osteosarcoma metastatic spreading via an IGF1Rbeta-dependent EMT-like process. BMC Cancer 19(1):62
73. Messerschmitt P et al (2008) Specific tyrosine kinases regulate human osteosarcoma cells in vitro. Clin Orthop Relat Res 466:2168–2175
74. Duan Z et al (2009) Insulin-like growth factor-I receptor tyrosine kinase inhibitor cyclolignan picropodophyllin inhibits proliferation and induces apoptosis in multidrug resistant osteosarcoma cell lines. Mol Cancer Ther 8(8):2122–2130
75. Girnita A et al (2004) Cyclolignans as inhibitors of the insulin-like growth factor-1 receptor and malignant cell growth. Cancer Res 64(1):236–242
76. Wu X et al (2013) Alternative cytotoxic effects of the postulated IGF-IR inhibitor picropodophyllin in vitro. Mol Cancer Ther 12(8):1526–1536
77. Waraky A et al (2014) Picropodophyllin causes mitotic arrest and catastrophe by depolymerizing microtubules via insulin-like growth factor-1 receptor-independent mechanism. Oncotarget 5(18):8379–8392
78. Allard JB, Duan C (2018) IGF-binding proteins: why do they exist and why are there so many? Front Endocrinol (Lausanne) 9:117
79. Yang R et al (2016) Transcriptional profiling identifies the signaling axes of IGF and transforming growth factor-b as involved in the pathogenesis of osteosarcoma. Clin Orthop Relat Res 474(1):178–189
80. Su Y et al (2011) Insulin-like growth factor binding protein 5 suppresses tumor growth and metastasis of human osteosarcoma. Oncogene 30(37):3907–3917
81. Kolb EA et al (2008) Initial testing (stage 1) of a monoclonal antibody (SCH 717454) against the IGF-1 receptor by the pediatric preclinical testing program. Pediatr Blood Cancer 50(6):1190–1197
82. Wang Y et al (2010) A fully human insulin-like growth factor-I receptor antibody SCH 717454 (Robatumumab) has antitumor activity as a single agent and in combination with cytotoxics in pediatric tumor xenografts. Mol Cancer Ther 9(2):410–418
83. Dong J et al (2010) Combination of two insulin-like growth factor-I receptor inhibitory antibodies targeting distinct epitopes leads to an enhanced antitumor response. Mol Cancer Ther 9(9):2593–2604
84. Kurmasheva RT et al (2009) The insulin-like growth factor-1 receptor-targeting antibody, CP-751,871, suppresses tumor-derived VEGF and synergizes with rapamycin in models of childhood sarcoma. Cancer Res 69(19):7662–7671
85. Kolb EA et al (2010) R1507, a fully human monoclonal antibody targeting IGF-1R, is effective alone and in combination with rapamycin in inhibiting growth of osteosarcoma xenografts. Pediatr Blood Cancer 55(1):67–75
86. Beltran PJ et al (2011) Efficacy of ganitumab (AMG 479), alone and in combination with rapamycin, in Ewing's and osteogenic sarcoma models. J Pharmacol Exp Ther 337(3):644–654
87. Pappo AS et al (2014) A phase 2 trial of R1507, a monoclonal antibody to the insulin-like growth factor-1 receptor (IGF-1R), in patients with recurrent or refractory rhabdomyosarcoma, osteosarcoma, synovial sarcoma, and other soft tissue sarcomas: results of a Sarcoma Alliance for Research Through Collaboration study. Cancer 120(16):2448–2456
88. Anderson PM et al (2016) A phase II study of clinical activity of SCH 717454 (robatumumab) in patients with relapsed osteosarcoma and Ewing sarcoma. Pediatr Blood Cancer 63(10):1761–1770
89. Schwartz GK et al (2013) Cixutumumab and temsirolimus for patients with bone and soft-tissue sarcoma: a multicentre, open-label, phase 2 trial. Lancet Oncol 14(4):371–382
90. Wagner LM et al (2015) Phase II study of cixutumumab in combination with temsirolimus in pediatric patients and young adults with recurrent or refractory sarcoma: a report from the

Children's Oncology Group. Pediatr Blood Cancer 62(3):440–444
91. Luk F et al (2011) IGF1R-targeted therapy and its enhancement of doxorubicin chemosensitivity in human osteosarcoma cell lines. Cancer Investig 29(8):521–532
92. Davaadelger B et al (2017) The IGF-1R/AKT pathway has opposing effects on Nutlin-3a-induced apoptosis. Cancer Biol Ther 18(11):895–903
93. Scotlandi K et al (2005) Antitumor activity of the insulin-like growth factor-I receptor kinase inhibitor NVP-AEW541 in musculoskeletal tumors. Cancer Res 65(9):3868–3876
94. Gvozdenovic A et al (2017) A bispecific antibody targeting IGF-IR and EGFR has tumor and metastasis suppressive activity in an orthotopic xenograft osteosarcoma mouse model. Am J Cancer Res 7(7):1435–1449
95. Pahl JH et al (2012) Anti-EGFR antibody cetuximab enhances the cytolytic activity of natural killer cells toward osteosarcoma. Clin Cancer Res 18(2):432–441
96. Lodhia KA, Tienchaiananda P, Haluska P (2015) Understanding the key to targeting the IGF axis in cancer: a biomarker assessment. Front Oncol 5:142
97. Asmane I et al (2012) Insulin-like growth factor type 1 receptor (IGF-1R) exclusive nuclear staining: a predictive biomarker for IGF-1R monoclonal antibody (Ab) therapy in sarcomas. Eur J Cancer 48(16):3027–3035
98. Cao Y et al (2014) Insulin-like growth factor 1 receptor and response to anti-IGF1R antibody therapy in osteosarcoma. PLoS One 9(8):e106249
99. Warsito D et al (2016) Nuclearly translocated insulin-like growth factor 1 receptor phosphorylates histone H3 at tyrosine 41 and induces SNAI2 expression via Brg1 chromatin remodeling protein. Oncotarget 7(27):42288–42302
100. Solomon-Zemler R, Sarfstein R, Werner H (2017) Nuclear insulin-like growth factor-1 receptor (IGF1R) displays proliferative and regulatory activities in non-malignant cells. PLoS One 12(9):e0185164
101. Aleksic T et al (2018) Nuclear IGF1R interacts with regulatory regions of chromatin to promote RNA polymerase II recruitment and gene expression associated with advanced tumor stage. Cancer Res 78(13):3497–3509
102. Takeuchi K (2019) Discovery stories of RET fusions in lung cancer: a mini-review. Front Physiol 10:216
103. Roskoski R Jr, Sadeghi-Nejad A (2018) Role of RET protein-tyrosine kinase inhibitors in the treatment RET-driven thyroid and lung cancers. Pharmacol Res 128:1–17
104. Gattelli A et al (2018) Chronic expression of wild-type Ret receptor in the mammary gland induces luminal tumors that are sensitive to Ret inhibition. Oncogene 37(29):4046–4054
105. Chen Y et al (2018) LncRNA MALAT1 promotes cancer metastasis in osteosarcoma via activation of the PI3K-Akt signaling pathway. Cell Physiol Biochem 51(3):1313–1326
106. Dong Y et al (2015) MALAT1 promotes the proliferation and metastasis of osteosarcoma cells by activating the PI3K/Akt pathway. Tumour Biol 36(3):1477–1486
107. Gobin B et al (2014) Imatinib mesylate exerts antiproliferative effects on osteosarcoma cells and inhibits the tumour growth in immunocompetent murine models. PLoS One 9(3):e90795
108. Tian S et al (2011) YN968D1 is a novel and selective inhibitor of vascular endothelial growth factor receptor-2 tyrosine kinase with potent activity in vitro and in vivo. Cancer Sci 102(7):1374–1380
109. Wilhelm SM et al (2011) Regorafenib (BAY 73-4506): a new oral multikinase inhibitor of angiogenic, stromal and oncogenic receptor tyrosine kinases with potent preclinical antitumor activity. Int J Cancer 129(1):245–255
110. Kitagawa D et al (2013) Activity-based kinase profiling of approved tyrosine kinase inhibitors. Genes Cells 18(2):110–122
111. Cha Y et al (2018) FGFR2 amplification is predictive of sensitivity to regorafenib in gastric and colorectal cancers in vitro. Mol Oncol 12(7):993–1003
112. Vasta JD et al (2018) Quantitative, wide-spectrum kinase profiling in live cells for assessing the effect of cellular ATP on target engagement. Cell Chem Biol 25(2):206–214.e11
113. Drilon A et al (2018) Targeting RET-driven cancers: lessons from evolving preclinical and clinical landscapes. Nat Rev Clin Oncol 15(3):151–167
114. Berlow N et al (2013) A new approach for prediction of tumor sensitivity to targeted drugs based on functional data. BMC Bioinformatics 14:239
115. Marley K et al (2015) Dasatinib modulates invasive and migratory properties of canine osteosarcoma and has therapeutic potential in affected dogs. Transl Oncol 8(4):231–238
116. Ahmadi M et al (2014) Hypoxia modulates the activity of a series of clinically approved tyrosine kinase inhibitors. Br J Pharmacol 171(1):224–236
117. Sevelda F et al (2015) EGFR is not a major driver for osteosarcoma cell growth in vitro but contributes to starvation and chemotherapy resistance. J Exp Clin Cancer Res 34:134
118. Collier C et al (2016) Opportunities for drug repurposing in osteosarcoma: a screen of FDA-approved oncology drugs in a micrometastatic model of disease in oral presentation at the annual meeting of the Orthopaedic Research Society
119. Yu D et al (2015) Identification of synergistic, clinically achievable, combination therapies for osteosarcoma. Sci Rep 5:16991
120. Sampson VB et al (2013) A review of targeted therapies evaluated by the pediatric preclinical testing program for osteosarcoma. Front Oncol 3:132
121. Kumar S et al (2011) Metronomic oral topotecan with pazopanib is an active antiangiogenic regimen

in mouse models of aggressive pediatric solid tumor. Clin Cancer Res 17(17):5656–5667
122. Sampson ER et al (2011) The orally bioavailable met inhibitor PF-2341066 inhibits osteosarcoma growth and osteolysis/matrix production in a xenograft model. J Bone Miner Res 26(6):1283–1294
123. Pan PJ, Liu YC, Hsu FT (2019) Protein kinase B and extracellular signal-regulated kinase inactivation is associated with regorafenib-induced inhibition of osteosarcoma progression in vitro and in vivo. J Clin Med 8(6):900
124. Pignochino Y et al (2009) Sorafenib blocks tumour growth, angiogenesis and metastatic potential in preclinical models of osteosarcoma through a mechanism potentially involving the inhibition of ERK1/2, MCL-1 and ezrin pathways. Mol Cancer 8:118
125. Pignochino Y et al (2013) The combination of sorafenib and everolimus abrogates mTORC1 and mTORC2 upregulation in osteosarcoma preclinical models. Clin Cancer Res 19(8):2117–2131
126. Tanaka T et al (2013) Dynamic analysis of lung metastasis by mouse osteosarcoma LM8: VEGF is a candidate for anti-metastasis therapy. Clin Exp Metastasis 30(4):369–379
127. Kumar RM et al (2015) Sunitinib malate (SU-11248) reduces tumour burden and lung metastasis in an intratibial human xenograft osteosarcoma mouse model. Am J Cancer Res 5(7):2156–2168
128. Hingorani P et al (2009) Inhibition of Src phosphorylation alters metastatic potential of osteosarcoma in vitro but not in vivo. Clin Cancer Res 15(10):3416–3422
129. Yamaguchi SI et al (2015) Synergistic antiproliferative effect of imatinib and adriamycin in platelet-derived growth factor receptor-expressing osteosarcoma cells. Cancer Sci 106(7):875–882
130. Jian C et al (2017) Co-targeting of DNA, RNA, and protein molecules provides optimal outcomes for treating osteosarcoma and pulmonary metastasis in spontaneous and experimental metastasis mouse models. Oncotarget 8(19):30742–30755
131. Xu J, Xie L, Guo W (2018) PDGF/PDGFR effects in osteosarcoma and the “add-on” strategy. Clin Sarcoma Res 8:15
132. Higuchi T et al (2019) Combination treatment with sorafenib and everolimus regresses a doxorubicin-resistant osteosarcoma in a PDOX mouse model. Anticancer Res 39(9):4781–4786
133. Higuchi T et al (2019) Sorafenib and palbociclib combination regresses a cisplatinum-resistant osteosarcoma in a PDOX mouse model. Anticancer Res 39(8):4079–4084
134. Liao Z et al (2019) Phase II trial of VEGFR2 inhibitor apatinib for metastatic sarcoma: focus on efficacy and safety. Exp Mol Med 51(3):24
135. Xie L et al (2019) Apatinib for advanced osteosarcoma after failure of standard multimodal therapy: an open label phase II clinical trial. Oncologist 24(7):e542–e550
136. Davis LE et al (2019) Randomized double-blind phase II study of regorafenib in patients with metastatic osteosarcoma. J Clin Oncol 37(16):1424–1431
137. Duffaud F et al (2019) Efficacy and safety of regorafenib in adult patients with metastatic osteosarcoma: a non-comparative, randomised, double-blind, placebo-controlled, phase 2 study. Lancet Oncol 20(1):120–133
138. Grignani G et al (2012) A phase II trial of sorafenib in relapsed and unresectable high-grade osteosarcoma after failure of standard multimodal therapy: an Italian Sarcoma Group study. Ann Oncol 23(2):508–516
139. Grignani G et al (2015) Sorafenib and everolimus for patients with unresectable high-grade osteosarcoma progressing after standard treatment: a non-randomised phase 2 clinical trial. Lancet Oncol 16(1):98–107
140. Anderson P (2009) Non-surgical treatment of pulmonary and extra-pulmonary metastases. Cancer Treat Res 152:203–215
141. Jaffe N (2014) Historical perspective on the introduction and use of chemotherapy for the treatment of osteosarcoma. Adv Exp Med Biol 804:1–30
142. Kallus S et al (2018) Nanoformulations of anticancer FGFR inhibitors with improved therapeutic index. Nanomedicine 14(8):2632–2643
143. Rovithi M et al (2019) Phase I dose-escalation study of once weekly or once every two weeks administration of high-dose sunitinib in patients with refractory solid tumors. J Clin Oncol 37(5):411–418
144. Rovithi M, Verheul HMW (2017) Pulsatile high-dose treatment with antiangiogenic tyrosine kinase inhibitors improves clinical antitumor activity. Angiogenesis 20(3):287–289
145. Zhou Y et al (2017) A case report of apatinib in treating osteosarcoma with pulmonary metastases. Medicine (Baltimore) 96(15):e6578
146. Zhu B et al (2018) Efficacy and safety of apatinib monotherapy in advanced bone and soft tissue sarcoma: an observational study. Cancer Biol Ther 19(3):198–204
147. Li F et al (2017) Efficacy and safety of apatinib in stage IV sarcomas: experience of a major sarcoma center in China. Oncotarget 8(38):64471–64480
148. Zheng K et al (2018) Efficacy and safety of apatinib in advance osteosarcoma with pulmonary metastases: a single-center observational study. Medicine (Baltimore) 97(31):e11734
149. Xie L et al (2018) Apatinib for advanced sarcoma: results from multiple institutions’ off-label use in China. BMC Cancer 18(1):396
150. Tian Z et al (2019) Efficacy and safety of apatinib in treatment of osteosarcoma after failed standard multimodal therapy: an observational study. Medicine (Baltimore) 98(19):e15650
151. Geller JI et al (2018) A study of axitinib, a VEGF receptor tyrosine kinase inhibitor, in children and adolescents with recurrent or refractory solid tumors: a Children’s Oncology Group phase 1

and pilot consortium trial (ADVL1315). Cancer 124(23):4548–4555
152. Chuk MK et al (2018) A phase 1 study of cabozantinib in children and adolescents with recurrent or refractory solid tumors, including CNS tumors: trial ADVL1211, a report from the Children's Oncology Group. Pediatr Blood Cancer 65(8):e27077
153. Fox E et al (2010) A phase 1 trial and pharmacokinetic study of cediranib, an orally bioavailable pan-vascular endothelial growth factor receptor inhibitor, in children and adolescents with refractory solid tumors. J Clin Oncol 28(35):5174–5181
154. Mosse YP et al (2013) Safety and activity of crizotinib for paediatric patients with refractory solid tumours or anaplastic large-cell lymphoma: a Children's Oncology Group phase 1 consortium study. Lancet Oncol 14(6):472–480
155. Aplenc R et al (2011) Pediatric phase I trial and pharmacokinetic study of dasatinib: a report from the Children's Oncology Group Phase I Consortium. J Clin Oncol 29(7):839–844
156. Schuetze SM et al (2016) SARC009: phase 2 study of dasatinib in patients with previously treated, high-grade, advanced sarcoma. Cancer 122(6):868–874
157. Bond M et al (2008) A phase II study of imatinib mesylate in children with refractory or relapsed solid tumors: a Children's Oncology Group study. Pediatr Blood Cancer 50(2):254–258
158. Chugh R et al (2009) Phase II multicenter trial of imatinib in 10 histologic subtypes of sarcoma using a bayesian hierarchical statistical model. J Clin Oncol 27(19):3148–3153
159. Penel-Page M et al (2015) Off-label use of targeted therapies in osteosarcomas: data from the French registry OUTC'S (Observatoire de l'Utilisation des Therapies Ciblees dans les Sarcomes). BMC Cancer 15:854
160. Elete KR et al (2018) Response to Pazopanib in patients with relapsed osteosarcoma. J PEDIATR Hematol Oncol. [Epub ahead of print] doi:10.1097/MPH.0000000000001375
161. Umeda K et al (2017) Pazopanib for second recurrence of osteosarcoma in pediatric patients. Pediatr Int 59(8):937–938
162. Safwat A et al (2014) Pazopanib in metastatic osteosarcoma: significant clinical response in three consecutive patients. Acta Oncol 53(10):1451–1454
163. Seto T et al (2019) Real-world experiences with pazopanib in patients with advanced soft tissue and bone sarcoma in Northern California. Med Sci (Basel) 7(3):48
164. Longhi A et al (2019) Pazopanib in relapsed osteosarcoma patients: report on 15 cases. Acta Oncol 58(1):124–128
165. Glade Bender JL et al (2013) Phase I pharmacokinetic and pharmacodynamic study of pazopanib in children with soft tissue sarcoma and other refractory solid tumors: a Children's Oncology Group Phase I Consortium report. J Clin Oncol 31(24):3034–3043
166. Mross K et al (2012) A phase I dose-escalation study of regorafenib (BAY 73-4506), an inhibitor of oncogenic, angiogenic, and stromal kinases, in patients with advanced solid tumors. Clin Cancer Res 18(9):2658–2667
167. Armstrong AE et al (2019) Prolonged response to sorafenib in a patient with refractory metastatic osteosarcoma and a somatic PDGFRA D846V mutation. Pediatr Blood Cancer 66(1):e27493
168. Raciborska A, Bilska K (2018) Sorafenib in patients with progressed and refractory bone tumors. Med Oncol 35(10):126
169. Cathomas R et al (2015) RANK ligand blockade with denosumab in combination with sorafenib in chemorefractory osteosarcoma: a possible step forward? Oncology 88(4):257–260
170. Widemann BC et al (2012) A phase I trial and pharmacokinetic study of sorafenib in children with refractory solid tumors or leukemias: a Children's Oncology Group Phase I Consortium report. Clin Cancer Res 18(21):6011–6022
171. Navid F et al (2013) Phase I and clinical pharmacology study of bevacizumab, sorafenib, and low-dose cyclophosphamide in children and young adults with refractory/recurrent solid tumors. Clin Cancer Res 19(1):236–246
172. Dubois SG et al (2011) Phase I and pharmacokinetic study of sunitinib in pediatric patients with refractory solid tumors: a Children's Oncology Group study. Clin Cancer Res 17(15):5113–5122

The Role of ALDH in the Metastatic Potential of Osteosarcoma Cells and Potential ALDH Targets

10

Rebekah Belayneh and Kurt Weiss

Abstract

Aldehyde dehydrogenases are a family of enzymes that oxidize aldehydes to carboxylic acids. These enzymes are important in cellular homeostasis during oxidative stress by the elimination of toxic aldehyde by-products from various cellular processes. In osteosarcoma, aldehyde dehydrogenase 1A1has been described as a cancer stem cell marker. Its activity has been found to correlate with metastatic potential and the metastatic phenotype. As such, a more complete understanding of aldehyde dehydrogenase in osteosarcoma will give us a deeper knowledge of its impact on osteosarcoma metastatic potential. Our hope is that this knowledge can be translated into novel antimetastatic therapeutic strategies and thus improve osteosarcoma prognoses.

Keywords

ALDH · Osteosarcoma · Metastasis · Disulfiram · mTOR · Copper · Retinal · Notch

R. Belayneh (✉) · K. Weiss
Department of Orthopaedic Surgery, University of Pittsburgh Medical Center, Pittsburgh, PA, USA

Musculoskeletal Oncology Lab, University of Pittsburgh, Pittsburgh, PA, USA

Introduction

Osteosarcoma (OS) is the most common primary malignancy of bone and usually occurs in the long bones of youths in the first and second decades of life [1]. This cancer's main diagnostic characteristic is the production of malignant osteoid by the tumor cells. Despite treatment with pre- and postoperative chemotherapy and wide surgical resection of the tumor, the overall survival of patients with OS without detectable metastasis is approximately 65–70% [2–6]. OS has a high propensity for metastasis, with the most common site of metastatic spread (>90%) being the lungs [7]. The prognosis for patients with detectable metastasis at the time of diagnosis is particularly poor, ranging from 15% to 30%. In turn, survival is ultimately dependent upon the presence or absence of pulmonary metastatic disease [2, 6, 8, 9].

The prognosis of patients with OS has not improved in the past several decades secondary to the lack of treatments that specifically targets OS metastatic biology. This problem remains unsolved due to our limited understanding of the mechanisms that are critical for the progression of metastatic disease. There is an unmet need to develop better understanding of OS metastasis in order to optimize current treatment strategies and develop novel approaches for treatment.

E. S. Kleinerman, R. Gorlick (eds.), *Current Advances in the Science of Osteosarcoma*, Advances in Experimental Medicine and Biology 1258, https://doi.org/10.1007/978-3-030-43085-6_10

Aldehyde dehydrogenase 1A1 has been studied by our group and others for its role in OS metastasis. It is a tetrameric enzyme that oxidizes aldehydes to carboxylic acids in the human body and enables cells to resist oxidative stress. There are 19 isozymes in the human ALDH family, and each possesses overlapping but unique functions. For example, ALDH1A1 mediates retinoic acid signaling, whereas ALDH2 is key in oxidizing acetaldehydes and has a role in alcohol metabolism. Many other isozymes of ALDH are important in oxidizing reactive aldehydes derived from lipid peroxidation and, in turn, help maintain cellular homeostasis [10].

Increased expression and activity of ALDH isozymes have been reported in many human cancers and have been associated with metastatic potential, regenerative capacity, drug resistance, and poor prognosis [10–14]. Clinical studies show that high ALDH1A1 (hereafter abbreviated simply as ALDH) activity is a predictor of poor survival in breast and ovarian cancer [15, 16]. On a cellular level, ALDH has been implicated as a cancer stem cell marker due to its high level of activity in cancer stem cells [16–21]. Cells with high levels of ALDH activity demonstrate enhanced tumorigenicity and invasion capacity [22]. ALDH activity has been found to correlate with clinical OS metastasis, and its inhibition in vitro diminishes the metastatic potential of OS cells. These findings suggest that ALDH is important to OS metastatic biology and may be a therapeutic target specific to OS cells with highly metastatic potential [23, 24].

ALDH's importance in murine OS cells and experimental metastasis has been established in previous studies [22–26]. K7M2 murine OS cells are highly metastatic to the lungs compared with the much less metastatic K12 OS cells, which were both derived from the same parental tumor [27]. Highly metastatic K7M2 cells demonstrated greater ALDH gene expression and activity than the less metastatic K12 cells. When sorted with fluorescence-activated cell sorting (FACS)according to ALDH activity, K7M2 cells demonstrated greater ALDH activity compared with less-metastatic K12 cells as demonstrated in Fig. 10.1 [22]. As demonstrated in Fig. 10.2, highly metastatic K7M2 cells express significantly more ALDH when compared to K12 cells. K7M2 cells were found to have a much stronger invasion capacity compared with K12 cells when tested through a semisolid Matrigel three-dimensional matrix, as demonstrated in Fig. 10.3 [22, 24]. ALDH may therefore represent a therapeutic target specific to OS cells with high metastatic potential.

ALDH in Other Molecular Pathways

Notch

The Notch signaling pathway is associated with ALDH activity and increased metastatic behavior in OS cells. Both ALDH and Notch are putative molecular targets for the treatment and prevention of OS metastasis. In investigating Notch signaling in OS metastasis, it was found that Notch signaling is upregulated in the highly metastatic murine K7M2 cells. Expression of Notch pathway genes was investigated in K7M2 and K12 cells, which demonstrated that *Notch1, Hes1,* and *ALDH* gene expressions were all upregulated in K7M2 cells. Notch signaling was inhibited with a gamma-secretase inhibitor, DAPT, to evaluate its effect on ALDH expression and activity. Expressions of *ALDH*, ALDH activity, and Notch signaling were significantly diminished after treatment with the gamma-secretase inhibitor DAPT as demonstrated in Fig. 10.4, suggesting crosstalk between the ALDH and Notch pathways as a function of OS metastatic potential.

mTOR

The mammalian target of rapamycin (mTOR) pathway has also been implicated in promoting metastatic potential in OS cells. It may also affect ALDH expression and activity. Rapamycin is an antimicrobial agent produced by *Streptomyces hygroscopius* that also exhibits potent immunosuppressive and antitumor properties, likely due to its ability to arrest the cell cycle in G1 [28]. Rapamycin inhibits ALDH gene expression and enzymatic

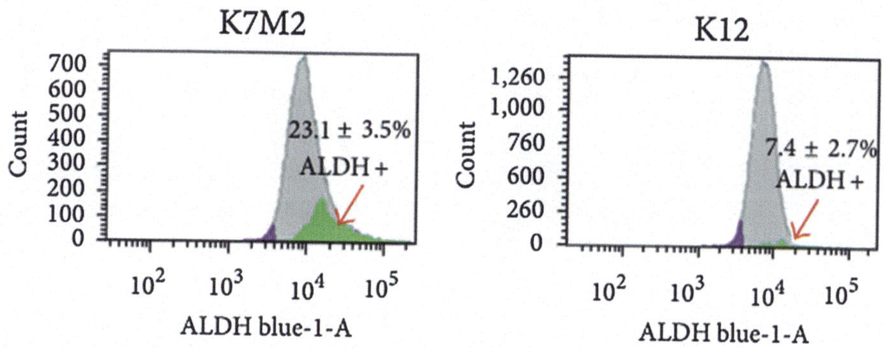

Fig. 10.1 ALDH activity was detected in K7M2 and K12 cells using FACS analysis and the relative amount of cells positive for ALDH is shown for each cell population. (Source: Mu et al. [24])

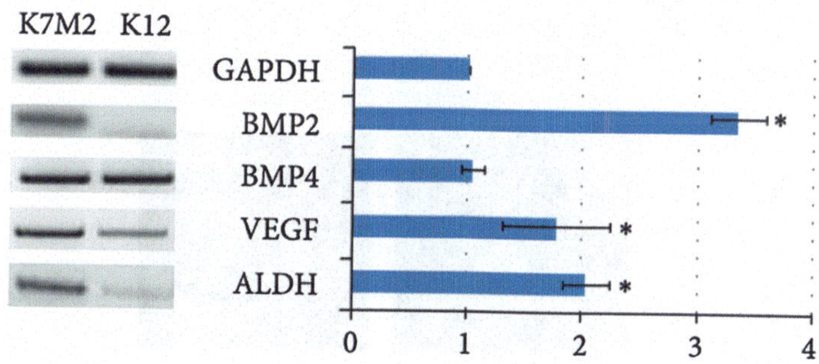

Fig. 10.2 RT-PCR was performed on K7M2 and K12 cells in order to quantitate the relative expression of BMP2, BMP4, VEGF, and ALDH-1A1. GAPDH serves as a loading control. It can be seen that the expression of ALDH-1A1 is significantly higher in the highly metastatic K7M2 cells as compared to the much less metastatic K12 cells. (Source: Mu et al. [24])

activity in K7M2 cells, reduces BMP2 and VEGF, and inhibits K7M2 proliferation, migration, and invasion in vitro. After treatment with rapamycin, the percentage of K7M2 cells with high ALDH activity diminished significantly compared with untreated cells and approached a level of activity more comparable to the less metastatic K12 cells. The population doubling time of both K7M2 and K12 cells was increased by rapamycin treatment with the effect being more profound on KM2 cells [24]. Previous studies have demonstrated that ALDH may function to neutralize oxidative stress and provide chemoresistance in cancer [29, 30]. By reducing ALDH expression and activity with rapamycin, K7M2 cells became more susceptible to apoptotic death when exposed to oxidative stress via hydrogen peroxide (Fig. 10.5) [24].

ALDH Inhibition with Disulfiram

Disulfiram is an ALDH inhibitor that has been used for many decades as a treatment for alcoholism [31]. It has also been shown to inhibit OS cell proliferation and metastasis in vitro. K7M2 cells treated with disulfiram, an irreversible ALDH inhibitor, demonstrated reduced ALDH activity and altered morphology [22, 24]. Disulfiram decreases the mTOR expression and activity of K7M2 cells. Inhibition of ALDH with disulfiram correlated with decreased mTOR expression and activity. This provides evidence for the interaction between ALDH, mTOR activity, and metastatic potential in murine OS cells [24, 25] (Fig. 10.6).

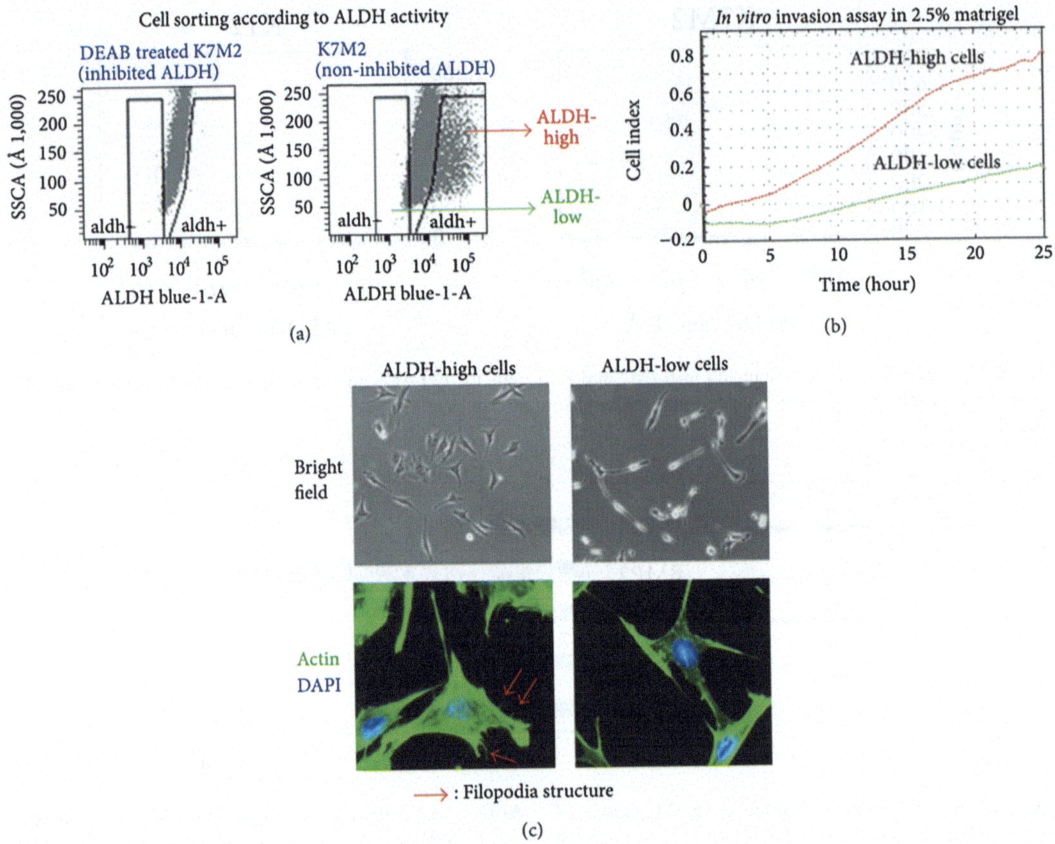

Fig. 10.3 Cell sorting of K7M2 cells via ALDH activity and differential features of ALDH-high cells and ALDH-low cells. (**a**) Highly metastatic K7M2 cells were suspended in Aldefluor buffer and sorted according to their enzymatic activity. Cells were treated with DEAB to block ALDH activity, and cells were deemed ALDH-high if their fluorescence was higher than that of the DEAB-treated controls. Cells were deemed ALDH-low if their fluorescence was lower than that of the DEAB-treated controls. (**b**) ALDH-high and ALDH-low K7M2 cell invasion was tracked over a period of 24 hours with live-cell imaging. The ALDH-high cells displayed much greater invasiveness with more cells invading through the Matrigel (2.5%). (**c**) Under bright field microscopy and under fluorescent microscopy (after staining for actin (green) and nuclei (blue) ALDH-high K7M2 cells displayed more filopodia than ALDH-low cells. Also the ALDH-high cells were more irregularly shaped and were more pleomorphic than ALDH-low K7M2 cells, which were more polygonal and reminiscent of less metastatic K12 cells. (Source: Mu et al. [24])

In Vivo Studies

The therapeutic efficacy of disulfiram has been evaluated in an immune-competent mouse model of metastatic OS. Therapeutic equivalence with doxorubicin in terms of the ability to reduce the burden of OS lung metastases was observed. Both disulfiram and doxorubicin were statistically superior to saline-treated animals. Interestingly, disulfiram and doxorubicin imparted different gene inductions within the primary tumor. As in the in vitro data mentioned above, disulfiram significantly reduced mTOR gene expression in experimental tumors compared with doxorubicin-treated animals. There were also decreases in *C-Myc* and *Akt* expressions with a concomitant increase in *Bad* expression. These data suggest that disulfiram and doxorubicin might be used in combination as they appear to work through different mechanisms [32].

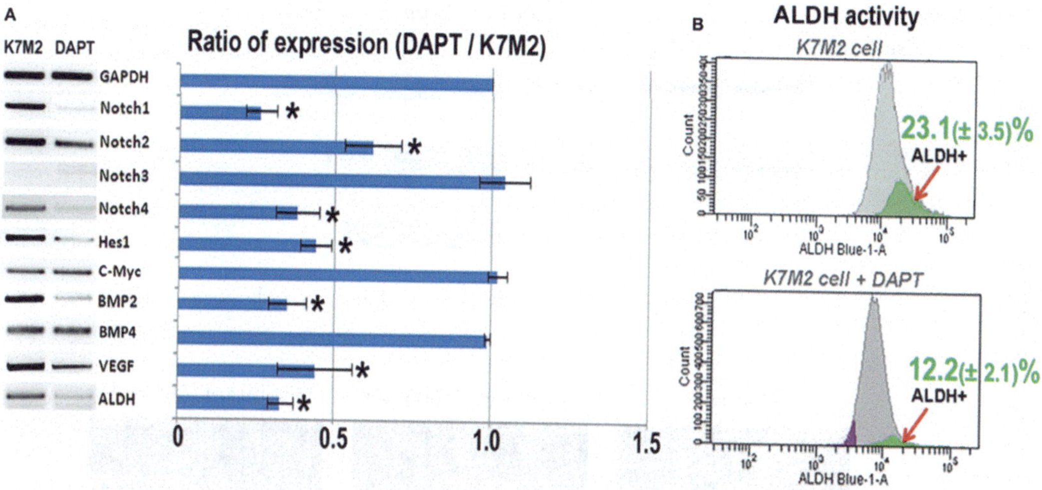

Fig. 10.4 Notch inhibition with DAPT reduces Notch signaling and *ALDH* expression. (**a**) RT-PCR was performed on cellular RNA extracted from K7M2 cells treated with DAPT or vehicle only (control) in order to quantitate the relative expression of Notch genes. In addition, ALDH expression was compared between DAPT-treated and DAPT-untreated K7M2 cells. GAPDH serves as a loading control for both lanes. Gene expression was normalized using GAPDH. (**b**) ALDH activity was detected in DAPT-treated and DAPT-untreated K7M2 cells using flow cytometry analysis, and the relative amount of cells positive for ALDH is shown for each cell population. * indicates the difference is significant comparing DAPT-treated with nontreated samples. (Source: Mu et al. [22])

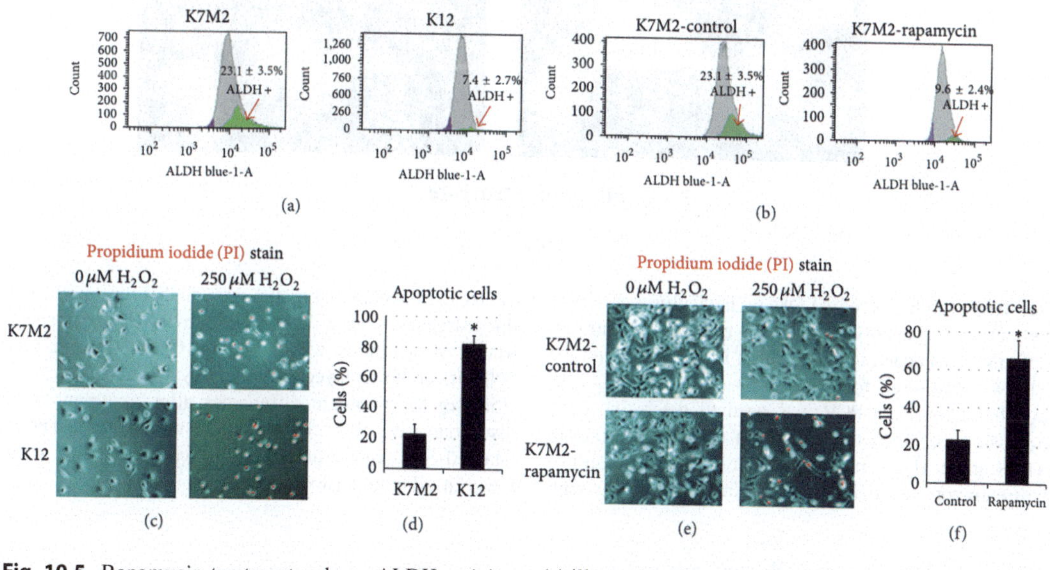

Fig. 10.5 Rapamycin treatment reduces ALDH activity and sensitizes K7M2 cells to oxidative stress. (**a**) ALDH activity was detected in K7M2 and K12 cells using FACS analysis and the relative amount of cells positive for ALDH is shown for each cell population. (**b**) K7M2 cells were treated with rapamycin or DMSO only (control) and analyzed by FACS as in (**a**). (**c**) K7M2 and K12 cells were treated with or without H_2O_2 (250 μM), and apoptosis was detected using PI staining. (**d**) A quantitative analysis of (**c**) illustrating the percentage of apoptotic cells after H_2O_2 treatment compared to untreated controls. (**e**) K7M2 cells were treated with or without H_2O_2, in the presence or absence (DMSO only control) of rapamycin, and apoptotic cells were detected as in (**c**). (**f**) A quantitative analysis of (**e**) illustrating the percentage of apoptotic cells after H2O2 compared to untreated controls. * indicates statistically significant differences ($P < 0.05$). (Source: Mu et al. [24])

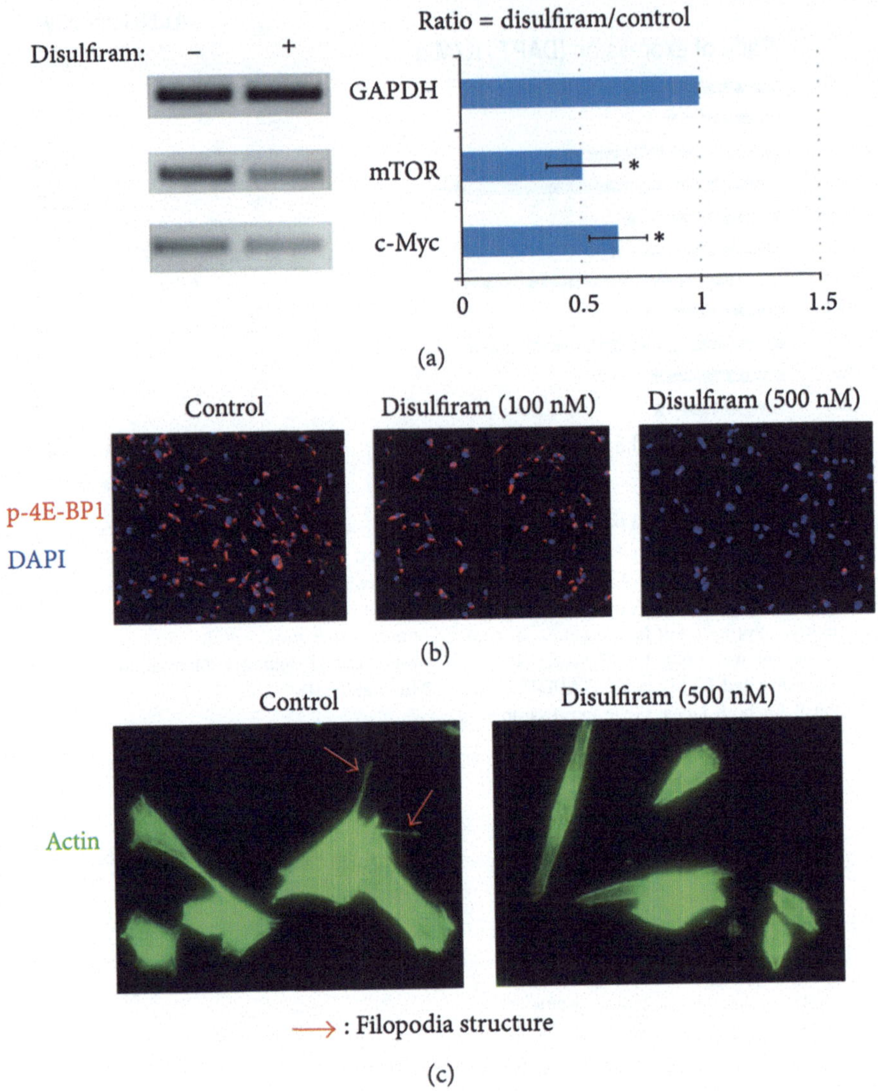

Fig. 10.6 Inhibition of ALDH with disulfiram inhibits the metastatic phenotype of K7M2 cells. (**a**) Disulfiram (250 μM) was added to ALDH-high K7M2 cells, and the cells were cultured for at least 24 hours in 10% FBS growth medium. RT-PCR was used to determine gene expression differences as a result of this treatment. Both mTOR and c-Myc expression were reduced as a result of the treatment with disulfiram. (**b**) Immunostaining with 4E-BP1 confirmed that treatment with disulfiram affected downstream targets in the mTOR pathway. As the concentration of disulfiram was increased (0 nm, 100 nm, and 500 nm), cells displayed progressively less mTOR expression. (**c**) Morphologic differences after treatment with disulfiram were also present, with disulfiram-treated cells (stained for actin) appearing less pleomorphic and with fewer invadopodia. (Source: Mu et al. [24])

Copper

Copper (Cu) is an essential micronutrient for physiologic redox reactions. It also plays an important role in oncogenic processes of invasion and metastasis [33, 34]. Many authors have observed that Cu potentiates the cytotoxic effects of disulfiram, a potent copper chelator, through incompletely understood mechanisms [24]. This combination is already being evaluated in several clinical trials. Based on these observations, the effect of Cu and disulfiram combination therapy was evaluated on OS cells in vitro. Commensurate with other observations, Cu was found to have a

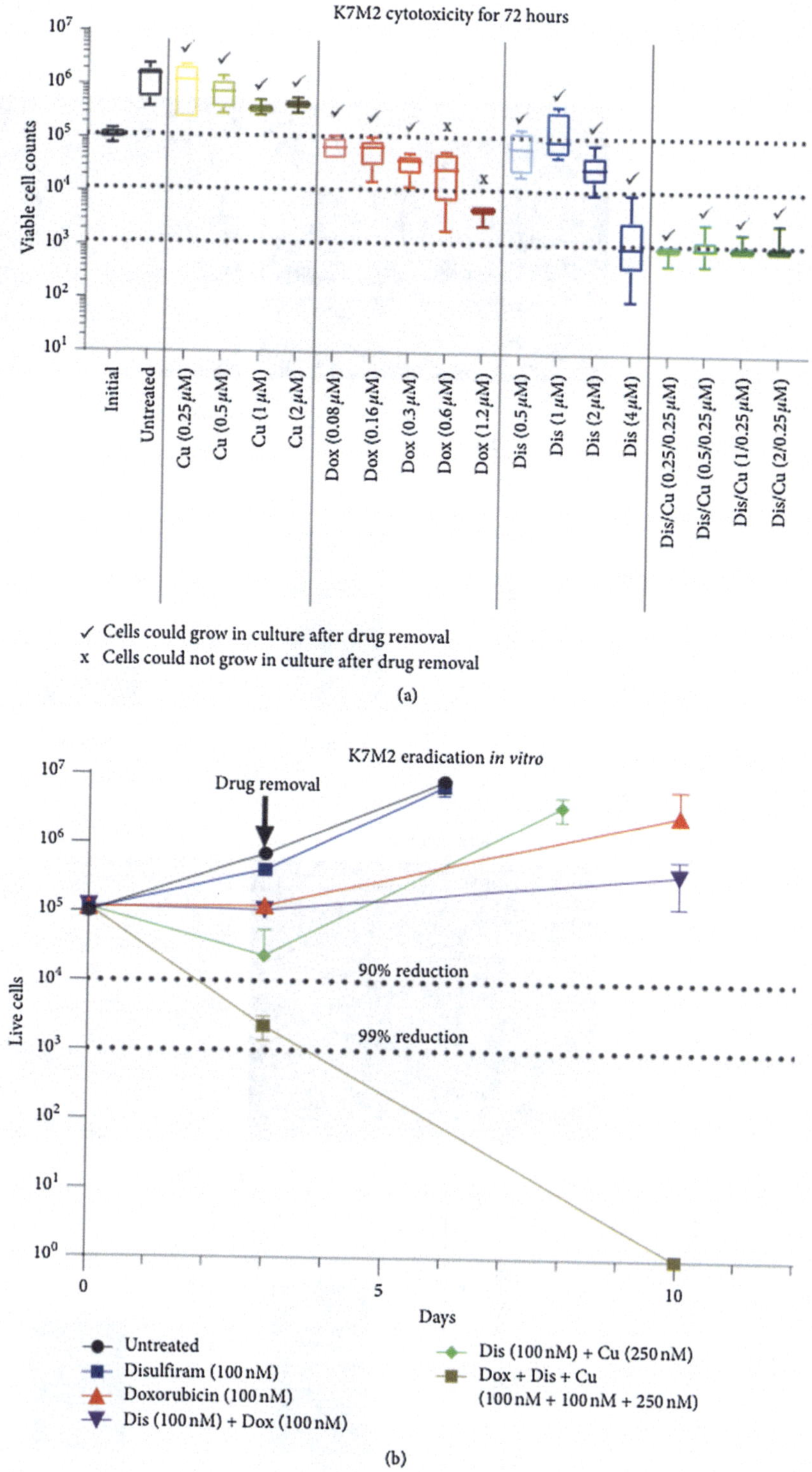

Fig. 10.7 Combination treatment of doxorubicin, disulfiram, and copper chloride effectively reduces viability of K7M2 cells and eliminates recovery in vitro. (**a**) K7M2 OS cells were treated with copper chloride (0.12–2 μM), disulfiram (0.5–4 μM), and disulfiram and copper chloride (0.12 and 0.25 μM) for 72 hours. After treatment, trypan blue exclusion staining was performed to obtain viable cell counts. Cells were cultured again in fresh media without drugs present and monitored for cellular growth. Disulfiram potentiated with copper chloride clearly killed most cells over 72 hours, but all could subsequently recover in culture. (**b**) Doxorubicin and disulfiram combination treatment also allowed for cellular recovery in vitro, but triple treatment with doxirubicin, disulfiram, and copper chloride resulted in over 90% killing over treatment duration and treated cells did not recover after drug removal. Experiments were performed using three independent cell cultures ($n = 3$). (Source: Mandell et al. [35])

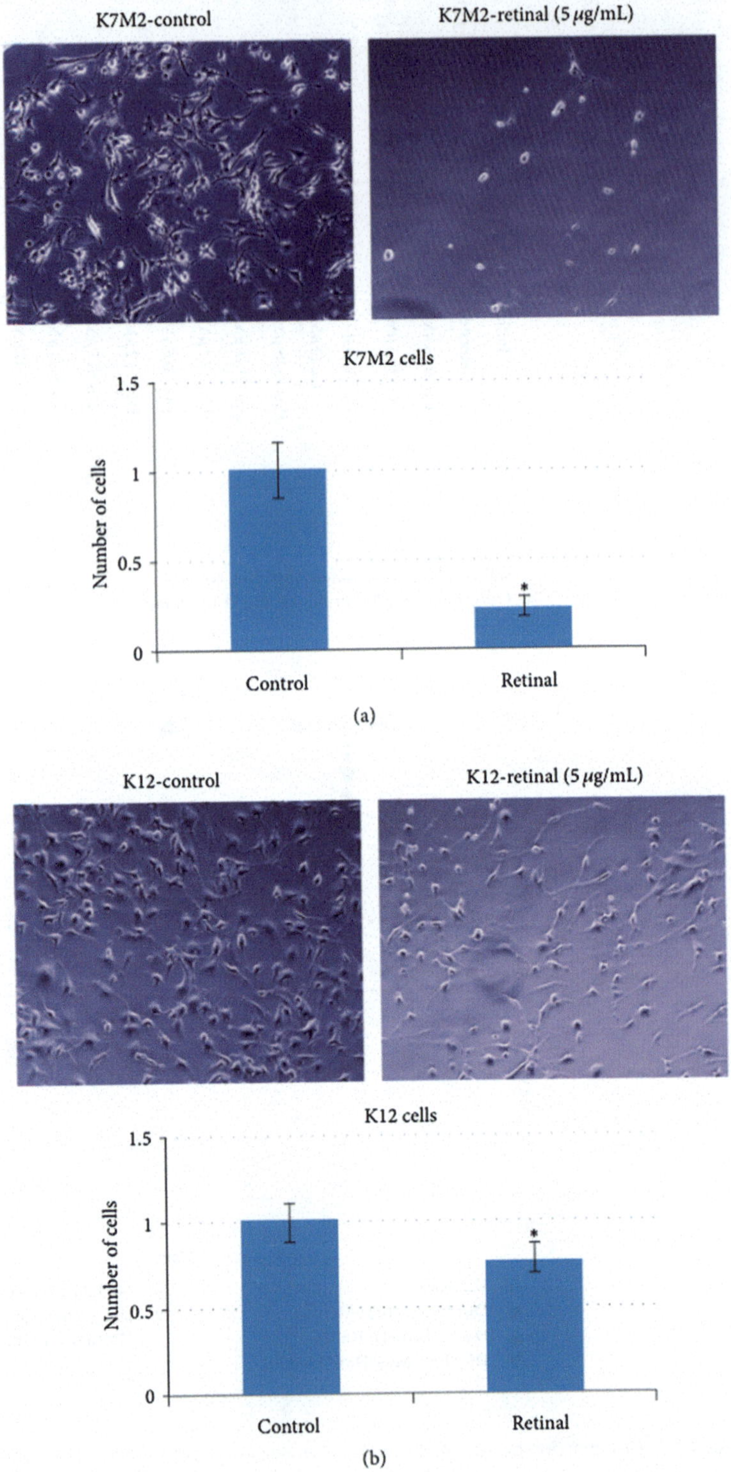

Fig. 10.8 Retinal treatment of K7M2 cells decreased cell proliferation and viability. (**a**, **b**) The cell proliferation and survival capacity of K7M2 cells was more dramatic with retinal treatment (5 μg/mL, 2 days), compared with K12 cells. (Source: Mu et al. [26])

dramatic effect on disulfiram cytotoxicity in K7M2 and K12 cells. The combination of disulfiram, Cu, and low-dose doxorubicin was also evaluated in vitro. This so-called triple treatment was the only experimental regimen from which K7M2 cells were fully eradicated and did not recover. These data suggest that combination therapy with these three agents should be evaluated in vivo as well [35] (Fig. 10.7).

Retinal

Retinal is another method by which ALDH can be targeted. Retinoic acid, a derivative of retinal, has antitumor properties including the induction of apoptosis as well as the inhibition of proliferation and differentiation of various cancer cells [36]. It has been shown to exert a tumor-suppressive effect on cells based on their interactions with cyclins and cyclin-dependent kinases (CDKs) to prevent cell cycle progression [37]. Many cancer cells have been shown to have abnormally low levels of various retinoids. Retinal, a precursor of retinoic acid, can be oxidized to retinoic acid by dehydrogenases, including ALDH [38]. Previous literature demonstrated that retinal targets ALDH-positive cancer stem cells and alters the phenotype of highly metastatic OS cells. Retinal preferentially affected the phenotypes of ALDH-high K7M2 cells in contrast with ALDH-low K12 cells, which could be mediated by the more efficient transformation of retinal to retinoic acid by ALDH in K7M2 cells. Retinal treatment of highly metastatic K7M2 cells decreased their proliferation, invasion capacity, and resistance to oxidative stress. Retinal also altered the expression of metastasis-related genes. These results indicate that retinal may be used to specifically target metastatic cancer stem cells [26] (Fig. 10.8).

Conclusions

These studies and examples support the concept that ALDH plays a key role in the metastatic biology of OS cells. ALDH inhibition thus represents a viable therapeutic strategy for the targeting of OS metastatic potential. OS is a disease that has not enjoyed any improvements in prognosis for several decades, and it is likely that current cytotoxic chemotherapy treatments have reached their zenith. The development of ALDH inhibition and other biologically intelligent treatments that specifically target OS metastases are essential to the improvement of OS prognosis.

References

1. Whelan J, McTiernan A, Cooper N, Wong YK, Francis M, Vernon S et al (2012) Incidence and survival of malignant bone sarcomas in England 1979-2007. Int J Cancer 131(4):E508–E517
2. Bielack SS, Kempf-Bielack B, Delling G, Exner GU, Flege S, Helmke K et al (2002) Prognostic factors in high-grade osteosarcoma of the extremities or trunk: an analysis of 1,702 patients treated on neoadjuvant cooperative osteosarcoma study group protocols. J Clin Oncol 20(3):776–790
3. Carrle D, Bielack SS (2006) Current strategies of chemotherapy in osteosarcoma. Int Orthop 30(6):445–451
4. Ferrari S, Smeland S, Mercuri M, Bertoni F, Longhi A, Ruggieri P et al (2005) Neoadjuvant chemotherapy with high-dose Ifosfamide, high-dose methotrexate, cisplatin, and doxorubicin for patients with localized osteosarcoma of the extremity: a joint study by the Italian and Scandinavian Sarcoma Groups. J Clin Oncol 23(34):8845–8852
5. Messerschmitt PJ, Garcia RM, Abdul-Karim FW, Greenfield EM, Getty PJ (2009) Osteosarcoma. J Am Acad Orthop Surg 17(8):515–527
6. Meyers PA, Schwartz CL, Krailo MD, Healey JH, Bernstein ML, Betcher D et al (2008) Osteosarcoma: the addition of muramyl tripeptide to chemotherapy improves overall survival--a report from the Children's Oncology Group. J Clin Oncol 26(4):633–638
7. Hegyi M, Semsei AF, Jakab Z, Antal I, Kiss J, Szendroi M et al (2011) Good prognosis of localized osteosarcoma in young patients treated with limb-salvage surgery and chemotherapy. Pediatr Blood Cancer 57(3):415–422
8. Meyers PA, Schwartz CL, Krailo M, Kleinerman ES, Betcher D, Bernstein ML et al (2005) Osteosarcoma: a randomized, prospective trial of the addition of ifosfamide and/or muramyl tripeptide to cisplatin, doxorubicin, and high-dose methotrexate. J Clin Oncol 23(9):2004–2011
9. Meyers PA, Gorlick R (1997) Osteosarcoma. Pediatr Clin N Am 44(4):973–989
10. Koppaka V, Thompson DC, Chen Y, Ellermann M, Nicolaou KC, Juvonen RO et al (2012) Aldehyde dehydrogenase inhibitors: a comprehensive review of the pharmacology, mechanism of action, substrate specificity, and clinical application. Pharmacol Rev 64(3):520–539

11. Charafe-Jauffret E, Ginestier C, Iovino F, Tarpin C, Diebel M, Esterni B et al (2010) Aldehyde dehydrogenase 1-positive cancer stem cells mediate metastasis and poor clinical outcome in inflammatory breast cancer. Clin Cancer Res 16(1):45–55
12. Cheung AMS, Wan TSK, Leung JCK, Chan LYY, Huang H, Kwong YL et al (2007) Aldehyde dehydrogenase activity in leukemic blasts defines a subgroup of acute myeloid leukemia with adverse prognosis and superior NOD/SCID engrafting potential. Leukemia 21(7):1423–1430
13. Cho H-J, Lee T-S, Park J-B, Park K-K, Choe J-Y, Sin D-I et al (2007) Disulfiram suppresses invasive ability of osteosarcoma cells via the inhibition of MMP-2 and MMP-9 expression. J Biochem Mol Biol 40(6):1069–1076
14. Honoki K, Fujii H, Kubo A, Kido A, Mori T, Tanaka Y et al (2010) Possible involvement of stem-like populations with elevated ALDH1 in sarcomas for chemotherapeutic drug resistance. Oncol Rep 24(2):501–505
15. Bortolomai I, Canevari S, Facetti I, De Cecco L, Castellano G, Zacchetti A et al (2010) Tumor initiating cells: development and critical characterization of a model derived from the A431 carcinoma cell line forming spheres in suspension. Cell Cycle 9(6):1194–1206
16. Ginestier C, Hur MH, Charafe-Jauffret E, Monville F, Dutcher J, Brown M et al (2007) ALDH1 is a marker of normal and malignant human mammary stem cells and a predictor of poor clinical outcome. Cell Stem Cell 1(5):555–567
17. Marcato P, Dean CA, Giacomantonio CA, Lee PWK (2011) Aldehyde dehydrogenase: its role as a cancer stem cell marker comes down to the specific isoform. Cell Cycle 10(9):1378–1384
18. Jiang F, Qiu Q, Khanna A, Todd NW, Deepak J, Xing L et al (2009) Aldehyde dehydrogenase 1 is a tumor stem cell-associated marker in lung cancer. Mol Cancer Res 7(3):330–338
19. Zhou F, Mu Y-D, Liang J, Liu Z-X, Zhou D, Ning W-L et al (2015) Aldehyde dehydrogenase 1: a specific cancer stem cell marker for human colorectal carcinoma. Mol Med Rep 11(5):3894–3899
20. Li H, Ma F, Wang H, Lin C, Fan Y, Zhang X et al (2013) Stem cell marker aldehyde dehydrogenase 1 (ALDH1)-expressing cells are enriched in triple-negative breast cancer. Int J Biol Markers 28(4):e357–e364
21. Huang EH, Hynes MJ, Zhang T, Ginestier C, Dontu G, Appelman H et al (2009) Aldehyde dehydrogenase 1 is a marker for normal and malignant human colonic stem cells (SC) and tracks SC overpopulation during colon tumorigenesis. Cancer Res 69(8):3382–3389
22. Mu X, Isaac C, Greco N, Huard J, Weiss K (2013) Notch signaling is associated with ALDH activity and an aggressive metastatic phenotype in murine osteosarcoma cells. Front Oncol 3:143
23. Greco N, Schott T, Mu X, Rothenberg A, Voigt C, McGough RL et al (2014) ALDH activity correlates with metastatic potential in primary sarcomas of bone. J Cancer Ther 5(4):331–338
24. Mu X, Isaac C, Schott T, Huard J, Weiss K (2013) Rapamycin inhibits ALDH activity, resistance to oxidative stress, and metastatic potential in murine osteosarcoma cells. Sarcoma 2013:480713
25. Yu S, Fourman MS, Mahjoub A, Mandell JB, Crasto JA, Greco NG et al (2017) Lung cells support osteosarcoma cell migration and survival. BMC Cancer 17(1):78
26. Mu X, Patel S, Mektepbayeva D, Mahjoub A, Huard J, Weiss K (2015) Retinal targets ALDH positive cancer stem cell and alters the phenotype of highly metastatic osteosarcoma cells. Sarcoma 2015:784954
27. Khanna C, Prehn J, Yeung C, Caylor J, Tsokos M, Helman L (2000) An orthotopic model of murine osteosarcoma with clonally related variants differing in pulmonary metastatic potential. Clin Exp Metastasis 18(3):261–271
28. Zhou H, Huang S (2010) mTOR signaling in cancer cell motility and tumor metastasis. Crit Rev Eukaryot Gene Expr 20(1):1–16
29. Magni M, Shammah S, Schiró R, Mellado W, Dalla-Favera R, Gianni AM (1996) Induction of cyclophosphamide-resistance by aldehyde-dehydrogenase gene transfer. Blood 87(3):1097–1103
30. Russo JE, Hilton J (1988) Characterization of cytosolic aldehyde dehydrogenase from cyclophosphamide resistant L1210 cells. Cancer Res 48(11):2963–2968
31. Wright C, Moore RD (1990) Disulfiram treatment of alcoholism. Am J Med 88(6):647–655
32. Crasto JA, Fourman MS, Morales-Restrepo A, Mahjoub A, Mandell JB, Ramnath K et al (2018) Disulfiram reduces metastatic osteosarcoma tumor burden in an immunocompetent Balb/c or-thotopic mouse model. Oncotarget 9(53):30163–30172
33. Denoyer D, Masaldan S, La Fontaine S, Cater MA (2015) Targeting copper in cancer therapy: "copper that cancer". Metallomics 7(11):1459–1476
34. Fouani L, Menezes SV, Paulson M, Richardson DR, Kovacevic Z (2017) Metals and metastasis: exploiting the role of metals in cancer metastasis to develop novel anti-metastatic agents. Pharmacol Res 115:275–287
35. Mandell JB, Lu F, Fisch M, Beumer JH, Guo J, Watters RJ et al (2019) Combination therapy with disulfiram, copper, and doxorubicin for osteosarcoma: in vitro support for a novel drug repurposing strategy. Sarcoma 2019:1320201
36. Noy N (2010) Between death and survival: retinoic acid in regulation of apoptosis. Annu Rev Nutr 30:201–217
37. Tang X-H, Gudas LJ (2011) Retinoids, retinoic acid receptors, and cancer. Annu Rev Pathol 6:345–364
38. Dräger UC, Wagner E, McCaffery P (1998) Aldehyde dehydrogenases in the generation of retinoic acid in the developing vertebrate: a central role of the eye. J Nutr 128(2 Suppl):463S–466S

Autophagy in Osteosarcoma

11

Grace Nehme and Nancy Gordon

Abstract

Osteosarcoma (OS) remains a difficult disease to treat. The standard chemotherapy regimen has not improved survival for the past three decades. Resistance to chemotherapy remains a challenge and constitutes a major concern to clinical investigators. Autophagy has been recognized as a survival mechanism implicated in resistance to chemotherapy. We previously demonstrated chemotherapy to induce autophagy in OS. However, whether induction of autophagy will lead to survival or death has been the focus of many laboratories. Autophagy is a very context-dependent process, and no specific biomarker has been identified to define whether the process will lead to survival or death. In the present chapter, we present some of the mechanisms involved in the process of autophagy and summarize some of the most recent work related to autophagy in OS and the challenges encountered with the use of old and new autophagy inhibitors.

Keywords

Autophagy · Osteosarcoma · Chemotherapy · Survival · Death

G. Nehme · N. Gordon (✉)
Department of Pediatric Research, MD Anderson Cancer Center, Houston, TX, USA
e-mail: ngordon@mdanderson.org

Introduction

The term autophagy derives from the Greek meaning "eating of self." It is a catabolic process by which cells self-degrade their own constituents to maintain homeostasis and allow regular turnover of cell components [1]. In mammals, three types of autophagy have been described: macroautophagy, microautophagy, and chaperone-mediated autophagy. Macroautophagy involves bulk degradation of cytosol and organelles, microautophagy engulfs only parts of the cytosol or organelles, and chaperone-mediated autophagy involves the degradation of specific cytosolic proteins [2]. In this chapter, we focus on macroautophagy (hereafter referred as autophagy), the most studied autophagy type.

Under stressful conditions such as hypoxia, starvation, and cytotoxicity, autophagy allows the recycling of cellular components to be used as a source of energy. Autophagy is implicated in various different biological functions. Not only it plays a role in cell survival but it is also implicated in metabolism and development.

Autophagy has been shown to play an important role in many diseases such as neurodegenerative diseases where defects in autophagy can result in neurodegeneration [3]. It is also associated with aging [4] and the development of autoimmune diseases [5, 6], metabolic disorders [7], and cancer. Deregulation of autophagy has been described in many cancers such as glioblastoma,

E. S. Kleinerman, R. Gorlick (eds.), *Current Advances in the Science of Osteosarcoma*, Advances in Experimental Medicine and Biology 1258, https://doi.org/10.1007/978-3-030-43085-6_11

melanoma, lymphoma, and other solid tumors [1, 8]. In cancer, autophagy plays a role at different levels of cancer progression [1]and it is not associated with a specific trigger.

Autophagy could promote cell survival by protecting malignant cells from unfavorable conditions but could also serve as a tumor suppressor by impairing malignant transformation and promoting malignant cell death through programmed cell death (PCD) type II [1, 9]. This dual role of autophagy has been demonstrated in many cancers including osteosarcoma (OS). Therefore, targeting autophagy has been the focus of many studies [3, 10–12].

The process of autophagy involves more than 30 autophagy-related genes (Atg) and includes several steps. As shown in Fig. 11.1, the autophagy process starts when a stressful signal (1) activates the Atg1 complex, comprised of Atg1, Atg13, Atg17, Atg29, and Atg31, which leads to the formation of a flat membrane cistern, the phagophore, via activation of the vesicle trafficking complex formed by vesicle-mediated vacuolar protein sorting 34 (Vps34), a phosphatidyl inositol 3 kinase (PI3K), and one of the first characterized autophagy proteins, Beclin1. Interaction of these complexes and other factors help to recluse proteins and lipids necessary for the autophagosome formation (2). Completion of the autophagosome formation happens during elongation, the next step in the autophagy process (3). This step is regulated by two ubiquitin-like systems: the first system involves the formation of the Atg12, Atg5, and Atg16 complex, which is mediated by the E1-like enzyme Atg7. The second system regulates the conjugation of the microtubule-associated protein 1 light chain 3 (LC3-I/Atg8) with phosphatidylethanolamine (PE). LC3 is first synthesized as an unprocessed form, proLC3, and subsequently converted to a proteolytically processed form, LC3-I. LC3-I is cleaved by the protease Atg4, modified into the PE-conjugated form, LC3-II, and translocated from the cytoplasm to the autophagosome membrane. LC3 is the only known marker of the autophagosome (4). It also acts in cargo recognition by directly interacting with sequestosome 1 (SQSTM1/p62) via a complex formed between the cargo and SQSTM1 also bound to the autophagosome membrane. At this stage, the lysosome fuses to the autophagosome, forming the autolysosome (5). As a final step, proteins are degraded in the autolysosome and amino acids are released into the cytoplasm. These final products can be used for protein synthesis or can be oxidized by the mitochondria electron transport chain to produce adenosine triphosphate (ATP) to use as source of energy for cell survival. All

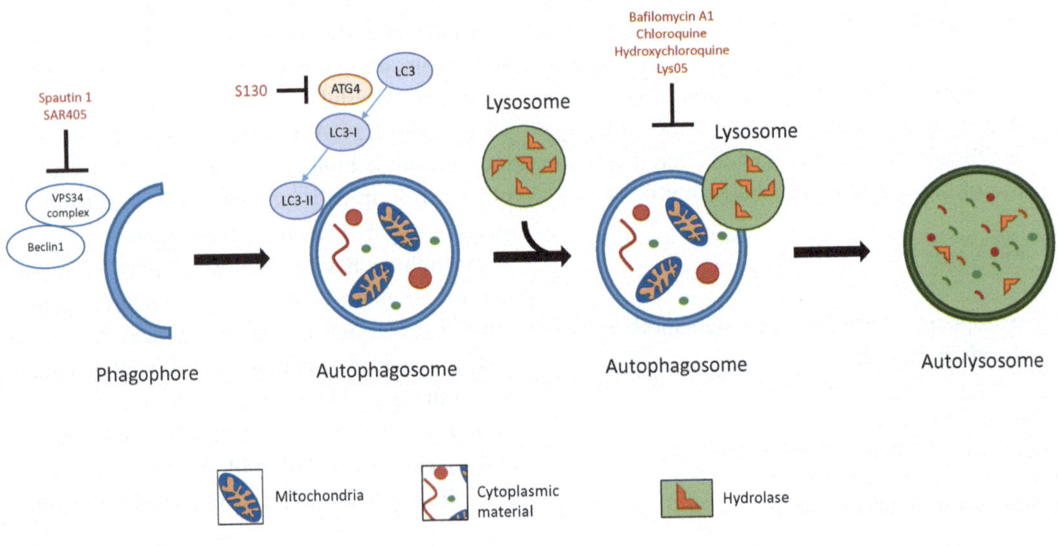

Fig. 11.1 The process of autophagy

proteins involved in the phagophore and autophagosome formation are released into the cytosol for reuse [13].

Regulation of Autophagy in Osteosarcoma

Autophagy is regulated through different mechanisms. The most studied mechanism involves the phosphoinositide 3 kinase (PI3K)/Akt/mammalian target of rapamycin (mTOR) pathway. In fact, the nutrient sensor PI3K is upstream of the mammalian target of rapamycin (mTOR) kinase, which negatively regulates autophagy. During normal nutrient conditions, the PI3K/AKT/mTOR pathway is activated, leading to inhibition of autophagy [14, 15]. However, during periods of nutrient deprivation, the PI3K/AKT/mTOR pathway is inhibited leading to autophagy induction [16].

Another important mechanism that regulates autophagy and tumorigenesis involves Beclin-1. Beclin-1 is part of a multiprotein complex formed by Vps34/class III PI3K. This complex initiates the formation of the phagophore. The interaction of Beclin-1 with Vps34 is modulated by anti-apoptotic molecules such as Bcl-2 and Bcl-xL. Under normal nutrient conditions, Beclin-1 is bound to Bcl-2 or Bcl-xL inhibiting autophagy. During starvation or stressful conditions, Beclin-1 is disrupted from Bcl-2/Bcl-xL through phosphorylation of the binding domain. The Beclin-1 complex can also be disrupted by other mechanisms that involve binding of the complex to the DAMP molecule or high-mobility group box 1 (HMGB1). The end result is induction of autophagy [17, 18].

In OS, these and other mechanisms are involved in autophagy regulation. Activation of the PI3K/AKT/mTOR signaling pathway has been demonstrated to inhibit autophagy in OS. The use of rapamycin, an mTOR inhibitor, induced autophagy and increased cell death in MG63 human osteosarcoma cells. Combination therapy rapamycin and cisplatin further enhanced cisplatin-induced cytotoxicity and stimulated autophagy [19]. Using a similar approach, arsenic trioxide in combination with radiation therapy induced autophagy and increased cytotoxicity in the HOS human OS cells through a mechanism that involves inhibition of the PI3K/AKT/mTOR signaling pathway [20]. Tumor-suppressing STF cDNA 3 (TSSC3) inhibition of the Src-mediated PI3K/AKT/mTOR signaling pathway induces autophagy and increases cytotoxicity of mineralized tissue-forming (MTF) osteoblasts and SaOS2 human OS cells [21]. Similarly, treatment of LM7, CCH-OS-D, and K7M3 metastatic OS cell lines with gemcitabine induces autophagy through a decrease in AKT and mTOR phosphorylation [12]. Furthermore, induction of Beclin-1 has also been shown to induce autophagy in OS. Panobinostat, a histone deacetylase inhibitor, suppresses Bcl-2 in SaOS2, U2-OS, and MG63 human OS cells and increases Beclin-1 expression leading to induction of autophagy and increased cytotoxicity [22]. Targeting MiR-100 inhibited mTOR, increased Beclin-1, and induced both autophagy and apoptosis in OS [23]. HMGB1-mediated autophagy induction leads to chemotherapy resistance in MG63, U2-OS, and SaOS2 human OS cells. Inhibition of both HMGB1 and autophagy led to increased drug sensitivity [24, 25]. A more recent study linked COP9 signalosome subunit 3 (COPS3), a protein-coding gene, to autophagy regulation and metastasis formation in OS [26].

Further, epigenetic alterations have been shown to play an important role in regulating the process of autophagy [13, 27–29]. Epigenetics involves the various mechanisms that allow for certain genes to be turned on and off under specific circumstances. Stable alterations in gene expression are essential for the development and differentiation of cells. Any abnormality in the regulatory process could lead to tumorigenesis. Several epigenetic mechanisms have been described that modulate gene expression such as DNA methylation, histone modifications, and nucleosome remodeling [29]. These mechanisms play important roles in gene transcription and regulation of gene expression. Several transcription factors that influence the process of autophagy have been identified. P53 and forkhead box O3 (FOXO3) were the first two transcription factors shown to induce autophagy [27]. Transcription factor EB (TFEB) is considered a key transcriptional regulator of autophagy as it activates the whole autophagy-lysosome pathway [30]. Under normal nutrient conditions, Zink Finger With KRAB and SCAN Domain 3 (ZKSCAN3) and Fork head transcription factor long isoform (FOXK) act as transcriptional repressors by inhibiting autophagy gene expression. The previous deleted reference should go as a number reference. Further, certain histone modifications can alter autophagy regulation [31] by having a direct effect on certain autophagy genes or by interacting with intermediates of the signal transduction pathway for autophagy. H4K16 acetylation and H3K9 dimethylation regulate core autophagy genes, whereas H3K27 trimethylation activates mTORC1 signaling leading to autophagy inhibition [27]. Bromodomain protein 4 (BRD4), a histone reader, links histone modifications to autophagy gene expression. BRD4 functions to inhibit autophagic activity under nutrient repletion status and knocking down BRD4 sustains autophagy during starvation status [27]. In pancreatic ductal adenocarcinoma, where autophagy has been described as a major resistance mechanisms to standard therapy, BRD4 was shown to be increased after gemcitabine treatment and contributed to drug resistance. Silencing BRD4 impaired cell viability and proliferation [32]. There is so far very limited knowledge on how epigenetic modifications can regulate autophagy in OS. Previous studies demonstrated that histone deacetylase inhibitors (HDACI) such as Trichostatin A inhibits the mTOR signaling pathway, enhances FOXO1 transcriptional activity, induces autophagy, and decreases cell death in human U2OS OS cells. Further inhibition of autophagy caused a marked enhancement of Trichostatin A-induced cell death in U2OS cells, suggesting potential efficacy of this combination for the treatment of OS [33].

Lastly, noncoding RNAs such as the small nucleolar RNA Host Gene 6 (SNHG6) can act as an oncogene in OS and induce autophagy through the regulation of Unc51-like autophagy activating kinase 1 (ULK1), a member of the preinitiation autophagy complex. Induction of autophagy through this mechanism decreases OS cell viability. Further silencing of the noncoding RNA SNHG6 inhibits OS cell growth and invasion [34].

In summary, various mechanisms are involved in the regulation of autophagy. None of them are specific to OS or any other disease process. Autophagy is a very context-dependent process, and its outcome might potentially be determined by the status and regulatory mechanisms triggered at the time the autophagy process is induced.

Autophagy and Tumorigenesis

Cell Survival Versus Cell Death

Autophagy exerts a dual role in tumorigenesis. It can either promote cell survival or cell death [35–37].

Autophagic cell death or programmed cell death (PCD) type II is described as a cell death mechanism that occurs in the presence of lysosomes. It differs from apoptosis (PCD type I) and necrosis (PCD type III) in that it lacks the chromatin condensation seen in apoptosis and swelling of the organelles seen in necrosis [38]. Autophagic cell death is caspase independent and can occur in the absence of proapoptotic proteins such as Bcl-2-associated X (Bax) and Bcl-2 homologous antagonist killer (Bak). In addition, during autophagic cell death, there is an increase in C-Jun N-terminal kinase (JNK), an essential cell death signaling molecule. However, insufficient JNK causes uncontrolled cell growth [39]. Certain chemotherapeutic agents can induce autophagy and lead to autophagic cell death. An example is obatoclax, a Bcl2 inhibitor, in acute lymphoblastic leukemia and Quinacrine in ovarian cancer [40, 41].

Alternatively, inability of cells to undergo autophagic cell death has been associated with tumorigenesis [42]. To this end, autophagy induction in cancer cells can also support tumor growth through various different mechanisms. It can induce cell survival during nutrient and oxygen shortage, promote chemotherapy resistance, and prevent apoptosis [43]. For example, in pancreatic cancer, under specific conditions, inhibition of autophagy causes tumor regression suggesting a potential contribution of autophagy in pancreatic tumor growth [44]. Indeed, induction of autophagy in pancreatic stellate cells within the tumor microenvironment was found to promote tumor growth [45]. Similarly, the role of autophagy in tumor growth has also been attributed to the tumor host autophagy status. In the face of an autophagy-competent host, autophagy leads to tumor growth. This is highlighted in a recent paper by Katheder et al. where dormant tumor cells from autophagy-deficient *Drosophila* reactivated tumor growth when implanted in an autophagy-competent host, emphasizing the potential role of host autophagy in tumorigenesis [46]. This duality has been described in various tumors including OS.

Dual Role of Autophagy in Osteosarcoma

As previously stated, autophagy has been described as a mechanism that is context dependent. Previous studies developed in our laboratory demonstrated autophagy to have a dual role in OS. Different OS cell lines and treatments were used. In the murine OS cell lines K7M3 and DLM8, we demonstrated that treatment of these cells with camptothecin(CPT) induced autophagy. However, inhibition of autophagy led to decrease CPT-induced cell death in DLM8 and increase in CPT-induced cell death in K7M3 OS cells [47]. Treatment of the two human OS cell lines, LM7 and CCH-OS-D, with the nucleoside analog, gemcitabine(GCB), also led to induction of autophagy. However, inhibition of autophagy in the LM7 cells caused increased cell death, whereas inhibition of autophagy in the CCH-OS-D cells led to an increase in cell survival confirming the dual effect of chemotherapy-induced autophagy in OS [12]. This duality is not species specific as the effect was seen in mouse (K7M3 and DLM8) and human (LM7 and CCH-OS-D) cells. It is not specific to any particular chemotherapy agent as different chemotherapeutic agents (CPT,GCB) with different mechanism of actions led to the same dual effect. There is still very limited understanding of the underlying mechanisms that define these responses. Many factors and pathways have been described as responsible for either increase in cell survival or death. However, this effect has so far been attributed to the specific context where autophagy takes place. Santiago O'Farril et al. are the first ones to describe the potential for a small heat shock protein to define autophagy outcome in OS. We describe this effect in the next section of the chapter.

Heat Shock Proteins and Autophagy

Heat shock proteins (HSPs) are a class of functionally related proteins whose expression is increased when cells are exposed to elevated temperatures and other types of stress. HSPs protect cells from stress-associated injury, are overexpressed in many malignancies, and are implicated in tumor cell proliferation, differentiation, invasion, and metastases. Santiago O'Farril et al. identified phosphorylated Hsp27(pHSP27) as a potential biomarker to determine whether autophagy induction will lead to survival or death in OS. Induction of pHSP27 following drug exposure with GCB correlated with the role of autophagy in drug sensitivity. Blocking autophagy in OS cells whose pHsp27 was increased following drug exposure with GCB resulted in enhanced drug sensitivity. However, blocking autophagy in OS cells where pHsp27 was decreased resulted in reduced cell sensitivity. These findings are the first to identify the potential of this heat shock protein to act as a biomarker to define the specific conditions where inhibition of autophagy will provide benefit [12]. Additionally, further studies demonstrated that positive expression of HSP27 and negative expression of LC3B in OS correlated with the worst 10-year overall survival, whereas negative HSP27 expression and positive LC3B expression had the best 10-year overall survival which suggested HSP27 as a negative prognostic marker in OS [48]. Other HSPs have also been described to play a role in autophagy induction. HSP90AA1 which belongs to the HSP90 family of HSP is upregulated in OS. It promotes autophagy and inhibits apoptosis leading to chemotherapy resistance [10]. The specific link between autophagy and heat shock proteins in OS is yet to be identified. However, these findings warrant future investigations on the potential role of HSPs in the modulation of autophagy in OS.

Autophagy Inhibition: From Drug Development to Challenges into Clinical Translation

Autophagy is a universal process present in every cell. Under physiologic conditions, autophagy is required to maintain tissue homeostasis. However, it can also contribute to the development and progression of certain diseases such as cancer. The development of autophagy inhibitors has become a challenge. Several drugs targeting autophagy have been described in the literature. Some compounds target the initial steps of the autophagy process, whereas others target autophagy at a later stage altering lysosomal functions [49]. Table 11.1 describes the different drugs that serve as autophagy inhibitors.

Early-stage inhibitors include pan-PI3K inhibitors such as 3-methyladenine (3-MA), which was first described in 1982 as a drug that acts to inhibit autophagy [50]. It was not until later when 3-MA was found to target both, class I PI3K and Vps34. 3-MA is nonspecific and poorly soluble which limits its potency [49]. More novel pan-PI3K inhibitors have been developed, but to date, none of those compounds have been shown to potently inhibit autophagy [49]. Another family of early stage inhibitors targets the Vps34 complex, a key structure in the autophagy process. Spautin-1 promotes the degradation of Vps34 complexes and causes cancer cell death under nutrient-deprived conditions. A preclinical study has shown synergistic effect of spautin-1 in combination with imatinib in the treatment of chronic myeloid leukemia [51]. SAR405, a pyrimidinone compound, was recently identified as a potent and selective catalytic inhibitor of Vps34, and it was shown to trigger an antiproliferative effect in renal cell carcinoma when combined with everolimus, an mTOR inhibitor, [49].

Late-stage autophagy inhibitors block the degradation of the autophagosome contents by the lysosomes. Bafilomycin A1 is a vacuolar-type H^+ ATPase inhibitor which blocks lysosomal proton transport thus inhibiting autophagic flux [49]. Inhibition of autophagy using bafilomycin A1 helped overcome chemotherapy resistance in gastric cancer cells [52].

There are also the so-called lysosomotropic agents used for the treatment of malaria. These include chloroquine (CQ) and hydroxychloroquine (HCQ). These agents disrupt lysosomal acidification and inhibit autolysosome formation. The major side effect of CQ is retinal toxicity. The addition of a hydroxyl group in HCQ ameliorates this effect by decreasing its ability to cross the blood-retinal barrier. CQ and HCQ are well tolerated. Efficacy of these agents in various

Table 11.1 Autophagy inhibitors

Compound	Target	Characteristics
3-Methyladenine	pan-PI3K inhibitors	Nonspecific, poorly soluble
Spautin-1	Vps34 inhibitors	Degrades VPS34 complexes and causes cancer cell death
SAR405	Vps34 inhibitors	Potent and selective catalytic inhibitor of Vps34
Bafilomycin A1	Blocks degradation of autophagosome contents	Inhibits autophagy flux to overcome chemotherapy resistance
Chloroquine	Inhibits autolysosome formation	Major side effect: retinal toxicity
Hydroxychloroquine	Inhibits autolysosome formation	Less retinal toxicity than CQ
Lys05	Inhibits autolysosome formation	More potent than CQ and HCQ
S130	ATG4 inhibitor	Potent inhibitor, causes cancer cells death in vitro

preclinical studies warranted their use in clinical trials. There are currently 31 active clinical trials using HCQ in combination with other drugs for the treatment of various malignancies. Temozolamide in combination with HCQ for the treatment of solid tumors and melanomas was tested in a Phase I clinical trial and demonstrated to be well tolerated with no associated toxicities [53]. However, an additional phase I/II trial that tested the same combination but with the addition of radiotherapy was used in patients with glioblastoma multiforme. The results demonstrated no improvement in survival at the chosen dose and severe myelosuppression at higher doses [54]. In vitro preclinical studies demonstrated effectiveness of the combination therapy HCQ + GCB in OS. A more recent Phase I study to explore the safety and tolerability of HCQ in combination with GCB and docetaxel (NCT03598595) in patients with recurrent or refractory OS was initiated and is ongoing.

Uncertainties remain with the use of chloroquine derivatives. A recent meta-analysis combined data from seven clinical trials using CQ and HCQ in combination with chemotherapy or radiation therapy and demonstrated that autophagy inhibitor-based therapy had a better treatment response than chemotherapy or radiation alone [55]. However, whether CQ/HCQ effectively inhibits autophagy in human tumors remains controversial. Potency at the tolerable doses remains suboptimal. Other derivatives are under development. Lys05, a bivalent analog of HCQ, has a tenfold greater potency than HCQ and demonstrated a better antitumor activity in preclinical models of glioblastoma, colon cancer, and melanoma [56, 57].

Additional autophagy inhibitors are under development. S130 targets the inhibition of ATG4. S130 tested in vitro demonstrated arrested growth of colorectal cancer cells and induced cell death [58].

In summary, here we describe the various autophagy inhibitors available and address their mechanism of action. Identification of an effective autophagy inhibitor remains a challenge. Further studies are needed to elucidate the best and more suitable autophagy inhibitor to use and in addition identify specific biomarkers of response.

Summary

The role of autophagy in OS remains unclear. Here, we describe autophagy as a mechanism that can either lead to survival or death in OS. We also point to some of the mechanisms implicated in the regulation of autophagy as it relates to OS. No one mechanism defines the outcome of autophagy in this disease. Furthermore, there isn't a well-identified biomarker to define the autophagy fate in OS whether it is induced by chemotherapy or other kinds of stress. We describe the potential for HSP27 to determine whether induction of autophagy will lead to survival or death, summarized the different autophagy inhibitors available, and point to the remaining challenges on the selection of one specific inhibitor. Better understanding of the mechanisms involved in the induction of autophagy in OS is necessary to define its role and select the most appropriate and effective agent to specifically target the process.

References

1. Aredia F, Ortiz LMG, Giansanti V, Scovassi AI (2012) Autophagy and cancer. Cell 1(3):520–534
2. Todde V, Veenhuis M, van der Klei IJ (2009) Autophagy: principles and significance in health and disease. Biochim Biophys Acta 1792(1):3–13
3. Towers CG, Thorburn A (2016) Therapeutic targeting of autophagy. EBioMedicine 14:15–23
4. Lapierre LR et al (2015) Transcriptional and epigenetic regulation of autophagy in aging. Autophagy 11(6):867–880
5. Muller S (2017) [Autophagy, autoimmunity and autoimmune diseases]. Med Sci (Paris) 33(3):319–327
6. Yin H et al (2018) The therapeutic and pathogenic role of autophagy in autoimmune diseases. Front Immunol 9:1512
7. Wang F, Jia J, Rodrigues B (2017) Autophagy, metabolic disease, and pathogenesis of heart dysfunction. Can J Cardiol 33(7):850–859
8. Levy JMM, Towers CG, Thorburn A (2017) Targeting autophagy in cancer. Nat Rev Cancer 17(9):528–542
9. Sun Y, Peng ZL (2009) Programmed cell death and cancer. Postgrad Med J 85(1001):134–140

10. Xiao X et al (2018) HSP90AA1-mediated autophagy promotes drug resistance in osteosarcoma. J Exp Clin Cancer Res 37(1):201
11. Chen Y et al (2018) Pelargonidin induces antitumor effects in human osteosarcoma cells via autophagy induction, loss of mitochondrial membrane potential, G2/M cell cycle arrest and downregulation of PI3K/AKT signalling pathway. J BUON 23(3):735–740
12. Santiago-O'Farrill JM et al (2018) Phosphorylated heat shock protein 27 as a potential biomarker to predict the role of chemotherapy-induced autophagy in osteosarcoma response to therapy. Oncotarget 9(2):1602–1616
13. Fullgrabe J, Klionsky DJ, Joseph B (2014) The return of the nucleus: transcriptional and epigenetic control of autophagy. Nat Rev Mol Cell Biol 15(1):65–74
14. Liu EY, Ryan KM (2012) Autophagy and cancer–issues we need to digest. J Cell Sci 125(Pt 10):2349–2358
15. Liu Q et al (2009) MTOR mediated anti-cancer drug discovery. Drug Discov Today Ther Strateg 6(2):47–55
16. Huber A, Rubio A, Loewith RJ (2009) Mammalian target of rapamycin signaling and cancer treatment. Cancer Chemother Rev 4(4):210–223
17. Gordy C, He YW (2012) The crosstalk between autophagy and apoptosis: where does this lead? Protein Cell 3(1):17–27
18. Kang R et al (2010) HMGB1: a novel Beclin 1-binding protein active in autophagy. Autophagy 6(8):1209–1211
19. Xie ZG, Xie Y, Dong QR (2013) Inhibition of the mammalian target of rapamycin leads to autophagy activation and cell death of MG63 osteosarcoma cells. Oncol Lett 6(5):1465–1469
20. Chiu HW et al (2011) Synergistic effects of arsenic trioxide and radiation in osteosarcoma cells through the induction of both autophagy and apoptosis. Radiat Res 175(5):547–560
21. Zhao GS et al (2018) TSSC3 promotes autophagy via inactivating the Src-mediated PI3K/Akt/mTOR pathway to suppress tumorigenesis and metastasis in osteosarcoma, and predicts a favorable prognosis. J Exp Clin Cancer Res 37(1):188
22. Wirries A et al (2018) Panobinostat mediated cell death: a novel therapeutic approach for osteosarcoma. Oncotarget 9(68):32997–33010
23. Yu Z et al (2018) MiR-100 up-regulation enhanced cell autophagy and apoptosis induced by cisplatin in osteosarcoma by targeting mTOR. Eur Rev Med Pharmacol Sci 22(18):5867–5873
24. Huang J et al (2012) Targeting HMGB1-mediated autophagy as a novel therapeutic strategy for osteosarcoma. Autophagy 8(2):275–277
25. Huang J et al (2012) HMGB1 promotes drug resistance in osteosarcoma. Cancer Res 72(1):230–238
26. Zhang F et al (2018) Novel oncogene COPS3 interacts with Beclin1 and Raf-1 to regulate metastasis of osteosarcoma through autophagy. J Exp Clin Cancer Res 37(1):135
27. Sakamaki JI et al (2018) Emerging roles of transcriptional programs in autophagy regulation. Transcription 9(2):131–136
28. Baek SH, Kim KI (2017) Epigenetic control of autophagy: nuclear events gain more attention. Mol Cell 65(5):781–785
29. Li B, Ye Z (2014) Epigenetic alterations in osteosarcoma: promising targets. Mol Biol Rep 41(5):3303–3315
30. Settembre C et al (2011) TFEB links autophagy to lysosomal biogenesis. Science 332(6036):1429–1433
31. Bao X et al (2018) Inhibition of LCMR1 and ATG12 by demethylation-activated miR-570-3p is involved in the anti-metastasis effects of metformin on human osteosarcoma. Cell Death Dis 9(6):611
32. Wang YH et al (2015) BRD4 promotes pancreatic ductal adenocarcinoma cell proliferation and enhances gemcitabine resistance. Oncol Rep 33(4):1699–1706
33. Bai Y et al (2019) Trichostatin A activates FOXO1 and induces autophagy in osteosarcoma. Arch Med Sci 15(1):204–213
34. Zhu X et al (2019) Silencing of SNHG6 induced cell autophagy by targeting miR-26a-5p/ULK1 signaling pathway in human osteosarcoma. Cancer Cell Int 19:82
35. Eskelinen EL (2011) The dual role of autophagy in cancer. Curr Opin Pharmacol 11(4):294–300
36. Yang ZJ et al (2011) The role of autophagy in cancer: therapeutic implications. Mol Cancer Ther 10(9):1533–1541
37. Yang SY, Winslet MC (2011) Dual role of autophagy in colon cancer cell survival. Ann Surg Oncol 18 Suppl 3:S239
38. Karch J et al (2017) Autophagic cell death is dependent on lysosomal membrane permeability through Bax and Bak. elife 6
39. Shimizu S et al (2014) Autophagic cell death and cancer. Int J Mol Sci 15(2):3145–3153
40. Fulda S, Kogel D (2015) Cell death by autophagy: emerging molecular mechanisms and implications for cancer therapy. Oncogene 34(40):5105–5113
41. Khurana A et al (2015) Quinacrine promotes autophagic cell death and chemosensitivity in ovarian cancer and attenuates tumor growth. Oncotarget 6(34):36354–36369
42. Zhang W et al (2015) Knockdown of autophagy-related protein 6, Beclin-1, decreases cell growth, invasion, and metastasis and has a positive effect on chemotherapy-induced cytotoxicity in osteosarcoma cells. Tumour Biol 36(4):2531–2539
43. Aunan JR, Cho WC, Soreide K (2017) The biology of aging and cancer: a brief overview of shared and divergent molecular hallmarks. Aging Dis 8(5):628–642
44. Yang S et al (2011) Pancreatic cancers require autophagy for tumor growth. Genes Dev 25(7):717–729
45. Endo S et al (2017) Autophagy is required for activation of pancreatic stellate cells, associated with pancreatic cancer progression and promotes growth

of pancreatic tumors in mice. Gastroenterology 152(6):1492–1506. e24
46. Katheder NS et al (2017) Microenvironmental autophagy promotes tumour growth. Nature 541(7637):417–420
47. Hollomon MG et al (2013) Knockdown of autophagy-related protein 5, ATG5, decreases oxidative stress and has an opposing effect on camptothecin-induced cytotoxicity in osteosarcoma cells. BMC Cancer 13:500
48. Livingston JA et al (2018) Analysis of HSP27 and the autophagy marker LC3B(+) puncta following preoperative chemotherapy identifies high-risk osteosarcoma patients. Mol Cancer Ther 17(6):1315–1323
49. Pasquier B (2016) Autophagy inhibitors. Cell Mol Life Sci 73(5):985–1001
50. Seglen PO, Gordon PB (1982) 3-Methyladenine: specific inhibitor of autophagic/lysosomal protein degradation in isolated rat hepatocytes. Proc Natl Acad Sci U S A 79(6):1889–1892
51. Shao S et al (2014) Spautin-1, a novel autophagy inhibitor, enhances imatinib-induced apoptosis in chronic myeloid leukemia. Int J Oncol 44(5):1661–1668
52. Li LQ et al (2016) Inhibition of autophagy by bafilomycin A1 promotes chemosensitivity of gastric cancer cells. Tumour Biol 37(1):653–659
53. Rangwala R et al (2014) Phase I trial of hydroxychloroquine with dose-intense temozolomide in patients with advanced solid tumors and melanoma. Autophagy 10(8):1369–1379
54. Rosenfeld MR et al (2014) A phase I/II trial of hydroxychloroquine in conjunction with radiation therapy and concurrent and adjuvant temozolomide in patients with newly diagnosed glioblastoma multiforme. Autophagy 10(8):1359–1368
55. Xu R et al (2018) The clinical value of using chloroquine or hydroxychloroquine as autophagy inhibitors in the treatment of cancers: a systematic review and meta-analysis. Medicine (Baltimore) 97(46):e12912
56. Amaravadi RK, Winkler JD (2012) Lys05: a new lysosomal autophagy inhibitor. Autophagy 8(9):1383–1384
57. McAfee Q et al (2012) Autophagy inhibitor Lys05 has single-agent antitumor activity and reproduces the phenotype of a genetic autophagy deficiency. Proc Natl Acad Sci U S A 109(21):8253–8258
58. Fu Y et al (2019) Discovery of a small molecule targeting autophagy via ATG4B inhibition and cell death of colorectal cancer cells in vitro and in vivo. Autophagy 15:295–311

12 The Fas/FasL Signaling Pathway: Its Role in the Metastatic Process and as a Target for Treating Osteosarcoma Lung Metastases

Nadya Koshkina, Yuanzheng Yang, and Eugenie S. Kleinerman

Abstract

Understanding how the tumor microenvironment participates in inhibiting or supporting tumor growth is critical for the development of novel therapies. Osteosarcoma (OS) metastasizes almost exclusively to the lung, an organ where Fas ligand (FasL) is constitutively expressed. This chapter focuses on our studies dedicated to the interaction of OS cells with the lung microenvironment. We will summarize our studies conducted over the past 20 years showing the importance of the Fas/FasL signaling pathway to the establishment and progression of OS metastases in the lung. We demonstrated that the $FasL^+$ lung microenvironment eliminates Fas-positive (Fas^+) OS cells that metastasize to the lungs, through apoptosis induced by Fas signaling following interaction of Fas on the tumor cell surface with FasL on the lung epithelial cells. Expression of the Fas receptor on OS cells inversely correlated with the ability of OS cells to form lung metastases. Blocking this pathway interferes with this process, allowing Fas^+ cells to grow in the lung. By contrast, upregulation of Fas on Fas^- OS cells inhibited their ability to metastasize to the lung. We demonstrated how the $FasL^+$ lung microenvironment can be leveraged for therapeutic intent through the upregulation of Fas expression. To this end, we demonstrated that the histone deacetylase inhibitor entinostat upregulated Fas expression on OS cells, reduced their ability to form lung metastases, and induced regression of established micrometastases. Fas expression in OS cells is regulated epigenetically by the microRNA miR-20a. We showed that expressions of Fas and miR-20a are inversely correlated, and that delivery of anti-miR-20a in vivo to mice with established osteosarcoma lung metastases resulted in upregulation of Fas and tumor regression. Therefore, targeting the Fas signaling pathway may present therapeutic opportunities, which target the lung microenvironment for elimination of OS lung metastases. We have also shown that in addition to being critically involved in the metastatic potential, the Fas signaling pathway may also contribute to the efficacy of chemotherapy. We demonstrated that the chemotherapeutic agent gemcitabine (GCB) increased Fas expression in both human and mouse OS cells in vitro. In vivo, aerosol GCB therapy induced upregulation of Fas expression and the regression of established osteosarcoma lung metastases. The therapeutic efficacy of GCB was contingent upon a $FasL^+$ lung microenvironment as aerosol GCB had no effect in FasL-deficient mice.

N. Koshkina (✉) · Y. Yang · E. S. Kleinerman
Department of Pediatrics-Research, The University of Texas MD Anderson Cancer Center, Children's Cancer Hospital, Houston, TX, USA

E. S. Kleinerman, R. Gorlick (eds.), *Current Advances in the Science of Osteosarcoma*, Advances in Experimental Medicine and Biology 1258, https://doi.org/10.1007/978-3-030-43085-6_12

Manipulation of Fas expression and the Fas pathway should be considered, as this concept may provide additional novel therapeutic approaches for treating patients with OS lung metastases.

Keywords

Osteosarcoma (OS) · Pulmonary metastasis · Fas · FasL · FADD (Fas-associated death domain) · FDN (Fas-associated death domain e negative) · C-FLIP · miR-20a · Methylation · Histone deacetylase inhibitors · Entinostat · Gemcitabine (GCB)

Introduction

The lungs remain the most common site of metastasis for osteosarcoma (OS) and the most common cause of death for patients with relapsed/refractory OS. Understanding of molecular mechanisms that allow OS cells to grow in the lung and become insensitive to chemotherapy and immunotherapy will allow the development of novel therapeutic approaches for the treatment of OS lung metastases. We have identified Fas expression and the Fas signaling pathway as one mechanism involved in the metastatic process to the lung [1]. In this chapter, we will summarize these findings and highlight possible approaches that can target Fas-regulated signaling to increase the vulnerability of OS cells to elimination by the $FasL^+$ lung microenvironment.

Fas Pathway Is Involved in Development of OS Metastases in the Lungs

Normal homeostasis in mammals is regulated by a programmed cell death mechanism called apoptosis. This mechanism involves several signaling molecules, which become overexpressed on the cell surface of deteriorating eukaryotic cells. This class of molecules, called death receptors, contains a cytoplasmic death domain (DD) and includes TNF, FAS/CD95, and TRAL proteins. All of these are activated upon binding with their ligands and trigger a similar cascade of downstream events leading to apoptosis [2]. Corruption of death receptor-mediated apoptosis pathway in cancer cells is mediated by several different mechanisms including overexpression of anti-apoptotic proteins and decreased expression of proapoptotic proteins [3, 4].

Ligands to the death receptor molecules, such as FasL, are commonly expressed or secreted by immune cells. Only a few types of epithelial cells express FasL, lung epithelial cells being one type [3]. Since pulmonary epithelial cells express FasL [5, 6], we hypothesized that malfunction of Fas-mediated apoptosis may play an important role in development of OS lung metastases. Our studies of Fas expression in human OS cells showed that the rate of growth of OS cells in the lungs of immunodeficient mice inversely correlated with the levels of Fas expression on their cell surface [7]. Supporting this concept, Nambu et al. reported that Fas protein was reduced or not expressed on the cell surface in primary lung adenocarcinoma [8]. Finally, interaction between Fas and FasL was found to be important for suppression of both melanoma and OS lung metastasis [7, 9, 10]. These findings led us to focus on the significance of Fas signaling in OS lung metastases.

Expression of Fas Receptor Correlates with OS Metastatic Potential in Animal Models and in Patient Specimens

Human SAOS-2 OS cells do not form lung metastases after intravenous injection into immunodeficient nude mice and have high levels of Fas receptor on their cell surface. In contrast, LM-6 and LM-7 cells derived from SAOS-2 cells by cycling through the mouse lung form lung metastases after intravenous (i.v.) administration and have significantly lower levels of Fas [5, 10]. Similarly, murine K7 OS cells form primary bone tumors when injected into the tibia of syngeneic immunocompetent mice but do not metastasize to the lungs. By contrast when the K7M3 subline (created by cycling K7 cells through the lung) was injected into the tibia, both primary tumor and lung metastases are induced [6]. Expression of Fas in the K7M3 cells was significantly decreased compared to the K7 cells, and all K7M3 lung metastases were Fas-negative (Fas^-). This suggested to us that only Fas^- cells were able to grow in the lung. We confirmed this by quantifying the percent of cells retained in the lung following i.v. injection using fluorescently labeled K7M3 cells. We demonstrated that retention of cells in the lung over time was increased when the Fas pathway was blocked through FAS-associated death domain (FADD) dominant-negative transfected cells [11].

We examined the pattern of Fas expression in OS lung metastases from patients in two different studies [10, 12]. Approximately 60% of samples were negative for Fas, while 40% were weakly positive for Fas. We were able to analyze a few samples of both the primary OS tumor in the bone and the corresponding lung metastases from the same patient. These specimens showed that the primary OS tumors were highly positive for Fas, whereas the corresponding lung metastases in the lungs showed negligible levels of Fas [5]. Similarly, in the K7 and K7M3 mouse models described above, staining of primary OS tumors from the bone revealed high levels of Fas with negligible Fas expression in the corresponding lung metastasis [6]. Taken together, we interpret these findings to mean that there are heterogeneities in terms of Fas expression in cells forming the primary bone tumor. However, the only cells that are able to form lung metastases are Fas^-. This also supports the concept that the $FasL^+$ lung microenvironment filters out the Fas^+ OS cells.

In another set of experiments, we demonstrated that conversion of highly metastatic LM7 Fas^-OS tumor cells to Fas^+ cells by transfection with the *Fas* gene or by stimulating Fas expression using IL-12 gene therapy dramatically reduced their metastatic potential in vivo [10, 13, 14].

These data support the concept that Fas^+Fas-positive OS tumor cells are eliminated from the $FasL^+$ lung microenvironment via induction of apoptosis induced by binding of the Fas^+ cells with the FasL expressed by pulmonary epithelial cells. In contrast, Fas^- OS tumor cells can evade this host defense mechanism and proliferate in the pulmonary environment.

Functional Fas, FasL, and Downstream Fas Signaling Are Important for OS Lung Metastasis Progression

In our studies, we reported that approximately 40% of pulmonary metastases specimens from patients with OS showed low expression of Fas receptor but appeared to be able to survive in the pulmonary microenvironment [12]. This phenomenon can have several explanations: (1) the Fas receptors in these cells could be mutated and therefore not functional; (2) the levels of Fas receptor may be not sufficient for induction of apoptosis; (3) there could be a dysfunction of FasL activity in the host defense mechanism; and finally (4) there could be inhibition of Fas signaling downstream of the Fas receptor.

We investigated the rates of *Fas* gene mutation in patients with OS. We intentionally did not study mutations of this gene in the tumor specimens, because OS tumors are heterogeneous. High chromosomal instability has been shown in OS tumors from patients with numerous structural variations and single-nucleotide variants with little consistency between the primary and metastatic lesions from the same patient. Genomic sequencing in primary and metastatic tumors also showed no consistent mutations [15, 16]. We examined mutations of the *Fas* gene in patient blood specimens to determine if there are stable or inherited mutations associated with OS risk and progression. In collaboration with E. Sturgis, who discovered two single-nucleotide polymorphism (SNP) mutations in the *Fas* promoter region as tumor risk and progression factor in patients with squamous cell carcinoma [17], we analyzed 123 pediatric patients with OS for *Fas* single-nucleotide polymorphisms: 2 of the promoter regions (−1377 G > A and −670 A > G) and 2 of the coding regions (exon 318,272 A

> G and exon 722,628 C > T). As a comparison group, we used blood specimens from 510 adults with no history of cancer due to the lack of blood bank specimens from healthy children. We found an increased risk of OS associated with the heterozygous genotype *Fas* exon 3 A > G, and this association was more pronounced in non-Hispanic whites. Additionally, the frequency of the variant allele (exon 3 G) was significantly higher in OS cases than in controls. We found no significant association between OS risk and the other Fas polymorphisms [18]. These studies included analysis of only a few SNP mutations in the *Fas* gene. Extended studies of other mutations in the *Fas* gene are required before this can serve as a specific risk and progression factors.

We still do not know what level of Fas expression is sufficient to trigger apoptosis in OS cells. At the minimum, there should be at least three Fas molecules, because FasL triggers trimerization of Fas receptors before the induction of downstream death-inducing events [19, 20]. However, we are skeptical that 3 Fas receptors will be sufficient. In addition, even abundant levels of Fas will not lead to tumor cell death without functional FasL expression in the pulmonary microenvironment. We demonstrated this by using FasL-deficient *gld* mice. When we injected these mice with highly Fas-positive K7 OS cells, which do not form lung metastases in mice with a functional FasL system, we found that the *gld* mice had visible and microscopic metastatic lesions in the lungs and that these tumors were Fas-positive [6]. Therefore, any changes leading to FasL inactivation in pulmonary microenvironment, such as genetic mutations or cleavage of the FasL molecule from the cell surface of pneumocytes or the formation of the inactive FasL-soluble form by external matrix metalloproteinase MMP-7, can facilitate OS lung metastases development [21].

Finally, many downstream reactions are triggered after Fas receptor activation and their inhibition can prevent apoptosis even after successful Fas-FasL binding. In our studies, we transfected Fas-positive nonmetastatic cells K7 OS cells with the Fas-associated death domain-negative (FDN) plasmid. Fas-associated death domain (FADD) is an essential adaptor molecule that binds the cytoplasmic part of the Fas-FasL complex with procaspases 8 and 10 to form the death-inducing signaling complex (DISC), as shown in Fig. 12.1. FDN molecule lacks the important part of FADD that binds with procaspases and, therefore, blocks the Fas signaling at the very beginning of cascade reactions. Transfection of Fas-positive nonmetastatic K7 OS cells made them insensitive to Fas-mediated apoptosis and when injected i.v. into mice led to the formation of Fas-positive metastases in the lungs [6, 11].

We also studied the role of the natural Fas inhibitor c-FLIP. C-FLIP prevents binding of the FADD molecule with procaspases [22]. Elevated in c-FLIP expression has been reported in numerous different tumors [23]. Overexpression of C-FLIP blocks Fas-mediated apoptosis, therefore protecting cells from FasL-induced cell death. Analysis of OS patient samples showed elevated levels of c-FLIP in some lung metastases [24]. Also, c-FLIP levels were found to be high in human Fas$^+$OS KRIB and CCH-OS-D lung metastases when compared with their primary bone tumor [24]. Thus, cell surface Fas expression, a functional Fas signaling pathway, and a FasL$^+$ microenviroment are all critical in the elimination of OS cells in the lung. Downregulation of cell surface Fas on OS cells, high expression of C-FLIP, or absence or alteration of FasL in the lung epithelium can individually prevent clearance of OS cells, thereby facilitating and contributing to the successful metastatic process in the lung. Taken together, these findings show the important role of Fas-mediated apoptosis in the metastatic potential of OS, suggesting that this pathway may represent a novel target for development of new therapeutic approaches to treat patients with lung metastases.

Fas Signaling Plays an Important Role in Response of OS to Therapy

In our early studies, when we investigated the efficacy of IL-12 gene therapy for the treatment of OS [13, 14, 25], we discovered that the efficacy

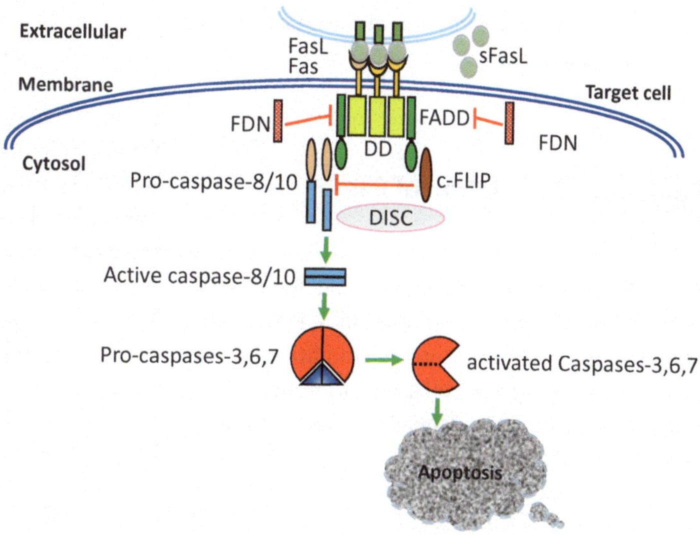

Fig. 12.1 Schema of Fas/FasL pathway. sFasL secreted FasL, DD cytoplasmic death domain, FADD Fas-associated death domain, FDN Fas-associated death domain dominant-negative, DISC death-inducing signaling complex, c-FLIP Cellular FADD-like IL-1β-converting enzyme(FLICE) inhibitory protein

of IL-12 therapy was accompanied by upregulation of the Fas receptor on OS cells and tumors [13]. Transfection of the LM6 metastatic OS cell line with the IL-12 gene resulted in the upregulation of Fas receptor on the cell surface more than two-fold without affecting FasL expression [14]. Analysis of different therapeutic agents used as a standard therapy for OS treatment in patients, such as 4-hydroperoxycyclophosphamide, cisplatin, adriamycin, and methotrexate, showed that only one of them, 4-hydroperoxycyclophosphamide, was able to upregulate FasL gene expression. The upregulation of Fas by IL-12 gene transfection together with the increased FasL by treated with 4-hydroperoxycyclophosphamide, led to increased cell death when compared with single agent exposure [26]. These studies showed that activation of Fas signaling by elevation of both Fas and FasL in OS cells can lead to enhanced cell death. Therefore, in all our following studies of anticancer agents, we investigated their ability to stimulate Fas and/or FasL expression. Numerous studies by other investigators also validated the role of Fas and FasL as mediators of anticancer therapy [8, 27, 28].

In the late 1990s to early 2000s, gemcitabine (GCB) was first successfully used for the first-line treatment of patients with pancreatic cancer, NSCL lung cancer, metastatic breast cancer, and in combination with other chemotherapeutic agents for treatment of different types of carcinomas. In our preclinical studies, GCB upregulated Fas expression in OS cells in vitro and enhanced FasL-mediated cell death [29, 30]. In mice and dogs with OS lung metastases, efficacy was demonstrated by administering GCB to the lungs via aerosol therapy [29–31]. Immunohistochemical analysis showed that treatment of OS lung metastases resulted in increased levels of Fas expression. It is important to note that the therapeutic effect of aerosol gemcitabine was dependent upon a FasL^{+} lung microenvironment as therapeutic activity was not seen in transgenic *gld* mice that do not express functional FasL [11]. These findings further confirmed our hypotheses that *an intact Fas signaling pathway and FasL expression in the tumor microenvironment are critical for response of OS lung metastases to chemotherapeutic agents.*

Regulation of the Fas Signaling Pathway

Having demonstrated that downregulation of Fas expression plays an important role in OS cell survival in the lung and that re-expression of Fas induces the regression of established OS lung metastases, we investigated how Fas expression is regulated in OS cells. The goal was to restore and/or increase the expression of Fas in order for the FasL+ lung microenvironment to assist with tumor regression.

Epigenetic Regulation of Fas Expression

Gene expression may be modulated by several different mechanisms from the DNA-RNA transcription, to posttranscriptional modification of protein. In eukaryotic cells, the accessibility of large regions of DNA depends on its chromatin structure, which can be altered epigenetically as a result of histone alteration directed by two major mechanisms, DNA methylation and acetylation/deacetylation.

Methylation of DNA is a common method of gene silencing. DNA is typically methylated by methyltransferase enzymes on cytosine nucleotides in a CpG islands. Analysis of the pattern of methylation in the region of interest on DNA can be performed by bisulfite mapping method. It was shown that Fas expression was regulated by its gene promoter methylation in some cancer cells and that this may contribute to the development of drug resistance [32–36]. However, PCR-based methylation and bisulfine-modified DNA sequencing analysis of various OS cell lines showed that 99.8% of CpG islands in the *Fas* promoter and first intron regions were unmethylated [36]. Moreover, the levels of methylation in these regions of the *Fas* gene were similar in the nonmetastatic SAOS-2 cells that express high levels of Fas receptor protein and the metastatic subline LM-7 that express low levels of Fas [37]. Treatment of these cells with the methylation agent 5-azadeoxycitidine did not change Fas protein expression on their cell surface. These findings indicate that suppression of *Fas* gene expression in OS cells is not mediated by DNA methylation.

Histone acetylation, regulated by the histone acetyl transferase enzyme, and its reverse process deacetylation, mediated by histone deacetylase (HDAC), are essential epigenetic regulatory mechanisms. Acetylated histones, octameric proteins that organize chromatin into nucleosomes and ultimately higher-order structures, represent a type of epigenetic marker within chromatin. Acetylation removes the positive charge on the histones, thereby decreasing the interaction of the N-termini of histones with the negatively charged phosphate groups of DNA. Consequently, the condensed chromatin is transformed into a more relaxed structure that is associated with greater levels of gene transcription. The use of agents that inhibit the deacetylation process and promote the acetylation status of the gene, so-called HDAC inhibitors, are widely used to promote expression of numerous genes, including Fas in various cancer cells [38–40]. Imai and colleagues were the first to describe the ability of the pan-HDAC inhibitor FR901228 to induce OS tumor regression via Fas signaling by inducing expression of FasL in vitro [41]. Watanabe et al. described the ability of FR901228 to sensitize OS tumor cells to death receptor–mediated apoptosis by suppression of cFLIP expression [42]. In our studies with DLM8 mouse OS cell that express low levels of Fas protein, we showed that treatment with entinostat, which inhibits only class type I HDACs, increased the acetylation status of the promoter region in the *Fas* gene and increased Fas mRNA and protein expression at subtoxic doses [43]. These results indicate that inhibiting HDAC type I is a mechanism for upregulating or restoring of the *Fas* gene expression in DLM8 cells. Our findings were supported by other investigators showing that HDAC inhibitors can stimulate Fas expression in U2OS cells [44, 45]. Interestingly, in LM7 cells, which also have low levels of Fas receptor expression, treatment with entinostat was followed by increased levels of Fas mRNA and protein. However, we were unable to detect increased Fas on the cell surface using flow cytometry [46]. Despite the fact that the Fas receptor expression on the tumor cell surface was not increased, entinostat-treated LM7 cells were more sensitive to FasL [43, 46]. In addition, there was a significant increase in soluble FasL(sFasL) binding, indicating that the increased Fas protein levels induced by entinostat were in the membrane compartments, which is not detectable using flow cytometry [46]. Membrane lipid raft platforms have been shown to be critical in Fas receptor signaling. Indeed, we demonstrated that treating OS cells with entinostat resulted in increased localization of the Fas receptor to the lipid raft microdomains, making it more accessible to FasL binding, thereby increasing FasL-mediated cell death [46]. Entinostat also decreased c-FLIP and FADD expression, proteins downstream of the Fas receptor, which inhibit Fas-mediated cytotoxicity [22, 43, 46]. Taken together these finding further substantiate the concept of identifying agents that increase cell surface Fas or its localization to the lipid rafts, or those that decrease c-FLIP and other proteins that block the Fas-signaling pathway, for the treatment of patients with OS lung metastases. This approach will sensitize OS cells to Fas-induced apoptosis, thereby harnessing the lung microenvironment in the therapeutic process.

Regulation of Fas Expression in OS Cells by miRNA

Several mechanisms can be involved in the regulation of Fas. We investigated the role of microRNA (miRNA) as one of the mechanisms controlling Fas expression. miRNAs are small (21–25 nucleotides) noncoding RNAs that negatively regulate gene expression [47, 48]. They are frequently dysregulated in a variety of cancers, including breast cancer, colon cancer, lung cancer, hepatocellular carcinoma, and OS [49–53]. miRNAs have been shown to regulate gene expression posttranscriptionally by targeting the 3′-untranslated region (3′-UTR) of specific mRNAs through binding to complimentary sequences in the 3′ untranslated regions (3′-UTR) or coding regions of the mRNA [47, 54] or by suppressing translation initiation through binding to complementary sequences in the 5′ untranslated regions (5′-UTR) of the mRNA [54]. Death receptor signaling proteins, such as TNFa, FADD, ribosome-inactivating protein (RIP), caspase-3, Bcl-2 interacting mediator of cell death protein (Bim), and p21 (CDKN1A), were found to be regulated by miRNAs [55–59].

Several miRNAs, including miR-21, miR-200c, let-7, miR-34, and miR-146a, have been reported to regulate Fas or FasL expression in different tumor cells [60–64]. We showed that a specific miRNA cluster, miR-17-92, is upregulated in Fas^{-} OS cells that metastasize to the lung [65]. Expression levels of several members of the miR-17-92 cluster including miR-20a and miR-19a were found to be higher in metastatic low Fas-expressing LM7 cells than in the parental SAOS-2 nonmetastatic high Fas-expressing cells. We demonstrated that miR-20a downregulated Fas expression and increased the metastatic potential of OS cells to the lung [65]. Overexpression of miR-20a in nonmetastatic Fas-positive SAOS-2 OS cells downregulated Fas expression and decreased their sensitivity to FasL. In contrast, inhibiting miR-20a in Fas-negative LM7 OS cells increased Fas expression and sensitivity to FasL. An inverse correlation between Fas and miR-20a expression in 8 cell lines derived from patient samples was also demonstrated. Furthermore, mice injected with LM7 cells stably transfected with an anti-miR-20a vector had fewer metastases than those transfected with control plasmid [65].

We showed that the regulation of Fas by miR-20a was not mediated by its binding to the 3'-UTR of Fas mRNA, which induces the degradation of Fas mRNA or the suppression of protein translation, but rather by an indirect effect on the Fas promoter activity [66]. Reporter assays using Fas promoter-driven luciferase expression showed that the activity of the Fas promoter was affected by miR-20a. Serial deletions of the *Fas* promoter region showed that a 90-bp region (−240 bp to −150 bp) on the Fas promoter was critical for Fas regulation by miR-20a. Targeting miR-20a by administering nanoparticle-formulated anti-miR-20a oligonucleotides significantly inhibited the growth of established OS lung metastases in mice, which correlated with the upregulation of Fas expression in the tumors.

We conclude that miR-20a encoded by the miRNA-17-92 cluster regulates Fas expression in OS cells and plays a critical role in the metastatic process of OS to the lung. These findings indicate that targeting miR-20a may be another novel therapeutic approach focusing on the Fas/FasL signaling pathway for patients with relapsed OS in the lung.

Summary

In this chapter, we have summarized our studies illustrating the importance of the organ microenvironment in the successful metastatic process for OS. The success or failure to form metastases in the lung depends on the ability of the metastasized OS cells to survive in the lung microenvironment, which is different than the bone microenvironment. Therefore, interactions between the tumor cells and the normal *bone* or *lung* cells can be different and therefore can influence the balance between tumor cell proliferation and tumor cell death in that particular organ microenvironment. OS cells with biologic characteristics that support adaptation to the lung microenvironment will be the ones that successfully grow and expand. While the ability to survive in the new microenvironment is not the only biologic criteria, it is the first step and arguably the most critical. Inducing vascular expansion to support the proliferating cells is another critical biologic characteristic, but without survival, this characteristic is secondary. OS cells that do not have the phenotype that is compatible with the lung microenvironment will be eliminated. We have focused on how the lung microenvironment supports or interferes with OS cell survival, as we are of the opinion that this presents a unique opportunity for developing new therapies particularly directed to treat patients with OS lung metastases who have not responded to chemotherapy. In short, we wish to change the therapeutic focus and target the lung microenvironment, thereby allowing the lung to contribute to killing and eradicating the established metastatic disease.

The studies detailed in this chapter show that the lung microenvironment, with its constitutive expression of FasL, can control the fate of OS cells that migrate to the lung, and determine whether these OS cells survive or are eliminated, and whether they can go on to form lung metastases. Expression of cell surface Fas or Fas protein in the lipid membranes, (where it is accessible to interaction with FasL on the lung epithelial cells), is one of the critical factors. OS cells that have downregulated their cell surface Fas or those with a blocked Fas signaling pathway (i.e., overexpression of c-FLIP) will be able to survive in the $FasL^+$ lung microenvironment. We showed that Fas expression on OS cells is epigenetically regulated, a process that is controlled by histone acetylation and a specific microRNA from the MiR 17-92 cluster, and not by *Fas* gene mutation or methylation. We demonstrated that interventions that induce re-expression of Fas using various therapies (including aerosol IL-2 and gemcitabine) resulted in regression of established lung metastases. This was independent of the agent's ability to directly induce tumor cell apoptosis and required a $FasL^+$ lung microenvironment. These compelling data underscore how incorporating and harnessing the lung microenvironment into the therapeutic strategy for patients with OS lung metastases can add to the successful eradication of the disease, particularly against tumor cells that show resistance to chemotherapy and other forms of salvage therapy. Since 90% of newly diagnosed patients have undetectable microscopic metastases, incorporating such an approach into the neoadjuvant treatment schema may also prove beneficial in increasing the disease-free and long-term survival rates.

References

1. Gordon N, Kleinerman ES (2009) The role of Fas/FasL in the metastatic potential of osteosarcoma and targeting this pathway for the treatment of osteosarcoma lung metastases. Cancer Treat Res 152:497–508
2. Guicciardi ME, Gores GJ (2009) Life and death by death receptors. FASEB J 23:1625–1637
3. Peter ME, Hadji A, Murmann AE et al (2015) The role of CD95 and CD95 ligand in cancer. Cell Death Differ 22(4):549–559
4. Pfeffer CM, Singh ATK (2018) Apoptosis: a target for anticancer therapy. Int J Mol Sci 19(2):E448
5. Jia SF, Worth LL, Kleinerman ES (1999) A nude mouse model of human osteosarcoma lung metastases for evaluating new therapeutic strategies. Clin Exp Metastasis 17:501–506

6. Koshkina NV, Khanna C, Mendoza A, Guan H, DeLauter L, Kleinerman ES (2005) Fas-negative osteosarcoma tumor cells are selected during metastasis to the lungs: the role of the Fas pathway in the metastatic process of osteosarcoma. Mol Cancer Res 5:991–999
7. Worth LL, Lafleur EA, Jia SF, Kleinerman ES (2002) Fas expression inversely correlates with metastatic potential in osteosarcoma cells. Oncol Rep 9:823–827
8. Nambu Y, Hughes SJ, Rehemtulla A, Hamstra D, Orringer MB, Beer DG (1998) Lack of cell surface Fas/APO-1 expression in pulmonary adenocarcinoma. J Clin Invest 101:1102–1110
9. Owen-Schaub LB, van Golen KL, Hill LL, Price JE (1998) Fas and Fas ligand interactions suppress melanoma lung metastasis. J Exp Med 188:1717–1723
10. Lafleur EA, Koshkina NV, Stewart J, Jia SF, Worth LL, Duan X, Kleinerman ES (2004) Increased Fas expression reduces the metastatic potential of human osteosarcoma cells. Clin Cancer Res 10:8114–8119
11. Gordon N, Koshkina NV, Jia SF, Khanna C, Mendoza A, Worth LL, Kleinerman ES (2007) Corruption of the Fas pathway delays the pulmonary clearance of murine osteosarcoma cells, enhances their metastatic potential, and reduces the effect of aerosol gemcitabine. Clin Cancer Res 13(15 Pt 1):4503–4510
12. Gordon N, Arndt CA, Hawkins DS, Doherty DK, Inwards CY, Munsell MF et al (2005) Fas expression in lung metastasis from osteosarcoma patients. J Pediatr Hematol Oncol 27:611–615
13. Worth LL, Jia SF, Zhou Z, Chen L, Kleinerman ES (2000) Intranasal therapy with an adenoviral vector containing the murine interleukin-12 gene eradicates osteosarcoma lung metastases. Clin Cancer Res 6(9):3713–3718
14. Lafleur EA, Jia SF, Worth LL, Zhou Z, Owen-Schaub LB, Kleinerman ES (2001) Interleukin (IL)-12 and IL-12 gene transfer up-regulate Fas expression in human osteosarcoma and breast cancer cells. Cancer Res 61(10):4066–4071
15. Sayles LC, Breese MR, Koehne AL, Leung SG, Lee AG, Liu HY, Spillinger A, Shah AT, Tanasa B, Straessler K, Hazard FK, Spunt SL, Marina N, Kim GE, Cho SJ, Avedian RS, Mohler DG, Kim MO, DuBois SG, Hawkins DS, Sweet-Cordero EA (2019) Genome-informed targeted therapy for osteosarcoma. Cancer Discov 9(1):46–63
16. Negri GL, Grande BM, Delaidelli A, El-Naggar A, Cochrane D, Lau CC, Triche TJ, Moore RA, Jones SJ, Montpetit A, Marra MA, Malkin D, Morin RD, Sorensen PH (2019) Integrative genomic analysis of matched primary and metastatic pediatric osteosarcoma. J Pathol 249(3):319–331. https://doi.org/10.1002/path.5319
17. Zhang Z, Wang LE, Sturgis EM, El-Naggar AK, Hong WK, Amos CI, Spitz MR, Wei Q (2006) Polymorphisms of FAS and FAS ligand genes involved in the death pathway and risk and progression of squamous cell carcinoma of the head and neck. Clin Cancer Res 12(18):5596–5602
18. Koshkina NV, Kleinerman ES, Li G, Zhao CC, Wei Q, Sturgis EM (2007) Exploratory analysis of Fas gene polymorphisms in pediatric osteosarcoma patients. J Pediatr Hematol Oncol 29:815–821
19. Chan FK (2007) Three is better than one: pre-ligand receptor assembly in the regulation of TNF receptor signaling. Cytokine 37:101–107
20. Fu Q, Fu TM, Cruz AC, Sengupta P, Thomas SK, Wang S, Siegel RM, Wu H, Chou JJ (2016) Structural basis and functional role of intramembrane trimerization of the Fas/CD95 death receptor. Mol Cell 61:602–613
21. Mitsiades N, Yu WH, Poulaki V, Tsokos M, Stamenkovic I (2001) Matrix metalloproteinase-7-mediated cleavage of Fas ligand protects tumor cells from chemotherapeutic drug cytotoxicity. Cancer Res 61(2):577–581
22. Rao-Bindal K, Koshkina NV, Stewart J, Kleinerman ES (2013) The histone deacetylase inhibitor, MS-275 (entinostat), downregulates c-FLIP, sensitizes osteosarcoma cells to FasL, and induces the regression of osteosarcoma lung metastases. Curr Cancer Drug Targets 13:411–422
23. Iyer AK, Azad N, Talbot S, Stehlik C, Lu B, Wang L, Rojanasakul Y (2011) Antioxidant c-FLIP inhibits Fas ligand-induced NF-kappaB activation in a phosphatidylinositol 3-kinase/Akt-dependent manner. J Immunol 187(6):3256–3266
24. Rao-Bindal K, Rao CK, Yu L, Kleinerman ES (2013) Expression of c-FLIP in pulmonary metastases in osteosarcoma patients and human xenografts. Pediatr Blood Cancer 60:575–579
25. Worth LL, Jia SF, An T, Kleinerman ES (1999) ImmTher, a lipophilic disaccharide derivative of muramyl dipeptide, up-regulates specific monocyte cytokine genes and activates monocyte-mediated tumoricidal activity. Cancer Immunol Immunother 48:312–320
26. Duan X, Zhou Z, Jia SF, Colvin M, Lafleur EA, Kleinerman ES (2004) Interleukin-12 enhances the sensitivity of human osteosarcoma cells to 4-hydroperoxycyclophosphamide by a mechanism involving the Fas/Fas-ligand pathway. Clin Cancer Res 10:777–783
27. Möller P, Koretz K, Leithäuser F, Brüderlein S, Henne C, Quentmeier A, Krammer PH (1994) Expression of APO-1 (CD95), a member of the NGF/TNF receptor superfamily, in normal and neoplastic colon epithelium. Int J Cancer 57(3):371–377
28. O'Brien DI, Nally K, Kelly RG, O'Connor TM, Shanahan F, O'Connell J (2005) Targeting the Fas/Fas ligand pathway in cancer. Expert Opin Ther Targets 9(5):1031–1044
29. Koshkina NV, Kleinerman ES (2005) Aerosol gemcitabine inhibits the growth of primary osteosarcoma and osteosarcoma lung metastases. Int J Cancer 116(3):458–463
30. Gordon N, Kleinerman ES (2010) Aerosol therapy for the treatment of osteosarcoma lung metastases: targeting the Fas/FasL pathway and rationale for the

use of gemcitabine. J Aerosol Med Pulm Drug Deliv 23(4):189–196
31. Rodriguez CO Jr (2014) Using canine osteosarcoma as a model to assess efficacy of novel therapies: can old dogs teach us new tricks? Adv Exp Med Biol 804:237–256
32. Petak I, Danam RP, Tillman DM, Vernes R, Howell SR, Berczi L, Kopper L, Brent TP, Houghton JA (2003) Hypermethylation of the gene promoter and enhancer region can regulate Fas expression and sensitivity in colon carcinoma. Cell Death Differ 2:211–217
33. Santourlidis S, Warskulat U, Florl AR, Maas S, Pulte T, Fischer J, Muller W, Schulz WA (2001) Hypermethylation of the tumor necrosis factor receptor superfamily 6 (APT1, Fas, CD95/Apo-1) gene promoter at rel/nuclear factor kappaB sites in prostatic carcinoma. Mol Carcinog 1:36–43
34. Butler LM, Dobrovic A, Bianco T, Cowled PA (2000) Promoter region methylation does not account for the frequent loss of expression of the Fas gene in colorectal carcinoma. Br J Cancer 1:131–135
35. Manoochehri M, Borhani N, Karbasi A, Koochaki A, Kazemi B (2016) Promoter hypermethylation and downregulation of the FAS gene may be involved in colorectal carcinogenesis. Oncol Lett 12(1):285–290
36. Cacan E (2016) Histone deacetylase-1-mediated suppression of FAS in chemoresistant ovarian cancer cells. Anticancer Res 36:2819–2826
37. Huang G, Koshkina NV, Kleinerman ES (2009) Fas expression in metastatic osteosarcoma cells is not regulated by CpG island methylation. Oncol Res 18(1):31–39
38. Wood TE, Dalili S, Simpson CD, Sukhai MA, Hurren R, Anyiwe K, Mao X, Suarez Saiz F, Gronda M, Eberhard Y, MacLean N, Ketela T, Reed JC, Moffat J, Minden MD, Batey RA, Schimmer AD (2010) Selective inhibition of histone deacetylases sensitizes malignant cells to death receptor ligands. Mol Cancer Ther 9:246–256
39. Kim HR, Kim EJ, Yang SH, Jeong ET, Park C, Lee JH, Youn MJ, So HS, Park R (2006) Trichostatin A induces apoptosis in lung cancer cells via simultaneous activation of the death receptor-mediated and mitochondrial pathway? Exp Mol Med 38:616–624
40. Zhang X, Jiang SJ, Shang B, Jiang HJ (2015) Effects of histone deacetylase inhibitor trichostatin A combined with cisplatin on apoptosis of A549 cell line. Thorac Cancer 6:202–208
41. Imai T, Adachi S, Nishijo K, Ohgushi M, Okada M, Yasumi T et al (2003) FR901228 induces tumor regression associated with induction of Fas ligand and activation of Fas signaling in human osteosarcoma cells. Oncogene 22:9231–9242
42. Watanabe K, Okamoto K, Yonehara S (2005) Sensitization of osteosarcoma cells to death receptor-mediated apoptosis by HDAC inhibitors through downregulation of cellular FLIP. Cell Death Differ 12:10–18
43. Koshkina NV, Rao-Bindal K, Kleinerman ES (2011) Effect of the histone deacetylase inhibitor SNDX-275 on Fas signaling in osteosarcoma cells and the feasibility of its topical application for the treatment of osteosarcoma lung metastases. Cancer 117:3457–3467
44. Escaffit F, Vaute O, Chevillard-Briet M, Segui B, Takami Y, Nakayama T, Trouche D (2007) Cleavage and cytoplasmic delocalization of histone deacetylase 3 are important for apoptosis progression. Mol Cell Biol 27:554–567
45. Senese S, Zaragoza K, Minardi S, Muradore I, Ronzoni S, Passafaro A, Bernard L, Draetta GF, Alcalay M, Seiser C, Chiocca S (2007) Role for histone deacetylase 1 in human tumor cell proliferation. Mol Cell Biol 27:4784–4795
46. Rao-Bindal K, Zhou Z, Kleinerman ES (2012) MS-275 sensitizes osteosarcoma cells to Fas ligand-induced cell death by increasing the localization of Fas in membrane lipid rafts. Cell Death Dis 3:e369
47. He L, Hannon GJ (2004) MicroRNAs: small RNAs with a big role in gene regulation. Nat Rev Gene 5:522–531
48. Gartel AL, Kandel ES (2008) miRNAs: little known mediators of oncogenesis. Semin Cancer Biol 18(2):103–110
49. Yu Z, Baserga R, Chen L, Wang C, Lisanti MP, Pestell RG (2010) microRNA, cell cycle, and human breast cancer. Am J Pathol 176(3):1058–1064
50. Pucci S, Mazzarelli P (2011) MicroRNA dysregulation in colon cancer microenvironment interactions: the importance of small things in metastases. Cancer Microenviron 4(2):155–162
51. Boeri M, Pastorino U, Sozzi G (2012) Role of microRNAs in lung cancer: microRNA signatures in cancer prognosis. Cancer J 18(3):268–274
52. Braconi C, Henry JC, Kogure T, Schmittgen T, Patel T (2011) The role of microRNAs in human liver cancers. Semin Oncol 38(6):752–763
53. Kobayashi E, Hornicek FJ, Duan Z (2012) MicroRNA involvement in osteosarcoma. Sarcoma 2012:359739
54. Shivdasani RA (2006) MicroRNAs: regulators of gene expression and cell differentiation. Blood 108:3646–3653
55. Lima RT, Busacca S, Almeida GM, Gaudino G, Fennell DA, Vasconcelos MH (2011) MicroRNA regulation of core apoptosis pathways in cancer. Eur J Cancer 47:163–174
56. Tili E, Michaille JJ, Cimino A, Costinean S, Dumitru CD, Adair B et al (2007) Modulation of miR-155 and miR-125b levels following lipopolysaccharide/TNF-alpha stimulation and their possible roles in regulating the response to endotoxin shock. J Immunol 179:5082–5089
57. Petrocca F, Visone R, Onelli MR, Shah MH, Nicoloso MS, de Martino I et al (2008) E2F1-regulated microRNAs impair TGFbeta-dependent cell-cycle arrest and apoptosis in gastric cancer. Cancer Cell 13:272–286
58. Inomata M, Tagawa H, Guo YM, Kameoka Y, Takahashi N, Sawada K (2009) MicroRNA-17-92

down-regulates expression of distinct targets in different B-cell lymphoma subtypes. Blood 113:396–402
59. Ovcharenko D, Kelnar K, Johnson C, Leng N, Brown D (2007) Genome-scale microRNA and small interfering RNA screens identify small RNA modulators of TRAIL-induced apoptosis pathway. Cancer Res 67:10782–10788
60. Sayed D, He M, Hong C, Gao S, Rane S et al (2010) MicroRNA-21 is a downstream effector of AKT that mediates its antiapoptotic effects via suppression of Fas ligand. J Biol Chem 285:20281–20290
61. Schickel R, Park SM, Murmann AE, Peter ME (2010) miR-200c regulates induction of apoptosis through CD95 by targeting FAP-1. Mol Cell 38(6):908–915
62. Geng L, Zhu B, Dai BH, Sui CJ, Xu F et al (2011) A let-7/Fas double-negative feedback loop regulates human colon carcinoma cells sensitivity to Fas-related apoptosis. Biochem Biophys Res Commun 408:494–499
63. Hau A, Ceppi P, Peter ME (2012) CD95 is part of a Let-7/p53/miR-34 regulatory network. PLoS One 7(11):e49636
64. Suzuki Y, Kim HW, Ashraf M, Haider H (2010) Diazoxide potentiates mesenchymalstem cell survival via NF-kappaB-dependent miR-146a expression by targeting Fas. Am J Physiol Heart Circ Physiol 299:H1077–H1082
65. Huang G, Nishimoto K, Zhou Z, Hughes D, Kleinerman ES (2012) miR-20a encoded by the miR-17-92 cluster increases the metastatic potential of osteosarcoma cells by regulating Fas expression. Cancer Res 72:908–916
66. Yang Y, Huang G, Zhou Z, Fewell JG, Kleinerman ES (2018) miR-20a regulates FAS expression in osteosarcoma cells by modulating FAS promoter activity and can be therapeutically targeted to inhibit lung metastases. Mol Cancer Ther 17(1):130–139

Exosomes: Dynamic Mediators of Extracellular Communication in the Tumor Microenvironment

13

Kerri Wolf-Dennen and Eugenie S. Kleinerman

Abstract

It is becoming increasingly recognized that the tumor microenvironment significantly contributes to the development, progression, and metastasis of cancer and also plays a role in response to treatment. The tumor microenvironment is a complex and heterogeneous niche comprised of stromal cells, cancer cells, blood vessels, areas of hypoxia and necrotic tissue, fibrosis, and extracellular matrix. Cellular communication takes place within the tumor microenvironment, both via cell to cell contact, and through extracellular mechanisms such as exosomal signaling. Exosomes are very small membrane-bound vesicles that have been shown to play key roles in the progression of cancer including modulation of the tumor microenvironment through the induction of angiogenesis, the transfer of genetic information that confers drug resistance, and increased cell migration, invasion, proliferation, and survival, as well as the modulation of immune cell interactions. The role of exosomes in several different cancers has been investigated. In the context of osteosarcoma, understanding how exosomes may modulate the tumor microenvironment to support metastatic growth particularly in the lung, the most common site of metastases, may identify novel therapeutic targets for relapsed patients.

K. Wolf-Dennen (✉) · E. S. Kleinerman
The University of Texas MD Anderson Cancer Center, Department of Pediatrics, Houston, TX, USA
e-mail: KKWolf@mdanderson.org

Keywords

Exosomes · Extracellular vesicles · Etracellular communication · Tumor microenvironment · Cell signaling · Osteosarcoma progression and metastasis

Introduction

Cell-to-cell communication within complex tissue microenvironments such as cancer is crucial to the sustained growth, invasion, and metastasis of cancer cells [1]. It is also becoming increasingly recognized that the tumor microenvironment is dynamic, heterogeneous, and composed of a wide variety of cells including tumor cells, innate and adaptive immune cells, fibroblasts, pericytes, endothelial cells, and mesenchymal stem cells [1, 2]. The tumor microenvironment is also comprised of extracellular matrix, collagen, fibrotic tissue, necrotic tissues, areas of hypoxia, and a diverse array of growth factors. Communication within the tumor microenvironment is complex as it can occur either intercellularly or extracellularly, which can be mediated by

E. S. Kleinerman, R. Gorlick (eds.), *Current Advances in the Science of Osteosarcoma*, Advances in Experimental Medicine and Biology 1258, https://doi.org/10.1007/978-3-030-43085-6_13

direct contact between cells or by the transfer of secreted molecules or extracellular vesicles such as exosomes [3–5].

Exosomes are very small, 30–150-nm vesicles that have been shown to be an integral part of intercellular communication [4]. Furthermore, they have been shown to be released by a wide variety of cells, including both normal cells and cancer cells in vitro and in vivo [4]. Exosomes can also be found in many body fluids, including blood, urine, semen, breast milk, saliva, amniotic fluid as well as ascites fluid [6]. Exosomes arise at the lipid raft domain of the plasma membrane, where endocytosis leads to the intracellular formation of an early endosome. These endosomes undergo a maturation process and become late endosomes, which then invaginate and give rise to multivesicular bodies. Then, these multivesicular bodies are either degraded within lysosomes or are released into the extracellular space as exosomes [4, 7]. In addition, exosomes have also been shown to be secreted by pathogens such as fungi, archaea, mycobacteria, and bacteria, as well as plant and animal cells, which suggests that the biogenesis and subsequent exosomal communication between cells is an important evolutionary conserved signaling mechanism [6].

Exosomes have been shown to contain a wide variety of constituents, such as proteins, mRNAs, miRNAs, as well as both single-stranded and double-stranded DNAs [8]. In addition, the generation of exosomes by inward budding ensures that the membrane proteins on the surface of an exosome preserve the same orientation and folding as those on the plasma membrane [7]. Exosomes have been shown to be a rich source of biomarkers and have also been shown to play a role in a wide variety of normal physiological processes such as tissue regeneration, angiogenesis, autophagy, blood coagulation, immunomodulation, stem cell differentiation, wound healing, pregnancy, as well as cancer progression [6, 8–11]. Exosomes have also been proposed to mediate cellular communication during the normal development of the nervous system and regeneration of normal neurons [8, 10]. Importantly, the discovery that the contents of an exosomes can be transferred to recipient cells and can mediate both cellular signaling and phenotypic changes within cells supports the idea that exosomes are dynamic mediators of intercellular communication, both locally and distantly [6].

The Role of Exosomes in Cancer

Exosomes have been detected within the tumor microenvironment, and emerging evidence suggests that exosomes play a role in regulating tumor proliferation, survival, angiogenesis, resistance, migration, invasion, lymphogenesis, cancer development, progression, and metastasis [9]. An increase in the rate of exosome secretion as well as differential cargo expression has been shown to be favorable in the development of cancer [12]. Within the tumor microenvironment, exosomes can mediate interactions between the following cell types: cancer cell to cancer cell, cancer cell to stromal cell, or stromal cell to cancer cell. For example, exosomes derived from a glioblastoma-astrocytoma cell line, which contain mutant EGFR variant III (EGFRvIII) expression can transform glioblastoma cells lacking EGFRvIII expression [13]. Breast cancer–derived exosomes can also transform nonmalignant mouse fibroblast cells, which will form tumors [14]. It has also been shown that exosomes can transfer oncogenic proteins or fusion gene mRNA, as well as oncogenic lncRNAs from cancer cells to other cells within the tumor microenvironment [6]. In addition, nonmalignant breast cells form tumors when exposed to exosomes from breast cancer cells or exosomes isolated from the serum of breast cancer patients [15]. Exosomes play many roles in the tumor microenvironment, and it is becoming increasingly recognized that they may play a much bigger role in tumor growth and metastasis than originally thought.

The Role of Exosomes in Drug Resistance

The ability of cancer cells to adapt and become resistant to radiotherapy, chemotherapy, or even immunotherapy is an unsolved problem to date. It is becoming increasingly understood that exosomes may contribute to resistance, specifically because of their ability to confer the resistant phenotype to nonresistant cancer cells in the tumor microenvironment using several different mechanisms [16]. Exosomes have been shown to package chemotherapeutic agents such as cisplatin from the cytosol in order to protect the surrounding cells from the cytotoxic effects of the drug [17]. Additionally, exosomes have also been shown to mediate resistance from a drug-resistant cell to a drug-sensitive cell simply by the transfer of exosomal contents [16]. For example, Adriamycin- and docetaxel-resistant breast cancer cells transferred miRNAs mediating resistance via exosomes to sensitive cell lines, resulting in the conference of resistance to both drugs [18]. Additionally, breast cancer cell exosomes have been found to carry HER2 and have also been shown to scavenge trastuzumab, thereby reducing its availability in circulation [19, 20]. In prostate cancer, exosomes have been shown to both carry and induce production of multidrug resistance protein (MDR-1/P-gp), which interferes with drug uptake, increasing cellular resistance to drugs such as docetaxel and anthracyclines [21]. Many studies have identified the exosomal transfer of miRNAs that mediate drug resistance, but more comprehensive genetic and proteomic investigations are needed to better understand the complex mechanisms responsible for exosome-mediated induction of resistance, how this is transferred between sensitive and resistant cells, and which signaling pathways are involved.

The Role of Exosomes in Angiogenesis and Lymphogenesis

Exosomes have also been shown to modulate the tumor microenvironment, which includes their ability to stimulate angiogenesis and lymphogenesis. Cancer cell–derived exosomes can carry TSPAN8 and integrin subunit α4 and have been shown to upregulate angiogenesis-related genes [22]. Additionally, exosomes have also been shown to carry soluble E-cadherin, which is a potent inducer of angiogenesis and were also shown to activate β-catenin and NF-kβ signaling [23]. It has been suggested that tumor-derived exosomes containing EGFR can activate endothelial cells to produce VEGF and upregulate VEGFR2 signaling [24]. Cancer cell–derived exosomes have also been shown to carry NOTCH ligand Delta-like 4 (DLL4), which resulted in increased vessel density and branching in vivo [25]. Within a hypoxic tumor microenvironment, exosomes have been shown to be especially important in the promotion of angiogenesis. For example, under hypoxic conditions, lung cancer cells produced more exosomes enriched in mir-23a, which resulted in accumulation of hypoxia-inducible factor-1-alpha (HIF1A) in endothelial cells and also targeted tight junction protein ZO1, thereby increasing vascular permeability [26]. Exosomes have also been shown to modulate the lymphatic system. As an example, exosomes released from melanoma cells have been shown to prepare the sentinel lymph node for tumor metastasis [27]. An exosome's ability to influence endothelial cells, modify both local and distant microenvironments, and promote angiogenesis can facilitate cancer progression and metastasis.

The Role of Exosomes in Cancer Cell Proliferation and Survival

Exosomes have also been shown to contain factors that modulate cancer cell proliferation and survival within the tumor microenvironment. Exosomes derived from chronic myeloid leukemia cells have been shown to contain the cytokine TGFβ-1, which promoted tumor growth through the activation of antiapoptotic pathways, as well as ERK and AKT [28]. Exosomes from cancer-associated fibroblasts have been shown to carry a signal recognition particle RNA that stimulated DDX58 signaling in breast cancer cells, which lead to increased chemoresistance as well as increased proliferation [29]. Gastric cancer–derived exosomes have been shown to increase proliferation in part due to PI3K/AKT activation [30]. In bladder cancer, exosomes have been suggested to activate AKT and ERK, thereby promoting proliferation and inhibiting apoptosis in recipient bladder cancer cells [31]. Colon cancer–derived exosomes containing mutant $KRAS^{G12D}$ were shown to induce anchorage-independent growth in colon cancer cells expressing the wild-type KRAS allele [32]. Exosomes have been associated with increased proliferation and decreased apoptosis in many cancer types, including but not limited to hepatocellular carcinoma, breast cancer, lung cancer, and osteosarcoma [6, 9, 11, 12]. Additionally, exosomes have also been associated with replicative immortality, a phenomenon that is associated with telomerase activation [33]. With each cell replication, cell telomeres are shortened until the cell reaches a nondividing stage or replicative senescence. Telomere length is controlled by the enzyme telomerase [33]. High expression of telomerase is often associated with cancer [33]. TERT, the catalytic subunit of telomerase, has been found to be contained within serum-derived exosomes from cancer patients [34]. Additionally, TERT mRNA (hTERT) has been found in cancer cell–derived exosomes, including pancreatic cancer and lung cancer [35]. When cancer cell–derived exosomes were added to fibroblasts, these contents were taken up and telomerase was activated, proliferation was increased, and cellular senescence was delayed, thereby extending the cell's lifespan [35]. These findings demonstrate that exosomes play a significant role in cancer cell proliferation and survival. Future studies must now focus on understanding the mechanisms by which cancer exosomes affect these functional changes.

The Role of Exosomes in Cancer Migration, Invasion, and Metastasis

Exosomes have also been shown to play a significant role in cancer cell migration, invasion, and metastasis. Exosomes can also control cell polarity and directional cell movement [11]. Fibrosarcoma cell–secreted exosomes that bound to cell surface integrin receptors facilitated the clustering of integrins, as well as the formation of a strong adhesion at the leading edge of the cell that promoted cellular migration [36]. Cancer-associated exosomes specifically loaded with WNT1 promoted the protrusion of breast cancer cells, as well as invasion and metastasis via autocrine activation of the WNT-planar cell polarity signaling pathway [37]. Exosomes have also been shown to modify the extracellular matrix (ECM) of the tumor microenvironment. Metastatic breast cancer–derived exosomes contained activated MMP2, which is a protease that degrades ECM and promotes invasion [38]. Additionally, exosomes have also been shown to unlock tight junctions of endothelial cells in breast cancer, which allows for extravasation into the surrounding vasculature and the promotion of metastasis [39]. Breast cancer–derived exosomes containing miR-181c were shown to compromise the blood–brain barrier by downregulating PDPK1 in endothelial cells, which resulted in the abnormal localization of actin and increasing metastasis to the brain [40]. Exosomes have also been shown to promote the development of a recipient cell's epithelial–mesenchymal transition (EMT) [16]. Factors associated with EMT such as TGFβ-1, β-catenin, and matrix metalloproteinases have been found to be contained within cancer cell exosomes [41–44]. In the exosomes of a highly metastatic lung cancer cell line, levels of vimentin were higher and conferred increased levels of migration, invasion, and proliferation when compared to the exosomes of its parental counterpart [45]. In another study, cancer-derived exosomes contained mir-21, which markedly increased the levels of vimentin and snail as well as decreased levels of E-cadherin, which are all markers associated with EMT [46, 47]. It is known that primary tumors release specific growth factors or cytokines, which promotes the development of metastasis. It has also been recently shown that exosomes can play a role in the development of the premetastatic niche. Cancer cell exosomes can condition lymph nodes or lungs tissues to become favorable to the metastatic colonization and subsequent outgrowth of melanoma cells [27, 48]. In prostate cancer, exosomes promoted osteoblast activity, which regulated the microenvironment of bone metastases [49, 50]. Additionally, exosomes from pancreatic cancer cells were shown to contain miR-494 and miR-542-3p, which resulted in the downregulation of cadherin-17 [48]. This led to increased levels of proteases, adhesion molecules, and other proteins in the lung and lymph node, which modulated the local microenvironment and made it more amenable to tumor metastasis [48]. Melanoma-derived exosomes have been shown to deliver the tyrosine kinase receptor MET to bone marrow progenitor cells, which in turn activates HGF-MET signaling, which subsequently promotes tumor metastasis to the bone [51]. The development of the metastatic niche has been observed in many cancers, such as pancreatic cancer, breast cancer, melanoma, lung cancer, and gastric cancer [11].

The Role of Exosomes in Cancer and Immune Cell Interactions

In recent years, it has been increasingly clear that cancer cell exosomes play a major role in the modulation of immune cell signaling and function within the tumor microenvironment. The first evidence that suggested that exosomes could modulate the immune system was the discovery that exosomes derived from B-cells that were transformed by Epstein–Bar virus contain MHCI and MHCII on their surface. It was later found that these exosomes were capable of antigen presentation and can activate CD4+ T cells, which showed that they can modulate immune cell function independently of direct cell-to-cell contact [52, 53]. Since that first discovery, exosomes have been shown to modulate multiple different types of immune cells, activate or suppress an immune-mediated tumor response, assist in tumor escape from immune surveillance, and activate various immunosuppressive pathways that support the continual growth of the tumor [54].

Exosomes have been shown to modulate innate immune cells that may be found in the tumor microenvironment, which include dendritic cells, natural killer cells, neutrophils, monocytes, and macrophages [54]. Additionally, they've also been shown to modulate the adaptive cells of the immune system, T cell and B cells [54]. Exosomes from tumors have been shown to transfer HSP70-80 as well as MHCI to dendritic cells, which in turn stimulates potent CD8+ antitumor effects [55, 56]. HSP70 on the exosomal surface has also been shown to stimulate natural killer cell migration and cytolytic activity, induce a stronger T helper cell response, and activate macrophages [57–59]. Although exosomes are capable of activating an antitumor response, they also have a very strong protumor effect. Tumor-derived exosomes have been shown to promote T-regulatory cells and have also been shown to inhibit both natural killer and T cell functions [54]. FAS ligand is a potent activator of cellular apoptosis and has been shown to be presented by tumor-derived exosomes, resulting in the apoptosis of activated T cells [60–62]. Melanoma cell exosomes stimulate high levels of reactive oxygen species, which in turn inhibits the activity of surrounding T cells [63]. Tumor-derived exosomes have also been shown to express PDL1, which supports tumor growth by inhibiting the activity of activated T cell function [64]. Pancreatic cancer exosomes have been shown to inhibit dendritic cell activity by downregulating TLR4 via miR-203 [65]. Exosomes derived from breast cancer cells were shown to be taken up by bone marrow myeloid progenitor cells, which impaired dendritic cell differentiation by the phosphorylation of STAT3 and the overexpression of IL6 [66]. Lung cancer exosomes have been shown to block the differentiation of dendritic cells by downregulating the expression of MHCII, CD86, and CD80 while also upregulating the expression of PD-L1 and CD11B [67]. Tumor-derived exosomes have also been shown to make adenosine from ATP by carrying the enzymatically active CD39 and CD73 and have been implicated in the suppression of activated B cells [68].

Tumor cell exosomes have been shown to modulate macrophage polarization. M1 macrophages are associated with antitumor functions as well as the stimulation of the local immune environment [69]. M2 macrophages are typically immunosuppressive and have been shown to be tumor promoting. Tumor cell exosomes have been shown to modulate this phenotype, resulting in an M2-like immunosuppressive phenotype [69]. Exosomes derived from glioblastoma stem cells not only induced an M2 phenotype, but also induced expression of PD-L1 on macrophages

[70]. Breast cancer–derived exosomes were able to modify macrophage polarization through gp130/STAT3 signaling [71]. Exosomes shed from colon cancer cells were shown to contain miR-1246, which induced the M2 macrophage phenotype with high levels of TGFβ expression [72]. Gastric cancer exosomes induced the expression of PD-L1 on the surface of macrophages, and impaired CD8+ T cell function via IL10 secretion [73]. Melanoma-, breast cancer-, and oral squamous cell carcinoma–derived exosomes have been shown to not only activate macrophages to an M1 phenotype but also facilitate the migration of cancer cells, metastasis, and immune escape, thus still playing a role in the progression of the tumor [74–76]. The role of exosomes in the tumor microenvironment can therefore be both tumor suppressive and tumor promoting. It will be important to investigate the mechanisms involved in this complex modification of cellular signaling, especially in the context of improving the efficiency of immunotherapy in cancer patients.

The Emerging Role of Exosomes in Osteosarcoma

The role of exosomes in osteosarcoma is still being elucidated and is not yet well understood. The first exosomes to be reported in the context of osteosarcoma found that extracellular vesicle–bound urokinase plasminogen activator (uPA) contributed to the progression of nonmetastatic osteosarcoma to a metastatic phenotype [77]. In a subsequent study, it was shown that human osteosarcoma cell exosomes can transfer resistance to doxorubicin by a mechanism involving the upregulation of MDR-1 and P-glycoprotein [78]. Proteomic analysis of osteosarcoma cell line–derived exosomes has suggested contents associated with tumor growth and metastasis [79]. Notch-activating factors have been observed inside osteosarcoma-derived exosomes and have been shown to mediate skeletal muscle atrophy in cancer cachexia [80]. Osteosarcoma cell–derived exosomes have also been shown to reduce the rate of T cell proliferation and activity, as well as promote a T regulatory phenotype [81]. Additionally, exosomes derived from bone marrow mesenchymal stem cells promoted cancer progression and tumor growth of human osteosarcoma cells, and increased proliferation, migration, invasion, and apoptotic resistance of osteosarcoma cells [82–85]. Metastatic osteosarcoma cell–derived exosomes have also been shown to contain miR-675, which promotes cell invasion and migration through CALN1 [86]. miR-1228 has also been shown to be contained in osteosarcoma-derived exosomes, which increased cell migration and invasion through SCA1 [87]. Extracellular vesicle–bound miR-25-3p promotes the capillary formation and invasion of vascular endothelial cells, thereby mediating angiogenesis and promoting tumor progression [88]. Osteosarcoma cell–derived exosomes have also been shown to promote osteoclast differentiation, bone resorption activity, tube formation of endothelial cells, and increase angiogenic markers [89]. Osteosarcoma-derived extracellular vesicles have been shown to induce a tumor-like phenotype in normal recipient fibroblasts [90]. Additionally, osteosarcoma exosomes that express a membrane associated form of TGFβ have been shown to educate mesenchymal stem cells to an inflammatory phenotype, which then leads to osteosarcoma progression [85]. Finally, extracellular vesicles secreted by highly metastatic clonal variants of osteosarcoma preferentially localize to the lungs. These vesicles are capable of inducing a metastatic phenotype in poorly metastatic clones [91]. The role of exosomes in osteosarcoma growth and metastasis, how exosomes from metastatic cells differ from the ones produced by nonmetastatic cells, and how metastatic-derived exosomes alter the organ microenvironment specifically in the lung to make it supportive of tumor growth needs to be elucidated.

Conclusions

The tumor microenvironment is a complex ecosystem composed of many different cells and components. The organ microenvironment for the primary tumor in the bone is not the same as the metastatic site, that is, the lungs. Exosomes play a key role in extracellular communication between cells. Therefore, future studies need to include a focus on exosomes from metastatic osteosarcoma cells and the role that exosomes play in the process of primary tumor development, progression, and metastasis and how exosomes modify the organ microenvironment to make it support growth at the distant site. A broader understanding of the role that exosomes play in extracellular communication of cancer cells may reveal novel therapeutic targets and opportunities and improve the efficiency of current therapies.

References

1. Hanahan D, Weinberg RA (2011) Hallmarks of cancer: the next generation. Cell 144(5):646–674
2. Kalluri R (2016) The biology and function of fibroblasts in cancer. Nat Rev Cancer 16(9):582–598
3. Chen W et al (2017) Tumor-related exosomes contribute to tumor-promoting microenvironment: an immunological perspective. J Immunol Res 2017:1073947
4. Keller S et al (2006) Exosomes: from biogenesis and secretion to biological function. Immunol Lett 107(2):102–108
5. Raposo G, Stoorvogel W (2013) Extracellular vesicles: exosomes, microvesicles, and friends. J Cell Biol 200(4):373–383
6. Xu R et al (2018) Extracellular vesicles in cancer – implications for future improvements in cancer care. Nat Rev Clin Oncol 15(10):617–638
7. Kowal J, Tkach M, Thery C (2014) Biogenesis and secretion of exosomes. Curr Opin Cell Biol 29:116–125
8. Vlassov AV et al (2012) Exosomes: current knowledge of their composition, biological functions, and diagnostic and therapeutic potentials. Biochim Biophys Acta 1820(7):940–948
9. Rajagopal C, Harikumar KB (2018) The origin and functions of exosomes in cancer. Front Oncol 8:66
10. Yanez-Mo M et al (2015) Biological properties of extracellular vesicles and their physiological functions. J Extracell Vesicles 4:27066
11. Zhang L, Yu D (2019) Exosomes in cancer development, metastasis, and immunity. Biochim Biophys Acta Rev Cancer 1871(2):455–468
12. Bebelman MP et al (2018) Biogenesis and function of extracellular vesicles in cancer. Pharmacol Ther 188:1–11
13. Al-Nedawi K et al (2008) Intercellular transfer of the oncogenic receptor EGFRvIII by microvesicles derived from tumour cells. Nat Cell Biol 10(5):619–624
14. Antonyak MA et al (2011) Cancer cell-derived microvesicles induce transformation by transferring tissue transglutaminase and fibronectin to recipient cells. Proc Natl Acad Sci U S A 108(12):4852–4857
15. Melo SA et al (2014) Cancer exosomes perform cell-independent microRNA biogenesis and promote tumorigenesis. Cancer Cell 26(5):707–721
16. Whiteside TL (2016) Tumor-derived exosomes and their role in cancer progression. Adv Clin Chem 74:103–141
17. Safaei R et al (2005) Abnormal lysosomal trafficking and enhanced exosomal export of cisplatin in drug-resistant human ovarian carcinoma cells. Mol Cancer Ther 4(10):1595–1604
18. Mrizak D et al (2015) Effect of nasopharyngeal carcinoma-derived exosomes on human regulatory T cells. J Natl Cancer Inst 107(1):363
19. Amorim M et al (2014) The overexpression of a single oncogene (ERBB2/HER2) alters the proteomic landscape of extracellular vesicles. Proteomics 14(12):1472–1479
20. Ciravolo V et al (2012) Potential role of HER2-overexpressing exosomes in countering trastuzumab-based therapy. J Cell Physiol 227(2):658–667
21. Corcoran C et al (2012) Docetaxel-resistance in prostate cancer: evaluating associated phenotypic changes and potential for resistance transfer via exosomes. PLoS One 7(12):e50999
22. Nazarenko I et al (2010) Cell surface tetraspanin Tspan8 contributes to molecular pathways of exosome-induced endothelial cell activation. Cancer Res 70(4):1668–1678
23. Tang MKS et al (2018) Soluble E-cadherin promotes tumor angiogenesis and localizes to exosome surface. Nat Commun 9(1):2270
24. Al-Nedawi K et al (2009) Endothelial expression of autocrine VEGF upon the uptake of tumor-derived microvesicles containing oncogenic EGFR. Proc Natl Acad Sci U S A 106(10):3794–3799
25. Sheldon H et al (2010) New mechanism for Notch signaling to endothelium at a distance by Delta-like 4 incorporation into exosomes. Blood 116(13):2385–2394
26. Hsu YL et al (2017) Hypoxic lung cancer-secreted exosomal miR-23a increased angiogenesis and vascular permeability by targeting prolyl hydroxylase and tight junction protein ZO-1. Oncogene 36(34):4929–4942
27. Hood JL, San RS, Wickline SA (2011) Exosomes released by melanoma cells prepare sentinel lymph nodes for tumor metastasis. Cancer Res 71(11):3792–3801
28. Raimondo S et al (2015) Chronic myeloid leukemia-derived exosomes promote tumor growth through an autocrine mechanism. Cell Commun Signal 13:8
29. Nabet BY et al (2017) Exosome RNA unshielding couples stromal activation to pattern recognition receptor signaling in cancer. Cell 170(2):352–366.e13
30. Qu JL et al (2009) Gastric cancer exosomes promote tumour cell proliferation through PI3K/Akt and MAPK/ERK activation. Dig Liver Dis 41(12):875–880

31. Yang L et al (2013) Bladder cancer cell-derived exosomes inhibit tumor cell apoptosis and induce cell proliferation in vitro. Mol Med Rep 8(4):1272–1278
32. Demory Beckler M et al (2013) Proteomic analysis of exosomes from mutant KRAS colon cancer cells identifies intercellular transfer of mutant KRAS. Mol Cell Proteomics 12(2):343–355
33. Al-Sowayan BS, Al-Shareeda AT, Al-Hujaily EM (2019) Exosomes, cancer's little army. Stem Cell Investig 6:9
34. Goldvaser H et al (2017) Characterisation of blood-derived exosomal hTERT mRNA secretion in cancer patients: a potential pan-cancer marker. Br J Cancer 117(3):353–357
35. Gutkin A et al (2016) Tumor cells derived exosomes contain hTERT mRNA and transform nonmalignant fibroblasts into telomerase positive cells. Oncotarget 7(37):59173–59188
36. Sung BH et al (2015) Directional cell movement through tissues is controlled by exosome secretion. Nat Commun 6:7164
37. Luga V et al (2012) Exosomes mediate stromal mobilization of autocrine Wnt-PCP signaling in breast cancer cell migration. Cell 151(7):1542–1556
38. Hendrix A et al (2010) Effect of the secretory small GTPase Rab27B on breast cancer growth, invasion, and metastasis. J Natl Cancer Inst 102(12):866–880
39. Zhou W et al (2014) Cancer-secreted miR-105 destroys vascular endothelial barriers to promote metastasis. Cancer Cell 25(4):501–515
40. Tominaga N et al (2015) Brain metastatic cancer cells release microRNA-181c-containing extracellular vesicles capable of destructing blood-brain barrier. Nat Commun 6:6716
41. Xiao D et al (2016) Melanoma cell-derived exosomes promote epithelial-mesenchymal transition in primary melanocytes through paracrine/autocrine signaling in the tumor microenvironment. Cancer Lett 376(2):318–327
42. Hakulinen J et al (2008) Secretion of active membrane type 1 matrix metalloproteinase (MMP-14) into extracellular space in microvesicular exosomes. J Cell Biochem 105(5):1211–1218
43. Chairoungdua A et al (2010) Exosome release of beta-catenin: a novel mechanism that antagonizes Wnt signaling. J Cell Biol 190(6):1079–1091
44. Webber JP et al (2015) Differentiation of tumour-promoting stromal myofibroblasts by cancer exosomes. Oncogene 34(3):290–302
45. Kidd ME, Shumaker DK, Ridge KM (2014) The role of vimentin intermediate filaments in the progression of lung cancer. Am J Respir Cell Mol Biol 50(1):1–6
46. Li L et al (2016) Exosomes derived from hypoxic oral squamous cell carcinoma cells deliver miR-21 to normoxic cells to elicit a prometastatic phenotype. Cancer Res 76(7):1770–1780
47. Wang Y et al (2013) The role of snail in EMT and tumorigenesis. Curr Cancer Drug Targets 13(9):963–972
48. Rana S, Malinowska K, Zoller M (2013) Exosomal tumor microRNA modulates premetastatic organ cells. Neoplasia 15(3):281–295
49. Ye Y et al (2017) Exosomal miR-141-3p regulates osteoblast activity to promote the osteoblastic metastasis of prostate cancer. Oncotarget 8(55):94834–94849
50. Karlsson T et al (2016) Tumor cell-derived exosomes from the prostate cancer cell line TRAMP-C1 impair osteoclast formation and differentiation. PLoS One 11(11):e0166284
51. Peinado H et al (2012) Melanoma exosomes educate bone marrow progenitor cells toward a pro-metastatic phenotype through MET. Nat Med 18(6):883–891
52. Zitvogel L et al (1998) Eradication of established murine tumors using a novel cell-free vaccine: dendritic cell-derived exosomes. Nat Med 4(5):594–600
53. Raposo G et al (1996) B lymphocytes secrete antigen-presenting vesicles. J Exp Med 183(3):1161–1172
54. Whiteside TL (2017) The effect of tumor-derived exosomes on immune regulation and cancer immunotherapy. Future Oncol 13(28):2583–2592
55. Altieri SL, Khan AN, Tomasi TB (2004) Exosomes from plasmacytoma cells as a tumor vaccine. J Immunother 27(4):282–288
56. Wolfers J et al (2001) Tumor-derived exosomes are a source of shared tumor rejection antigens for CTL cross-priming. Nat Med 7(3):297–303
57. Cho JA et al (2009) MHC independent anti-tumor immune responses induced by Hsp70-enriched exosomes generate tumor regression in murine models. Cancer Lett 275(2):256–265
58. Vega VL et al (2008) Hsp70 translocates into the plasma membrane after stress and is released into the extracellular environment in a membrane-associated form that activates macrophages. J Immunol 180(6):4299–4307
59. Gastpar R et al (2005) Heat shock protein 70 surface-positive tumor exosomes stimulate migratory and cytolytic activity of natural killer cells. Cancer Res 65(12):5238–5247
60. Klibi J et al (2009) Blood diffusion and Th1-suppressive effects of galectin-9-containing exosomes released by Epstein-Barr virus-infected nasopharyngeal carcinoma cells. Blood 113(9):1957–1966
61. Abusamra AJ et al (2005) Tumor exosomes expressing Fas ligand mediate CD8+ T-cell apoptosis. Blood Cells Mol Dis 35(2):169–173
62. Iero M et al (2008) Tumour-released exosomes and their implications in cancer immunity. Cell Death Differ 15(1):80–88
63. Soderberg A et al (2007) Redox-signaling transmitted in trans to neighboring cells by melanoma-derived TNF-containing exosomes. Free Radic Biol Med 43(1):90–99
64. Wieckowski EU et al (2009) Tumor-derived microvesicles promote regulatory T cell expansion and induce apoptosis in tumor-reactive activated CD8+ T lymphocytes. J Immunol 183(6):3720–3730
65. Fanini F, Fabbri M (2017) Cancer-derived exosomic microRNAs shape the immune system within the

tumor microenvironment: state of the art. Semin Cell Dev Biol 67:23–28
66. Yu S et al (2007) Tumor exosomes inhibit differentiation of bone marrow dendritic cells. J Immunol 178(11):6867–6875
67. Ning Y et al (2018) Tumor exosomes block dendritic cells maturation to decrease the T cell immune response. Immunol Lett 199:36–43
68. Schuler PJ et al (2014) Human CD4+ CD39+ regulatory T cells produce adenosine upon co-expression of surface CD73 or contact with CD73+ exosomes or CD73+ cells. Clin Exp Immunol 177(2):531–543
69. Biswas SK, Mantovani A (2010) Macrophage plasticity and interaction with lymphocyte subsets: cancer as a paradigm. Nat Immunol 11(10):889–896
70. Gabrusiewicz K et al (2018) Glioblastoma stem cell-derived exosomes induce M2 macrophages and PD-L1 expression on human monocytes. Onco Targets Ther 7(4):e141 2909
71. Ham S et al (2018) Breast cancer-derived exosomes alter macrophage polarization via gp130/STAT3 signaling. Front Immunol 9:871
72. Cooks T et al (2018) Mutant p53 cancers reprogram macrophages to tumor supporting macrophages via exosomal miR-1246. Nat Commun 9(1):771
73. Wang F et al (2018) Tumor-derived exosomes induce PD1(+) macrophage population in human gastric cancer that promotes disease progression. Oncogenesis 7(5):41
74. Chow A et al (2014) Macrophage immunomodulation by breast cancer-derived exosomes requires Toll-like receptor 2-mediated activation of NF-kappaB. Sci Rep 4:5750
75. Marton A et al (2012) Melanoma cell-derived exosomes alter macrophage and dendritic cell functions in vitro. Immunol Lett 148(1):34–38
76. Xiao M et al (2018) M1-like tumor-associated macrophages activated by exosome-transferred THBS1 promote malignant migration in oral squamous cell carcinoma. J Exp Clin Cancer Res 37(1):143
77. Endo-Munoz L et al (2015) Progression of osteosarcoma from a non-metastatic to a metastatic phenotype is causally associated with activation of an autocrine and paracrine uPA axis. PLoS One 10(8):e0133592
78. Torreggiani E et al (2016) Multimodal transfer of MDR by exosomes in human osteosarcoma. Int J Oncol 49(1):189–196
79. Jerez S et al (2017) Proteomic analysis of exosomes and exosome-free conditioned media from human osteosarcoma cell lines reveals secretion of proteins related to tumor progression. J Cell Biochem 118(2):351–360
80. Mu X et al (2016) Notch signaling mediates skeletal muscle atrophy in cancer cachexia caused by osteosarcoma. Sarcoma 2016:3758162
81. Troyer RM et al (2017) Exosomes from Osteosarcoma and normal osteoblast differ in proteomic cargo and immunomodulatory effects on T cells. Exp Cell Res 358(2):369–376
82. Lin S et al (2019) Microvesicles derived from human bone marrow mesenchymal stem cells promote U2OS cell growth under hypoxia: the role of PI3K/AKT and HIF-1alpha. Hum Cell 32(1):64–74
83. Qi J et al (2017) Exosomes derived from human bone marrow mesenchymal stem cells promote tumor growth through hedgehog signaling pathway. Cell Physiol Biochem 42(6):2242–2254
84. Qin F et al (2020) Bone marrow-derived mesenchymal stem cell-derived exosomal microRNA-208a promotes osteosarcoma cell proliferation, migration, and invasion. J Cell Physiol 235(5):4734–4745
85. Baglio SR et al (2017) Blocking tumor-educated MSC paracrine activity halts osteosarcoma progression. Clin Cancer Res 23(14):3721–3733
86. Gong L et al (2018) Exosomal miR-675 from metastatic osteosarcoma promotes cell migration and invasion by targeting CALN1. Biochem Biophys Res Commun 500(2):170–176
87. Wang JW et al (2019) Exosomal miR-1228 from cancer-associated fibroblasts promotes cell migration and invasion of osteosarcoma by directly targeting SCAI. Oncol Res 27(9):979–986
88. Yoshida A et al (2018) Clinical and functional significance of intracellular and extracellular microRNA-25-3p in osteosarcoma. Acta Med Okayama 72(2):165–174
89. Raimondi L et al (2019) Osteosarcoma cell-derived exosomes affect tumor microenvironment by specific packaging of microRNAs. Carcinogenesis, 2019(7):1–12.
90. Urciuoli E et al (2018) Osteosarcoma-derived extracellular vesicles induce a tumor-like phenotype in normal recipient cells. J Cell Physiol 233(8):6158–6172
91. Macklin R et al (2016) Extracellular vesicles secreted by highly metastatic clonal variants of osteosarcoma preferentially localize to the lungs and induce metastatic behaviour in poorly metastatic clones. Oncotarget 7(28):43570–43587

14 Comparative Immunology and Immunotherapy of Canine Osteosarcoma

Nicola J. Mason

Abstract

Approximately 800 people are diagnosed with osteosarcoma (OSA) per year in the USA. Although 70% of patients with localized OSA are cured with multiagent chemotherapy and surgical resection, the prognosis for patients with metastatic or relapsed disease is guarded. The small number of patients diagnosed annually contributes to an incomplete understanding of disease pathogenesis, and challenges in performing appropriately powered clinical trials and detecting correlative biomarkers of response. While mouse models of OSA are becoming increasingly sophisticated, they generally fail to accurately recapitulate tumor heterogeneity, tumor microenvironment (TME), systemic immune dysfunction, and the clinical features of tumor recurrence, metastases, and chemoresistance, which influence outcome. Pet dogs spontaneously develop OSA with an incidence that is 30–50 times higher than humans. Canine OSA parallels the human disease in its clinical presentation, biological behavior, genetic complexity, and therapeutic management. However, despite therapy, most dogs die from metastatic disease within 1 year of diagnosis. Since OSA occurs in immune-competent dogs, immune factors that sculpt tumor immunogenicity and influence responses to immune modulation are in effect. In both species, immune modulation has shown beneficial effects on patient outcome and work is now underway to identify the most effective immunotherapies, combination of immunotherapies, and correlative biomarkers that will further improve clinical response. In this chapter, the immune landscape of canine OSA and the immunotherapeutic strategies used to modulate antitumor immunity in dogs with the disease will be reviewed. From this immunological viewpoint, the value of employing dogs with spontaneous OSA to accelerate and inform the translation of immunotherapies into the human clinic will be underscored.

Keywords

Canine osteosarcoma · Immunotherapy · Tumor microenvironment · Animal model · Immune modulation · Comparative oncology

N. J. Mason (✉)
Department of Clinical Sciences and Advanced Medicine, School of Veterinary Medicine, University of Pennsylvania, Philadelphia, PA, USA

Department of Pathobiology, School of Veterinary Medicine, University of Pennsylvania, Philadelphia, PA, USA

Parker Institute for Cancer Immunotherapy, University of Pennsylvania, Philadelphia, PA, USA
e-mail: nmason@vet.upenn.edu

E. S. Kleinerman, R. Gorlick (eds.), *Current Advances in the Science of Osteosarcoma*, Advances in Experimental Medicine and Biology 1258, https://doi.org/10.1007/978-3-030-43085-6_14

Introduction

OSA affects approximately 800 people per year in the USA, and, as such, it is subject to the challenges that orphan diseases present for therapeutic advances. The relatively small number of patients contributes to an incomplete understanding of the disease pathogenesis, challenges to performing appropriately powered, randomized, controlled clinical trials, and identifying correlative biomarkers of response that might direct patient stratification and improve outcomes. Furthermore, in comparison to more common cancers, funding opportunities and dollars for basic and clinical research are limited. These factors have contributed to the lack of significant advances in the treatment of OSA for more than 30 years [1]. While mouse models of OSA are becoming increasingly sophisticated, they often fail to recapitulate tumor heterogeneity and the immunosuppressive microenvironment and systemic immune dysfunction that are frequently encountered in cancer patients [2–4]. These shortcomings are emphasized when using murine models to evaluate the safety and efficacy of immunotherapeutic agents, that act on the immune system to augment antitumor immunity and prevent metastatic disease. Metastasis to the lung, bone, and soft tissues is the principal cause of death in OSA patients, and this natural progression of the disease is also poorly modeled in murine systems.

Pet dogs spontaneously develop OSA with an incidence that is 30–50 times higher than humans (~45,000 cases/year in the USA) and has a lifetime risk of up to 10% in predisposed breeds [5]. OSA arising spontaneously in large breed dogs parallels the human disease in its clinical presentation, biological behavior, genetic complexity, and therapeutic management [6, 7]. However, despite therapy, most dogs will develop metastatic disease, which is ultimately responsible for the death of the canine patient within 1 year of diagnosis [6]. As such, dogs with spontaneous OSA have been recognized as a valuable comparative model of pediatric OSA in which to investigate disease pathogenesis and evaluate therapies to prevent and treat metastatic disease [8]. Over 75% of canine OSA occurs in the long bones with the metaphyseal region of the distal radius, proximal humerus and distal femur being the most common sites affected [9]. Commonly affected sites in humans are the distal femur and proximal tibia, with the proximal humerus and distal radius, both non-weight-bearing bones in humans compared to dogs, being less frequently affected [10, 11]. In contrast to OSA in humans, middle- to older-aged adult, skeletally mature dogs are most commonly affected although there is a second, smaller peak incidence at 12–24 months of age [12]. Similar to pediatric OSA, the majority of canine OSAs are high grade, and elevated serum levels of alkaline phosphatase and metastatic disease are poor prognostic indicators in both species [13, 14]. In both species, metastatic disease occurs in the lungs, bone, and soft tissue. Unlike pediatric patients, standard of care in dogs consists of surgical resection followed by four to six cycles of adjuvant carboplatin and/or doxorubicin chemotherapy [15, 16]. However, although improved survival is seen with the addition of chemotherapy after primary tumor removal [15, 17], up to 90% of dogs develop metastatic disease despite standard of care with overall survival times ranging from 8–12 months [18]. Indeed, as many as 25% of dogs receiving chemotherapy will develop gross metastatic disease within 14 weeks of amputation, underscoring the aggressive nature of the disease in dogs and suggesting that neoadjuvant therapies might be worthy of investigation [19]. Multiple chemotherapeutic strategies have been investigated in dogs with OSA over the last 30 years; however, none have significantly improved disease-free intervals or median survival times [16, 19–22].

Given the similarities identified between OSA in dogs and humans, researchers from diverse scientific disciplines have taken advantage of the canine translational "model" to inform diagnostics, prognostics, and therapeutics for both species. For example, taking advantage of the reduced genetic heterogeneity that occurs within specific dog breeds, molecular biologists have utilized genome-wide gene-expression profiling, exomic profiling, comparative genomic hybridization, and comparative transcriptomics to identify molecular subtypes, recurrent driver/suppressor gene mutations, and signatures that are predictive of outcome in dogs with OSA [23–27]. These signatures have been successfully applied to tumor samples from a more genetically diverse human patient population where they are also predictive of outcome [23, 24]. Surgeons and bioengineers have taken advantage of the canine OSA patient to develop limb-sparing techniques that have been applied to human patients [28]. The similar body size, genetic make-up, metabolism, and drug distribution kinetics between species have resulted in the canine OSA patient being a valuable asset for medical oncologists in evaluating safety and determining optimal dosing schedules of novel cytotoxic agents, small molecule inhibitors, and immune modulating agents to prevent metastatic disease. Finally, radiation oncologists have explored different radiation types and dosing schedules to optimize management of nonresectable lesions in the dog and to induce immunogenic cell death (ICD), which may augment immunotherapeutic strategies aimed at preventing metastatic disease [29].

While the canine OSA patient has already contributed much to our understanding of disease biology and the development of surgical and chemotherapeutic strategies to manage human OSA patients, perhaps, its greatest contribution will be realized in the development of safe and effective immunotherapies or combination immunotherapies to prevent metastatic disease. OSA is an immune-responsive tumor, and William Coley's observations in the late 1800s that concurrent bacterial infections increased patient survival provided some of the first evidence of this concept [30, 31]. Similar observations have been made more recently in canine OSA patients that experience surgical site infections following limb-sparing surgeries [32–34]. Although rare, spontaneous regression of primary OSA has also been reported in both species and is considered to be immunologically mediated [35–37]. Conversely, tumor-mediated suppression of innate and adaptive immunity occurs with many different types of neoplasia including OSA and this contributes to disease progression, metastases, and therapeutic resistance [38, 39]. Given the recent unprecedented success of immunotherapies including chimeric antigen receptor (CAR), T cells, and checkpoint inhibitors for the treatment of hematological and solid tumors with high mutational load, and the known immune responsiveness of OSA, there is increasing interest in evaluating immunotherapeutic strategies to treat OSA [40, 41]. Unlike immune-compromised rodent models of OSA that employ subcutaneous or orthotopic implantation of human tumor tissue or cell lines for research purposes, dogs that spontaneously develop OSA are immune competent, making them much better suited to evaluate therapies that act on the immune system to promote antitumor immunity. Furthermore, the spontaneity of tumor development means that tumor heterogeneity is preserved in dogs [42, 43] and the complex interplay that exists between the developing tumor and the immune system that sculpts tumor immunogenicity and directs the development of an immunosuppressive microenvironment is expected to be intact. Similar to pediatric OSA patients, dogs with OSA also exhibit systemic immune dysfunction that may serve as a significant barrier that needs to be overcome to improve response toimmunotherapies [38]. Since standard of care for dogs with OSA is surgical resection followed by cytotoxic chemotherapy, most canine OSA tissues are from chemotherapy-naïve primary samples, which may provide a more accurate assessment of tumor, tumor microenvironment (TME), and immune infiltrates than pediatric OSA samples taken at resection after multiagent chemotherapy. Finally, since owners of canine cancer patients may choose not to pursue standard of care, due to cost, patient size, or concerns surrounding quality-of-life issues, novel immunotherapies can be used at an earlier stage of disease compared to pediatric patients, increasing the likelihood of a favorable response that is not adversely affected by prolonged chemotherapy or advanced disease status.

Many questions now face immunotherapists aiming to improve the outcome of patients with OSA [44]. These include what is the immune status of the patient and of the tumor microenvironment (TME); how will these factors influence the clinical and immunological responses to immunotherapy; which tumor targets are relevant and safe; how can immunotherapies be rationally combined to broaden and augment antitumor immunity and provide a permissive TME to optimize antitumor effect; what are accurate biomarkers of response; and can they be employed to improve outcome via patient stratification? These clinically relevant questions can only be answered in patients with spontaneous tumors that exhibit tumor heterogeneity, recapitulate the tumor microenvironment, have intact and functional innate and adaptive immune responses, and either are known to develop metastatic disease with high frequency or already have metastatic disease.

Canine OSA patients present a spontaneous, immune competent "model" system that may be used to address a number of these questions and accelerate our understanding of OSA pathogenesis. Furthermore, they provide a valuable parallel patient population in which to evaluate the safety and efficacy of combination immunotherapeutic strategies and identify correlative biomarkers of clinical responses [45]. In this chapter, the immune landscape of canine OSA will be reviewed and compared with human OSA. Furthermore, the immunotherapeutic strategies that have been employed to modulate antitumor immunity in dogs with OSA will be presented. The review will examine the evidence that supports the use of canine patients to evaluate immunotherapeutic strategies, accelerate their translation into the human clinic, and identify correlative biomarkers that will assist in patient stratification for human clinical trials.

The Immunology of Canine OSA

Mutational Burden

The genetics of canine OSA have been well studied and are reviewed in detail elsewhere in this book. However, given that a tumor's somatic mutational load is the best predictor of neoepitope burden and, in turn, neoepitope burden is the most predictive measure of immune therapeutic response, it is worth mentioning here what is known about the mutational status of canine OSA [46]. Similar to humans, canine OSA exhibits considerable genomic instability from a karyotypic standpoint and shares microaberrations in commonly mutated genes such as p53 and Rb [47, 48]. However, OSA arising in humans and dogs generally exhibits a low point-mutation burden (median 1.98 mutations per Mb canine DNA) with a trend toward higher mutational loads in metastatic lesions [49–51]. Although the mutational burden of OSA is comparably low across the spectrum of evaluated human cancers, it is high in relation to other pediatric cancers, and nonsynonymous mutations may serve as potential neoantigens for tumor-specific T cells that may be augmented by immunotherapeutic strategies in both species [51, 52].

Immune Landscape

Understanding the factors that influence the immune responsiveness of OSA and identifying correlative biomarkers that predict this response are key to improving the outcome of human and canine patients with this disease. Here we provide a comparative overview of what is known about the immunological landscape of canine OSA, identifying key players that may be manipulated by immunotherapeutic strategies to enhance patient response.

One of the most comprehensive studies that investigated the comparative immunological landscape of OSA utilized RNAseq to evaluate transcriptional profiles from primary appendicular OSA of humans, dogs, and genetically engineered mouse models to identify shared transcriptional profiles that influenced tumor development and progression [24]. Three highly conserved gene clusters were identified across species that were enriched in cell cycle transcripts, immune transcripts associated with monocytes, and transcripts associated with T cells. In humans and dogs, increased expression of transcripts associated with cell cycle correlated with poor patient outcome. In humans, patients whose tumors showed loss of immune cell transcripts had the shortest survival time, suggesting this may serve as a prognostic biomarker for metastatic disease. However, the lack of immune transcripts was not significantly correlated with survival times in the dog. The authors postulate that this may be due to the aggressive nature of canine OSA, with dogs not surviving long enough for the role of immune activation to be recognized. Taking advantage of the reduced genetic heterogeneity seen within dog breeds, the same investigators performed genome-wide gene-expression profiling, which separated OSA tumor samples into two different molecular subgroups distinguished by expression of G2/M transition, DNA damage checkpoints, and microenvironment-interaction signatures. These different subtypes had different metastatic potentials that correlated with the presence or absence of immune cell infiltrates within the stroma [23].

Monocytes/Macrophages

During tumor development, circulating monocytes/macrophages traffic into tumors where they are co-opted by the tumor microenvironment, shifting from a classical proinflammatory type 1 (M1) phenotype to an anti-inflammatory, protumorigenic type 2 (M2) phenotype [53]. Accumulations of M2 macrophages in tumors such as breast and cervical cancer have been associated with a poor clinical outcome [53, 54]. Buddingh et al. used gene-expression analysis and IHC to show that high-grade human OSA samples contained both type 1 (CD14/HLA-DRα$^+$) and type 2 (CD14/CD163$^+$) TAMs and that the presence of TAMs was associated with reduced metastases and improved survival [55, 56]. Similar findings were reported by Gomez-Bruchet who analyzed pretreatment biopsy samples from patients enrolled on the French phase 3 trial (OS 2006) and demonstrated that patients with core biopsies showing >50% of cells as CD163$^+$ TAM experienced improved overall survival [56]. Finally, recent evidence in a murine xenograft model of metastatic OSA showed that the beneficial effects of PD-1 antagonism on pulmonary metastases were associated with increased infiltration by M1 macrophages and a reduction in M2 macrophages and depletion of macrophages in this model system negated the therapeutic effect of the checkpoint inhibitor [57]. Indeed, it is thought that the balance between M1 and M2 macrophages, which is controlled by the tumor cells themselves, plays a key role in determining the outcome of T cell responses within the tumor, with recent evidence suggesting that this outcome is dictated by PD-1/PD-L1 interactions. These findings underscore the complexity of immune interactions with the tumor and suggest that therapeutic strategies that influence the M1/M2 balance and promote a predominantly proinflammatory milieu may enhance antitumor T cell responses to control metastases and promote a more favorable outcome.

Using quantitative IHC to determine the presence of CD204$^+$ macrophages, CD3$^+$ T cells, and FOXP3$^+$ (forkhead box P3) cells in primary tumors of 24 dogs with appendicular OSA, Withers et al. reported that the only prognostic subset was CD204$^+$ cells, with dogs with high levels of CD204$^+$ infiltrate experiencing prolonged disease-free intervals (DFI) [58]. Dogs with proximal humeral OSA, a location that is generally associated with a poor prognosis, tended to have lower CD204$^+$ infiltrates compared to all other tumor locations and experienced shorter median survival times (MST) [58, 59]. In the same study, the authors demonstrated that tumors that contained high numbers of CD204$^+$ TAMs also had greater lymphocytic infiltrates and patients with lymphocytic infiltrates above the top quartile showed a statistically significant prolongation of survival [58]. It is worth noting that CD204 expression is commonly associated with an M2 phenotype; however, the presence of lymphocytic infiltrates that correlate with improved survival might suggest that functionally, these TAMs are more consistent with a proinflammatory subset. As the canine reagent toolbox expands, further investigation into the phenotype and functional properties of these TAMs will ensue. These studies suggest that at a basic level, canine and pediatric OSAs share comparable immune infiltrate features and suggest dogs with OSA are relevant to further investigations into agents that manipulate the TME to promote effective antitumor immunity.

Several studies have demonstrated that high-circulating monocyte counts (>400 cells/μl in dogs, but still within the normal range) are associated with shorter DFI in dogs and in pediatric patients with appendicular and axial OSA [60–63]. Recently, the mechanistic basis for this has been investigated in dogs. Researchers found that circulating monocytes from dogs with OSA had reduced expression of cell adhesion molecules and chemokine receptors including CD62L, CCR7, CCR2, and CXCR2 [64]. They also exhibited decreased chemotactic function. These findings are consistent with the idea of tumor-mediated monocyte dysfunction in which monocytes from OSA patients have a reduced ability to traffic into tumor sites and initiate an antitumor immune response. This idea is further supported by the finding that canine OSA patients that express higher levels of CCR2 on circulating monocytes, enabling them to migrate into areas of inflammation in response to chemoattractant proteins, have improved survival [64]. Interestingly, when monocyte counts were high in these dogs, the cells tended to express higher levels of CD14 and lower levels of CD16 compared with patients with lower monocyte counts. In humans, this macrophage subset (CD14hi, CD16int) denotes a proinflammatory macrophage phenotype, that is MHCII high, produces TNF-α, IL-1β, and IL-6, and is a potent T cell stimulator [65]; however, the functional attributes of a CD14hi, CD16int/lo subset have not been explored in canines. Consistent with systemic monocyte dysfunction, stimulation of circulating monocytes from canine OSA patients with LPS led to the production of significantly more TNF-α and PGE2 than monocytes from healthy dogs. TNF-α is classically proinflammatory; however, it also promotes PGE2 production and can exhibit protumorigenic effects in part via IL-34 production in the TME [66, 67]. PGE2 plays an important direct role in immune dysfunction through multiple mechanisms in patients with cancer [68]. It inhibits the function of neutrophils, monocytes, and macrophages; disrupts cross-talk between DCs and T cells; skews T cells to a type 2 protumorigenic phenotype; and promotes the accumulation of regulatory T cells (Tregs) and myeloid-derived suppressor cell (MDSCs) [68]. Canine OSA cell lines and tumor tissues have also been shown to produce PGE2 [69, 70]. Millanta et al. confirmed these findings by IHC showing that 93% of canine OSA tissues expressed COX-2, 85% expressed microsomal PGE2 synthase-1, and 89% of tumors expressed the PGE2 receptor [71]. In similar studies, Wasserman et al. showed that myeloid cells exposed to tumor-derived soluble factors from OSA cell lines had reduced phagocytic activity, downregulated MHCII and CD80 expression reducing their capacity to activate antigen-specific $CD4^+$ T cells, and suppressed responding effector cell proliferation [72]. Although not confirmed, it is possible that tumor-derived exosomes exert this immunosuppressive influence and contribute to the broad, tumor-mediated immune dysfunction seen in OSA patients. Similar immunosuppressive leukocyte profiles have been identified in pediatric sarcoma patients [61]. Together these data suggest that as in human patients, canine OSA avoids the immune response by adversely affecting the function and chemotactic capabilities of monocytes/macrophages [73]. Further investigations into the phenotype and function of different macrophage subsets are required in healthy and tumor-bearing dogs, but the current data suggests that canines with OSA can serve as a clinically relevant, patient population in which to investigate the biological and therapeutic effects of agents that modulate monocyte/macrophage subsets in the oncology clinic such as L-MTP-PE [74] and All-trans retinoic acid (ATRA) [75].

Myeloid-Derived Suppressor Cells

Myeloid-derived suppressor cells (MDSCs) are immature myeloid cells that are produced in the bone marrow and traffic to tumor microenvironments under the influence of certain chemokines [76]. They are potent suppressors of T cell responses through a variety of different mechanisms and have the capacity to differentiate into TAMs within the tumor microenvironment [76,

77]. MDSCs play an important role in tumor progression and metastases, and their presence has been shown to predict response to immunotherapy and correlate with poor clinical outcome in a number of different solid tumor types [76]. Recently, MDSCs that resemble fibroblasts and have T cell suppressor capabilities have been described in pediatric sarcoma patients, although no OSA patients were included in the dataset [78]. Canine MDSCs have recently been characterized into monocytic and granulocytic subsets both phenotypically and functionally, and both subsets were shown to be increased in the peripheral blood of dogs with hematological and solid tumors compared to healthy controls [79]. Several earlier studies evaluated the presence and function of circulating MDSC in dogs with cancer [80, 81]. Sherger et al. identified a functionally immunosuppressive subset of MDSCs ($CD11b^{lo}$ $CADO48^{lo}$) that were increased in the peripheral blood of dogs with different cancer types including OSA [81]. Similarly, Goulart et al. found a significantly higher percentage of $CD11b^{+}CD14^{-}MHCII^{-}$ granulocytic MDSCs in the peripheral blood of dogs with advanced or metastatic cancers, including OSA. This group further showed that these cells expressed hallmark features of human MDSC including ARG1, iNOS2, TGF-β, and IL-10, which mediate suppressor activity against T cells [80]. Although our understanding of the role that MDSCs play either directly or indirectly in OSA remains rudimentary, these results suggest that they contribute to the systemic and local immune dysfunctions in both species that must be overcome to improve the clinical response to immunotherapies.

Regulatory T Cells

Regulatory T cells are thought to play a central role in suppressing antitumor immunity and contributing to poor outcomes in human cancer patients. As such, their presence has been evaluated in dogs with various cancers including OSA [82, 83]. Utilizing a combination of an anti-canine CD4 antibody and a cross-reactive anti-mouse FOXP3 antibody [82], Biller et al. showed that prior to amputation, dogs with OSA had significantly higher numbers of circulating Tregs and reduced numbers of $CD8^{+}$ T cells compared to their healthy counterparts, resulting in a low CD8:Treg ratio that was predictive of shorter overall survival [38]. These aberrations in cell numbers and CD8:Treg ratio remained unchanged for at least 24 hours after amputation. Percentages of Tregs in the draining and distant lymph nodes of dogs with OSA and healthy controls were comparable [38]. Recently, proteins such as TGF-β [84], alpha fetoprotein, and heat shock proteins (HSP) within exosomes released from cultured canine OSA cell lines have been shown to suppress T cell proliferation, decrease CD25 expression on T cells, and direct a regulatory T cell phenotype, providing a potential mechanism for tumor-associated immune suppression in canine OSA patients [39]. However, in a follow-up study, Rissetto et al. used CD4, FOXP3, and CD25 to identify canine Tregs and found no difference in the percentage of Tregs in the peripheral blood or the draining lymph node of dogs with appendicular OSA when compared to healthy control dogs [83, 85]. Both studies evaluated samples from a small number of canine OSA patients, and evaluation of a larger cohort of dogs will be required to confirm the presence and predictive value of circulating Tregs in canine OSA.

T Cells

Tumor-infiltrating lymphocytes are identified in the majority of OSA biopsy samples, and multiple studies suggest that the presence of cytotoxic T cells controls the development of metastatic disease. Recently, Scott et al. reported that the presence of T cell infiltrates in human primary appendicular OSA predicted increased survival [24]. This supported previous findings from a multi-institutional European study that revealed a high ratio of intratumoral $CD8^{+}$:$FOXP3^{+}$ cells (>3.08) was predictive of improved survival [86]. Furthermore, Lussier et al. demonstrated that tumor-infiltrating cytotoxic T cells express PD-1 and that PD-1/PD-L1 blockade increases CTL activity, leading to a decrease in tumor burden and improved survival in a mouse model of OSA [87]. In dogs with OSA, RNAseq [24], IHC [58], flow cytometry [38], and histomorphometry [88] have been used to evaluate the presence of tumor-infiltrating lymphocytes. Histomorphometric and IHC studies on treatment of naïve, primary appendicular canine tumors showed that seven out of ten dogs had mild inflammation with a median of 8% of nucleated cells in the tumor being CD3⁺ T cells [88]. These cells were found in areas of necrosis and fibrosis as well as in viable tumor. Withers et al. used IHC and showed accumulations of B and T lymphocytes that resembled tertiary lymphoid structures (TLSs) in some canine patients [59]. Interestingly, neither RNAseq nor IHC data showed a correlation between T cells in canine primary appendicular tumors and overall survival [24, 58]. This is in contrast to the data obtained from human primary appendicular OSA tumors [24]. This discrepancy between species may arise due to the more rapid progression of OSA in the dog and the lack of time available to mount an immune response and/or the fact that overall survival time in dogs is highly influenced by the owners' perception of quality of life and their capability of paying for treatment, leading to earlier euthanasia of the canine patient and highly variable overall survival times. In dogs, the comparative lack of reagents that enable T cell subset identification by IHC and flow cytometry makes it challenging to define the T cell subsets within TILs that might influence outcome. Furthermore, no studies have yet addressed whether TILs present in canine OSA are tumor specific.

Checkpoint Molecule Expression

The expression of immune checkpoint molecules such as PD-L1 on OSA cell lines and tumor tissues has been investigated as another mechanism by which OSA can inhibit immune function within the TME. Shen et al. reported that using IHC, 23.7% and 50% of human OSA tissue samples expressed high and intermediate levels of PD-L1, respectively, and that PD-L1 expression levels correlated with metastatic disease and poor overall prognosis [89, 90]. These findings suggest that OSA cells actively participate in antitumor immunity and that checkpoint blockade in the form of anti-PD-L1 and anti-PD-1 therapies may have therapeutic benefit in pediatric OSA. However, clinical trial results with PD-1/PD-L1 and CTLA4 inhibitors used as monotherapies have been disappointing in pediatric OSA, and combination therapies aimed at inducing antitumor immunity together with checkpoint blockade may represent important areas of research moving forward [91–93]. With the advent of canine-specific or cross-reactive antibodies that recognize key checkpoint molecules, the role that they play in restricting antitumor immune responses and the benefit of checkpoint inhibition in dogs with different cancers including OSA is beginning to be explored. A recent study using a murine anti-canine PD-L1 antibody demonstrated expression of PD-L1 on the surface of three different canine OSA cell lines, and expression was increased following treatment with recombinant canine (rc) IFN-γ or supernatants from mitogen-stimulated T cells [94]. Using IHC, Maekawa et al. demonstrated that PD-L1 was also expressed in primary canine OSA, suggesting that as in pediatric OSA, strategies to inhibit PD-1:PD-L1 interaction might have a beneficial effect [95]. However, unlike pediatric OSA, no studies have yet been performed in canine primary or metastatic OSA lesions to determine whether PD-L1 is positively correlated with the amount of TILs or whether it serves as a prognostic indicator [87, 96, 97]. Circulating canine monocytes did not express PD-L1 but did upregulate its expression following treatment with rcIFN-γ [98]. Similar results were obtained using canine monocyte-derived macrophages [94], suggesting that these mononuclear cells may contribute to systemic and tumor-associated T cell suppression. Functional studies using checkpoint inhibitors have shown that blockade of the PD-1/PD-L1 axis promotes CTL responses and enhances IFN-γ production in vitro and in vivo, leading to reduced metastatic tumor burden in murine models [87]. These findings support the notion that this key checkpoint axis is intact and open to manipulation to enhance antitumor immunity in dogs with OSA and other tumors [95, 99]. With the development of canine anti-PD-1 and PD-L1-blocking antibodies, it is likely that pet dogs with spontaneous OSA will serve as valuable subjects in which to evaluate the effectiveness of combination vaccine or cellular therapies with checkpoint inhibition and to identifycorrelative biomarkers that predict response.

Metastatic Lesions

While several studies have been performed that compare the genetic makeup of paired primary appendicular OSA with metastatic lesions, studies evaluating the immunological landscape of metastatic lesions are rare [49, 100]. In human patients, immune infiltrates have been identified in primary and metastatic OSA lesions although lymphocytic infiltrate in metastatic lesions was shown to be higher than in the paired primary samples [101]. Withers et al. used IHC to evaluate $CD3^{+}$ T cells, $FOXP3^{+}$ cells, B cells, and $CD204^{+}$ macrophages in 21 paired primary and metastatic canine OSA samples [59]. They showed positive correlations of $CD3^{+}$ T cells and $FOXP3^{+}$ cells between primary and metastatic samples and that metastatic lesions had significantly more $CD3^{+}$, $PAX\text{-}5^{+}$, and $CD204^{+}$ cells compared with the primary tumor. In human patients, $CD3^{+}$ T cells were also higher in metastatic lesions compared with the primary tumor, but T cell subsets in the primary and metastatic lesions were the same [101]. Although B cells were the least prevalent immune cells in OSA lesions, they were observed to form clusters at the edge of half of the primary and 1/3 of the metastatic lesions, a feature that is reminiscent of tertiary lymphoid structures (TLSs) and has rarely been reported in human OSA.Unfortunately, the lack of canine-specific reagents to identify FDCs, follicular helper T cells, and chemokines makes further interrogation of these structures challenging. Indeed, additional geospatial molecular studies will be required to further define the immune cell types and their function within primary and metastatic lesions, to determine whether they have tumor-promoting or antitumor activity and perhaps to provide additional biomarkers of response to immunotherapies.

In summary, to the extent to which the immune status of canine patients and the immune landscape of their primary and metastatic OSA lesions have been explored, remarkable similarities have been identified with the human disease. These findings suggest that dogs with OSA will be valuable in investigating the benefit of microenvironment modulators such as macrophage activators, inhibitors of Tregs and MDSCs and their suppressive factors, and checkpoint inhibitors. More work is required to better understand the TME in both human and canine OSA patients particularly to identify biomarkers that may predict the patient's ability to favorably respond to immunomodulatory agents and immunotherapies. As the diagnostic reagent toolbox continues to develop for canine tissues and new technologies including geospatial gene-expression analysis are adopted, it is anticipated that our understanding of the immune microenvironment will improve, and this will guide the rational selection of immune therapies and combination immune therapies that aim to improve outcome for human and canine patients with OSA.

Immunotherapy of Canine OSA

The evidence outlined above indicates that the innate and adaptive arms of the immune system play a role in controlling OSA progression and that tumor-associated local and systemic immune dysfunction enables tumor progression. Therefore, therapeutic strategies aimed at augmenting antitumor activity and reversing immune dysfunction have the potential to improve patient outcome. The studies that have been performed to evaluate the safety and efficacy of immunotherapies aimed at improving the prognosis of humans and pet dogs with OSA are outlined below.

Immune Modulatory Agents

Coley's Toxins

In the late 1800s, William Coley made the seminal observation that patients with bone sarcoma suffering from concurrent streptococcal infections had prolonged durable remission times, suggesting that nonspecific immune activation was able to delay metastatic disease. Coley's efforts to recapitulate the favorable effects of natural streptococcal infection on patient outcome resulted in the development of Coley's toxins. This *Streptococcus/Serratia* concoction of either live or heat-killed bacteria was administered repeatedly to patients with sarcoma or following surgical resection of their sarcoma with favorable outcomes documented in a number of cases [30, 31]. Although the mechanism of improved overall response was unknown at the time, the adjuvant effects of bacterial components particularly on macrophages, promoting a permissive milieu that supports effective antitumor immunity, appear central to the effect [102].

BCG

Similar to Coley's toxins, Bacillus Calmette-Guerin (BCG), a live, attenuated strain of *Mycobacterium bovis*, promotes antitumor immunity and is FDA approved for first-line use in patients with high-risk nonmuscle invasive bladder cancer. Its therapeutic effect is thought to be mediated by T cells, NK cells, granulocytes, macrophages, and dendritic cells, plus a potential direct effect on the bladder cancer cells themselves [103]. Bech-Nielsen and colleagues treated dogs with spontaneous OSA after amputation with q2 weekly flank injections of BCG and noted a significant increase in survival extending from 13 weeks (control group $n = 5$) to 40 weeks in the vaccinated group ($n = 6$) [104]. The Kaplan-Meier survival curve reflected those of many medical immunotherapy trials performed today, with an elevation of the tail of the curve representing a greater proportion of vaccinated patients experiencing prolonged, durable remissions. A similar study by Owen and Bostock reported prolonged survival in dogs ($n = 6$) with appendicular OSA who underwent amputation followed by intravenous injections of 10^7–10^8 BCG organisms 1, 2, 4, and 8 weeks postoperatively [105]. Observations of a transient, low-grade fever within hours of administration suggested innate immune activation. Follow-up studies in healthy dogs showed that intravenous BCG mediated an increase in NK cell cytotoxicity and enhanced pulmonary macrophage activation, which likely played a role in controlling micrometastatic disease [106, 107]. These studies were performed in the absence of adjuvant chemotherapy. Almost 100 years after Coley made his seminal observation that bacterial infections improve patient outcome, similar observations were reported for dogs that had undergone limb salvage surgery for the treatment of appendicular OSA [32–34, 108]. Bacterial infections of *Staphylococcus* spp., *Pseudomonas* spp., and/or *Streptococcus* spp. were reported [33]. Similar to the studies using BCG, the mechanisms resulting in decreased pulmonary metastases and prolonged survival associated with bacterial infections are thought to be mediated by macrophages and NK cells, a concept supported in part by the finding that the survival benefit associated with osteomyelitis in murine OSA models is lost if monocytes/macrophages are depleted [102, 109].

Muramyl Tripeptides

The role of monocyte and macrophage activation in delaying or preventing metastatic disease in humans and dogs with OSA has been further underscored by favorable clinical responses in both species to liposome-encapsulated muramyl tripeptide-phosphatidylethanolamine (L-MTP-PE) – a mycobacterial wall extract. In a randomized, double-blinded, clinical trial, L-MTP-PE was administered intravenously twice a week for 8 weeks to 14 dogs with appendicular OSA after amputation [110]. Thirteen amputated dogs received empty liposomes as placebo controls. L-MTP-PE produced a transient, low-grade fever but was otherwise well tolerated. Median metastasis-free interval and median survival time for dogs receiving L-MTP-PE was 168 and 222 days, respectively, and 58 and 77 days for the placebo group. Follow-up studies demonstrated a similar beneficial effect of L-MTP-PE when administered after adjuvant cisplatin chemotherapy [111]. Here the MST of placebo dogs ($n = 14$) was 9.8 months compared with 14.4 months for dogs receiving L-MTP-PE. Interestingly, the survival benefit of L-MTP-PE was lost when treatment was administered concurrently with cisplatin [111], suggesting that concurrent cisplatin may either obviate the effects of L-MTP-PE or pre-treatment with cisplatin is required for the effects of L-MTP-PE to be realized. In vitro studies have revealed that L-MTP-PE is a potent activator of canine monocytes and macrophages, increasing their production of TNF-α and IL-6 and enhancing their cytostatic capabilities against tumor cells [112, 113]. Furthermore, pulmonary alveolar macrophages taken from canine patients treated with L-MTP-PE plus doxorubicin showed greater cytotoxic activity against OSA cells when compared to dogs treated with either agent alone [112]. In contrast to the in vivo results suggesting that cisplatin suppresses the beneficial effects of L-MTP-PE, monocyte cytotoxicity and TNF-α production were increased in dogs with splenic hemangiosarcoma treated with doxorubicin (a known inducer of ICD) plus L-MTP-PE [114]. Thus, it appears that different chemotherapies differentially influence the immunomodulatory and antitumor activities of L-MTP-PE. Additional work is required to identify the optimal combination and order of chemotherapy and immunomodulatory agents to achieve the most beneficial outcome. Understanding this order, which may depend upon the specific agents involved, remains an important challenge in the field of cancer immunotherapy today.

Cytokines

Given the pivotal role that innate and adaptive immune responses play in controlling tumor progression, several groups have investigated whether administration of IL-2, a potent T and NK cell growth factor, can augment antitumor immunity and delay progression or induce regression of pulmonary metastases [115]. Since high-dose systemic administration of IL-2 has a narrow therapeutic index, Khanna et al. explored the effects of aerosolized IL-2 liposomes on local and systemic immune effectors of normal healthy dogs [116]. The study showed that inhalation of human IL-2 liposomes significantly increased the number and activation status of leucocytes in bronchoalveolar lavage (BAL) fluid and skewed their composition toward lymphocytes and eosinophils rather than monocytes and macrophages, demonstrating biological activity of the administeredIL-2 [116]. Significant activation of systemic immune effectors was not observed and the aerosolized IL-2 was well tolerated, providing much needed safety data. To determine the clinical effects of aerosolized IL-2 in canine patients with metastatic OSA, four dogs with metastatic disease received aerosolized IL-2 liposomes two and three times a day for 30 days [117]. Two out of four dogs had complete and durable regression of metastases. Similar effects were observed on the composition of BAL cells with a statistically significant increase in lymphocytes, eosinophils, and macrophages after treatment. Cytolytic activity of BAL cells was also increased after 2 weeks of treatment, an effect that was attributed to NK cell activity and also possibly eosinophilic cytotoxicity [118]. However, cytotoxic activity declined thereafter, which may have been caused in part by the recorded development of antibodies against human IL-2. To circumvent formation of antidrug antibodies, cationic liposome–DNA complexes encoding canine IL-2 were delivered via intravenous infusion to 20 dogs with metastatic pulmonary OSA [119]. IL-2 expression was identified in the lung tissue and systemic immune activation in the form of transient fever, lymphopenia and thrombocytopenia, upregulation of costimulatory molecules and MHCII on monocytes, and increased NK cell cytotoxicity occurred. Overall survival of treated dogs was significantly increased compared with historical controls matched for disease stage. Furthermore, three dogs showed partial or complete regression of pulmonary metastases. These effects are most likely to be associated with a combination of local IL-2 production and innate immune responses induced by liposomes. In this study, the effects of IL-2 production on the local environment including leukocyte composition within BAL fluid were not evaluated.

Losartan

Losartan is a type I angiotensin II receptor antagonist. It has immunomodulatory effects on monocytes and macrophages and reduces pulmonary metastatic tumor burden in several mouse models of metastatic cancer (CT26 and 4T1) in part through inhibition of monocyte recruitment to the TME and a reduction in granulocytic MDSCs [120]. Losartan acts similarly to block CCL2-mediated migration in canine monocytes [121]. A clinical trial in dogs with metastatic OSA showed that a combination of high-dose losartan and the tyrosine kinase inhibitor toceranib was well tolerated in dogs and showed reduced monocyte trafficking to the metastatic lesions and exhibited antitumor activity (Steve Dow, personal communication). The results of these canine studies have supported the initiation of a pediatric trial (NCT03900793) for patients with relapsed/refractory OSA investigating the value of losartan in combination with sunitinib. The results of this trial are eagerly awaited.

Bisphosphonates

Bisphosphonates have been employed in the palliative setting to alleviate pain and reduce bone resorption by inhibiting osteoclast function. In addition to their effects on osteoclast apoptosis and inhibition of osteoclastogenesis, recent studies have indicated that bisphosphonates such as zoledronate and pamidronate have immunomodulatory functions through effects on innate and adaptive immune responses [122, 123]. In vitro, zoledronate inhibits regulatory T cell expansion, migration, and immunosuppressive activity [122]. In an HER2/neu (ErbB-2) transgenic mouse model, zoledronate switched tumor-associated macrophages from an M2 to M1 phenotype, reduced infiltration of macrophages into mammary tumors, and reduced VEGF concentrations and vascularization of the tumor [124]. However, in a murine model of OSA, where canine OSA cells were implanted orthotopically, zoledronate administered alone or following amputation did not reduce the incidence of pulmonary metastases [125]. The immunomodulatory effects of bisphosphonates have not yet been evaluated in vivo in the dog; however, given the common clinical use of these agents in canines with OSA, their effects on enabling antitumor immunity that is induced or augmented by other immunomodulatory agents, vaccines, or adoptive cellular therapies could be readily evaluated.

Active Vaccination

As nonspecific immune activation has shown moderate clinical benefit in delaying or preventing metastatic disease, attempts have been made to further improve outcomes by combining innate immune activation with tumor-specific adaptive immune responses using bacterial or viral vectors that supply TAA in the context of immune activation or provide additional cytokine support for T cell responses. Antigens identified and specifically targeted for therapeutic gain in human OSA patients include the epidermal growth cell factor receptor HER2/neu [126], GD2 and GD3 antigens [127, 128], TP-3-PAP [129, 130], and IGF-1R [84].

Bacterial Vaccines

Listeria monocytogenes

Listeria monocytogenes is a facultative, aerobic, intracellular bacteria that is a potent stimulator of innate and adaptive immunity. Through its ability to secrete the pore-forming lysin listeriolysin O (LLO), the bacteria can escape the phagosome and access the class I processing machinery of antigen-presenting cells [131]. As such, attenuated strains of recombinant *Listeria*, modified to express TAA fused to a truncated LLO, have been used in mouse models and in human patients to deliver antigens to APCs and generate tumor-specific $CD4^+$ and $CD8^+$ T cell responses [132]. The potent tumor-specific T cell responses generated break peripheral tolerance and lead to tumor regression [132, 133]. The epidermal growth factor receptor HER2/neu is expressed by 40–60% of primary OSA samples and serves as a relevant target for T cell therapies in humans and dogs with OSA [126, 134–136]. As such, active vaccination strategies to prime and expand HER2/neu specific T cells may be employed effectively to prevent or treat metastatic OSA. The ability of a recombinant chimeric huHER2/neu-expressing *Listeria* to prevent metastatic disease when administered in the setting of minimal residual disease after amputation and chemotherapy was evaluated in an 18-dog prospective clinical trial [137]. Vaccinated dogs experienced a median DFI of 650 days and OS of 956 days compared with an OS of 423 days for a historical, HER2/neu positive control group that received the same standard of care treatment without vaccination.

This promising result has led to a larger prospective, controlled, national clinical trial conducted through the Comparative Oncology Trials Consortium. Tumor tissue and serial plasma, serum, and PBMC samples are being prospectively collected during this trial to evaluate immune responses and identify correlative predictive biomarkers.

Salmonella Typhimurium

Anaerobic bacteria that preferentially home to and accumulate in the hypoxic microenvironment of tumors have been used to promote antitumor immunity [138]. A highly attenuated *Salmonella typhimurium* (VNP20009) that only induces low levels of TNF-α and is dependent upon external purines for growth was administered intravenously every week or every 2 weeks to dogs with different tumor types in a phase I basket clinical trial to evaluate toxicity and early antitumor efficacy [139]. The majority of patients had soft tissue sarcomas or carcinomas, while 4 out of 41 treated dogs had OSA. CRs were seen in 10% of patients and 10% of patients experienced stable disease. One dog with metastatic OSA showed a partial response for 68 days. Fever, nausea, vomiting, and diarrhea were common side effects. Although immunological endpoints were not addressed in this study, the antitumor responses may have been associated with the organism's direct tumoricidal activity, innate and adaptive immune activation, depletion of nutrients, and/or alteration of the TME. A second phase I study evaluated the safety of orally administered, attenuated *Salmonella typhimurium* as a vector to deliver IL-2 to 19 dogs with appendicular OSA. Dogs received oral dosing once 10 days prior to amputation and then after surgery concurrently with each of five doses of adjuvant doxorubicin for a total of six doses [140]. *Salmonella* was safe and well tolerated, and treated dogs experienced longer DFI but not OS when compared to two comparable historical control groups. An inflammatory leucogram (lymphocytosis and monocytosis) was seen in 18 out of 19 dogs 10 days after the first *Salmonella* administration, suggesting it was biologically active. *Salmonella* was not detected in any tumor tissue cultured after amputation, suggesting that any observed beneficial effects were more likely associated with antitumor immunity rather than a direct tumoricidal activity of the vector. Randomized, placebo-controlled, prospective trials are warranted to determine the true value of this approach.

DNA Vaccines

Alternative approaches to induce HER2/neu-specific T cell responses have been explored in dogs using DNA encoding the extracellular and transmembrane domains of human HER2/neu and electroporation as a priming strategy followed by a boost with an adenovirus 6 vector expressing the same HER2/neu construct [141]. This regime induced HER2/neu-specific IFN-γ responses and HER2/neu-specific IgG responses, although the adenoviral vector was found to be highly immunogenic, limiting the efficacy of any subsequent attempts to boost immunity using this serovar. Although no studies have yet been published using this approach in dogs with OSA to assess therapeutic effectiveness, the approach has been shown to be safe and induces durable HER2/neu-specific T cell responses in healthy dogs. Overcoming vector immunogenicity to enable effective booster treatments will be important to take this approach further clinically.

Oncolytic Viruses

Defects in antiviral defense mechanisms in tumor cells provide an ideal opportunity for oncolytic viruses (OVs) to be used therapeutically to selectively infect and destroy tumor cells. Tumor lysis and ICD results in induction of systemic polyclonal T cell responses that aim to control both primary and metastatic diseases. Many OVs also exert immunomodulatory effects on the microenvironment through their ability to induce the release of pathogen-associated molecular patterns (PAMPs) from tumor cells and promote the production of type I interferons [142]. Le Boeuf and colleagues demonstrated the ability of the oncolytic rhabdovirus, Maraba (MG1), to infect and kill both canine sarcoma cell lines and human sarcoma explants and confirmed these cytotoxic effects in a murine sarcoma model [143].

Similarly, Naik et al. evaluated VSV expressing IFNβ in dogs with different tumor types, including one dog with axial OSA and metastatic disease. All dogs tolerated intravenous oncolytic viral therapy well and the one dog with OSA showed stabilization of primary and metastatic disease for 6 months [144]. Laborda et al. utilized a locally delivered, hyaluronidase-armed, oncolytic adenovirus in a total of six dogs, including two dogs with OSA [145]. No adverse side effects occurred that could be directly attributed to the adenoviral therapy and partial responses were seen in two dogs, although neither had OSA. Although experience with oncolytic viral therapy in dogs with OSA is limited, these studies lay the groundwork for further evaluation of this approach either as a monotherapy or in combination with immunomodulatory agents or immune checkpoint inhibition to augment clinical effect.

FasL-Mediated Inflammation

The death receptor Fas (CD95) is expressed by many different tumor types and its engagement by FasL (CD95L) triggers apoptosis, leading to the hypothesis that FasL may represent a promising cancer therapeutic [146]. Both innate and adaptive immune responses are induced by intratumoral delivery of FasL, effects that are mediated via apoptosis of Fas^+ macrophages and the resulting influx of neutrophils that are ultimately responsible for tumor cell death. Subsequent recruitment and activation of APCs promotes a systemic antitumor immune response that aims to control metastatic spread. However, controversy surrounds its use in part, because Fas-signaling has also been shown to be required for tumor cell survival [147]. Furthermore, systemic administration of FasL results in lethal hepatotoxicity in mouse models [148]. To mitigate these risks while evaluating the effects of neoadjuvant FasL in dogs with OSA, Modiano et al. delivered a single intratumoral dose of adenovirus expressing canine FasL (Ad-FasL) to 56 dogs [88]. Ten days later, dogs underwent standard amputation and adjuvant carboplatin chemotherapy. Ad-FasL was generally well tolerated, with adverse effects associated with transient increases in transaminases and creatine phosphokinase. Adenoviral delivery of FasL induced a potent inflammatory response with increased lymphocytic infiltration within the tumor compared to dogs who did not receive Ad-FasL. Furthermore, dogs with high inflammation scores within the treated tumor experienced improved overall survival [88]. Dogs with reduced Fas expression on their tumors had greater inflammation scores supporting the notion that the improved survival effects of FasL are primarily associated with its induction of inflammation rather than direct Fas-mediated tumor apoptosis.

Cell-Based Therapies

Tall 104 Cells

The earliest recorded use of adoptive T cell therapy for dogs with OSA was in 1999, when Daniela Santoli's group at the Wistar Institute evaluated the safety and efficacy of the human cytotoxic T cell line, TALL104 cells, to prevent metastatic disease. Dogs that had undergone amputation and adjuvant cisplatin chemotherapy received 1×10^8 γ-irradiated cells/kg systemically daily for 5 days and then every month for up to 9 months. Only mild and transient grade 1 and 2 related GI toxicities were reported. Although the overall median survival was 11.5 months and the median DFI was 9.8 months, the Kaplan-Meier survival curve demonstrated uncharacteristic long-term survival of some patients. These effects were speculated to be in part mediated through enhanced endogenous NK cell activity that occurred as a direct result of TALL104 administration. Perhaps unsurprisingly, the unmodified, xenogenic, adoptively transferred cells did not persist and antibody responses against them were detected in all treated dogs and cellular immune responses against them were detected in 80% of treated dogs. Although performed 20 years ago, these studies have set the stage for evaluating genetic modifications of human T cells that will enable them to cross xenogenic barriers and persist to mediate antitumor immunity in canine patients. Employment of a comparative approach in these endeavors aims to provide greater clarity surrounding the modifications that will be required for successful allogeneic adoptive T cell therapy in human patients (N. Mason, personal communication).

Polyclonal Activated T Cells

Isolation, ex vivo expansion, and reinfusion of autologous tumor-infiltrating lymphocytes have proven effective in the treatment of immunogenic tumors such as melanoma; however, this strategy is underexplored in OSA patients [149]. Instead, veterinary researchers are currently evaluating active vaccination of canine patients with appendicular OSA using an autologous tumor lysate vaccine to prime circulating T cells. Primed, tumor-specific T cells are harvested by apheresis and polyclonally expanded ex vivo using a proprietary cocktail, before being adoptively transferred back into the patient after amputation. Adjuvant cytotoxic chemotherapy is not employed in the protocol (J. Bryan, personal communication). Early results suggest the procedure is safe, but outcome data is yet to be published.

NK Cell Therapies

NK cells play a fundamental role in tumor surveillance and elimination, and, as such, efforts have been made to employ autologous or allogeneic NK cells in adoptive transfer strategies to treat or prevent metastatic disease in human cancer patients. Activation of NK cells is MHC independent and mediated via receptors that recognize cell surface proteins that are upregulated on stressed target cells or are non-self proteins [150]. In addition to direct killing, NK cell activity is augmented in response to antibodies, cytokines, and immunomodulatory agents including chemotherapy and radiation therapy. As such, strategic combinations of adoptive NK cell transfer with immunomodulatory agents, sensitizing chemotherapy, or radiation therapy are being put forward for clinical evaluation. Advances in the use of adoptive immunotherapy (AI) with NK cells in dogs have previously been hampered by the lack of specific, validated antibodies to identify canine NK cells. A CD5 low, $CD8^{+}$, $CD3^{+/-}$ subset has been described that cytologically displays features consistent with NK cell morphology, expresses high levels of message for NK cell receptors, and displays cytotoxic activity against MHC null, thyroid adenocarcinoma cells [151]. More recently, an anticanine NKp46 mAb was generated, and $NKp46^{+}CD3^{-}$ canine cells showed cytolytic activity against canine OSA cell lines. Furthermore, these cells were effectively expanded ~20,000-fold over 3 weeks in coculture with irradiated K562 feeder cells that express hu4-1BBL and membrane-bound huIL-21 and huIL-2 [152]. Cytolytic activity of expanded $CD5dimCD3^{-}NKp46^{+}$ cells was significantly increased in vitro against allogeneic OSA cell lines after their treatment with γ-radiation [153]. Furthermore, radiation of canine OSA xenografts in NSG mice significantly increased homing of ex vivo expanded adoptively transferred canine NK cells and tumor killing. In contrast to human OSA treatments, radiation therapy is commonly employed in canine patients that do not undergo amputation to alleviate pain [154–156]. Canter et al. combined palliative radiation with intratumoral delivery of ex vivo expanded autologous NK cells once a week for 2 weeks after palliative radiation [153]. Limited toxicity was observed with this approach and posttreatment biopsies demonstrated persistence of labeled NK cells within the tumor for at least 1 week. Five out of ten dogs remained metastases-free at 6 months, and one dog showed resolution of a suspicious pulmonary nodule following treatment. Overall survival times were favorable compared with historical controls. Follow-up studies are now planned to evaluate the effects of the NK cell activating cytokine IL-15 as monotherapy and then in combination with autologous NK cell transfer in patients with metastatic OSA (R. Rebhun, personal communication). Taken together, this work described the successful isolation, activation, expansion, and transfer of canine NK cells and illustrates the enhancing effects of RT on NK cell cytotoxicity. Furthermore, it sets the stage for future studies evaluating AI with NK cells alone or in combination with the sensitizing effects of RT and supportive cytokines (IL-2/IL-15).

CAR T Cell Technology

Several groups have described protocols for generating canine CAR T cells either via RNA transfection with a first-generation CD20-targeting CAR or a second-generation IL-13Rα2-targeting CAR construct [157, 158] or transduction with an RD114 pseudotyped retroviral vector containing a second-generation HER2-targeting CAR construct [136]. Second-generation canine CAR T cells expressing the humanized, cross-reactive anti-IL-13Rα2 scFv (Hu08) produced IFN-γ when cocultured with three different canine OSA cell lines expressing IL-13Rα2 [158]. Furthermore, lentiviral transduced human CAR T cells expressing the same scFv effectively inhibited tumor growth when administered intravenously to NOD/SCID mice bearing established canine MC-KOSA xenografts [158]. Similarly, Mata et al. demonstrated that second-generation canine CAR T cells expressing the cross-reactive antiHER2/neu (FRP5) scFv and canine intracellular signaling domains secrete IFN-γ and effectively kill HER2-positive canine OSA cell lines in an antigen-specific manner. Furthermore, adoptive transfer of HER2-redirected T cells into SCID mice with established intraperitoneal OSA xenografts leads to tumor regression [159]. Similar tumor regression occurred following adoptive transfer of HER2-specific CART cells into mice with established OSA pulmonary metastases [159]. The same group of investigators showed low levels of HER2 expression on the surface of $CD133^+$ OSA tumor-initiating cells (TICs) and that HER2-specific CAR T cells specifically killed TICs in established orthotopic OSA tumors [160]. These data suggest that HER2-targeted CAR T cells may be valuable in targeting micrometastases to prevent metastatic disease. Given that up to 95% of canine patients will have micrometastatic disease at the time of initial presentation, they again represent a valuable patient population in which to evaluate CAR T cell strategies to prevent or treat metastatic disease.

Together this work sets the stage for evaluating both IL-13Rα2- and HER2-targeting CARs in pet dogs with OSA either alone or in combination with other immunotherapies such as checkpoint inhibitors or immunomodulatory agents for the treatment of both primary and metastatic disease. Furthermore, the identification of anticanine or cross-reactive antibodies against GD2 (e.g., 14G2a), IL-11Rα, and FAP will enable additional canine CARs to be constructed against these OSA-associated cell surface targets and then evaluated in dogs for their safety and ability to improve outcome [161].

In all cases, it remains to be seen whether adoptive cell transfer alone will be sufficient to control or prevent metastatic disease or whether combination with immunomodulatory agents that influence the TME will be required to achieve optimal clinical results. Indeed, as our understanding of the immune landscape of OSA increases and validated biomarkers emerge that predict immune responsiveness, it is anticipated that rational combinations of agents that augment tumor-specific T cells with agents that reverse systemic immune dysfunction and immune suppression within the TME will lead to improved patient outcome for both species.

Additional Strategies for Induction of ICD

It is now apparent that standard-of-care cancer therapeutic modalities such as certain chemotherapeutic agents and radiation therapy can induce ICD of the tumor, which is valuable in broadening antitumor immune responses initiated or augmented by immunotherapeutic strategies [29, 162]. In the last decade in particular, radiation has been shown to promote antitumor immune responses via increasing expression of target antigens, activating dendritic cells and inducing tumor-specific $CD8^+$ T cell responses, promoting $CD4^+$ and $CD8^+$ T cell infiltration into tumors, upregulating MHCI, and downregulating immunosuppressive molecules within the TME including arginase-I, CTLA4, PD-1, PD-L1, IDO, FOXP3, TGF-β, and IL-10 [163, 164]. These local effects translate into systemic antitumor immunity and are responsible for abscopal effects that have been reported following RT therapy. Although radiation therapy is infrequently employed in the treatment of pediatric OSA, its combined use with hyperthermia therapy, checkpoint inhibitors, locally delivered cytokines, vaccination, and adoptive cellular therapies is being actively pursued in other cancer types [165–168]. Conversely, both coarse fraction external beam radiation and megadose stereotactic radiosurgery are commonly used in canine patients with OSA that do not undergo primary tumor removal [169]. Palliative radiation is employed often as monotherapy, providing pain relief in up to 74% of dogs for 2–3 months [154, 155, 170]. Thus, dogs with OSA provide a readily available model system in which to explore the immunogenic effects of RT on the primary tumor immunome and to evaluate the safety and therapeutic effectiveness of its combination with vaccines, cellular therapies, and immunomodulatory agents to control primary disease and prevent metastatic disease [29, 171].

In summary, compelling evidence exists to indicate that OSA is an immune responsive tumor and that therapies aimed at initiating, enabling, and broadening antitumor immunity hold great promise for preventing and treating metastatic disease and improving patient outcome. A number of challenges lie ahead, not least of which is the design and implementation of rational combinations of immunotherapies and immunomodulatory agents that will promote tumor-specific adaptive T cell responses and enable them to function effectively within the TME. It is likely that not all patients will require the same immune modulation regime, and identifying biomarkers that can predict each patient's requirement enabling therapy to be tailored to their needs, may lead to the improvement in patient survival that the field has been searching for over the last Four decades. Given the remarkable similarity between canine and pediatric OSA, particularly as it relates to the local and systemic immune landscape, and the large number of pet dogs diagnosed with OSA per year, it seems that we have a remarkable opportunity to address some of these key challenges in the veterinary setting, leading to improved outcome for both patient populations.

References

1. Siegel R, Naishadham D, Jemal A (2013) Cancer statistics, 2013. CA Cancer J Clin 63:11–30
2. Guijarro MV, Ghivizzani SC, Gibbs CP (2014) Animal models in osteosarcoma. Front Oncol 4:189
3. Sanmamed MF, Chester C, Melero I, Kohrt H (2016) Defining the optimal murine models to investigate immune checkpoint blockers and their combination with other immunotherapies. Ann Oncol 27:1190–1198
4. Rangarajan A, Weinberg RA (2003) Opinion: comparative biology of mouse versus human cells: modelling human cancer in mice. Nat Rev Cancer 3:952–959
5. Anfinsen KP, Grotmol T, Bruland OS, Jonasdottir TJ (2011) Breed-specific incidence rates of canine primary bone tumors—a population based survey of dogs in Norway. Can J Vet Res 75:209–215
6. Fenger JM, London CA, Kisseberth WC (2014) Canine osteosarcoma: a naturally occurring disease to inform pediatric oncology. ILAR J 55:69–85
7. Fan TM, Khanna C (2015) Comparative aspects of osteosarcoma pathogenesis in humans and dogs. Vet Sci 2:210–230
8. Owen LN (1967) Comparative aspects of bone tumours in man and dog. Proc R Soc Med 60:1309–1310
9. Morello E, Martano M, Buracco P (2011) Biology, diagnosis and treatment of canine appendicular osteosarcoma: similarities and differences with human osteosarcoma. Vet J 189:268–277
10. Ottaviani G, Jaffe N (2009) The epidemiology of osteosarcoma. Cancer Treat Res 152:3–13
11. Unni K (1996) Dahlin's bone tumors: general aspects and data on 11,087 cases, 5th edn. Lippincott-Raven, Philadelphia
12. Misdorp W, Hart AA (1979) Some prognostic and epidemiologic factors in canine osteosarcoma. J Natl Cancer Inst 62:537–545
13. Garzotto CK, Berg J, Hoffmann WE, Rand WM (2000) Prognostic significance of serum alkaline phosphatase activity in canine appendicular osteosarcoma. J Vet Intern Med 14:587–592
14. Hillers KR, Dernell WS, Lafferty MH, Withrow SJ, Lana SE (2005) Incidence and prognostic importance of lymph node metastases in dogs with appendicular osteosarcoma: 228 cases (1986-2003). J Am Vet Med Assoc 226:1364–1367
15. Phillips B, Powers BE, Dernell WS et al (2009) Use of single-agent carboplatin as adjuvant or neoadjuvant therapy in conjunction with amputation for appendicular osteosarcoma in dogs. J Am Anim Hosp Assoc 45:33–38
16. Kent MS, Strom A, London CA, Seguin B (2004) Alternating carboplatin and doxorubicin as adjunctive chemotherapy to amputation or limb-sparing surgery in the treatment of appendicular osteosarcoma in dogs. J Vet Intern Med 18:540–544

17. Chun R, Garrett LD, Henry C, Wall M, Smith A, Azene NM (2005) Toxicity and efficacy of cisplatin and doxorubicin combination chemotherapy for the treatment of canine osteosarcoma. J Am Anim Hosp Assoc 41:382–387
18. Withrow SJ, Powers BE, Straw RC, Wilkins RM (1991) Comparative aspects of osteosarcoma. Dog versus man. Clin Orthop Relat Res 270:159–168
19. London CA, Gardner HL, Mathie T et al (2015) Impact of toceranib/piroxicam/cyclophosphamide maintenance therapy on outcome of dogs with appendicular osteosarcoma following amputation and carboplatin chemotherapy: a multi-institutional study. PLoS One 10:e0124889
20. McMahon M, Mathie T, Stingle N, Romansik E, Vail D, London C (2011) Adjuvant carboplatin and gemcitabine combination chemotherapy postamputation in canine appendicular osteosarcoma. J Vet Intern Med 25:511–517
21. Selmic LE, Burton JH, Thamm DH, Withrow SJ, Lana SE (2014) Comparison of carboplatin and doxorubicin-based chemotherapy protocols in 470 dogs after amputation for treatment of appendicular osteosarcoma. J Vet Intern Med 28:554–563
22. Berg J, Weinstein MJ, Springfield DS, Rand WM (1995) Results of surgery and doxorubicin chemotherapy in dogs with osteosarcoma. J Am Vet Med Assoc 206:1555–1560
23. Scott MC, Sarver AL, Gavin KJ et al (2011) Molecular subtypes of osteosarcoma identified by reducing tumor heterogeneity through an interspecies comparative approach. Bone 49:356–367
24. Scott MC, Temiz NA, Sarver AE et al (2018) Comparative transcriptome analysis quantifies immune cell transcript levels, metastatic progression, and survival in osteosarcoma. Cancer Res 78:326–337
25. Shao YW, Wood GA, Lu J et al (2019) Cross-species genomics identifies DLG2 as a tumor suppressor in osteosarcoma. Oncogene 38:291–298
26. Sakthikumar S, Elvers I, Kim J et al (2018) SETD2 is recurrently mutated in whole-exome sequenced canine osteosarcoma. Cancer Res 78:3421–3431
27. Davis LE, Jeng S, Svalina MN et al (2017) Integration of genomic, transcriptomic and functional profiles of aggressive osteosarcomas across multiple species. Oncotarget 8:76241–76256
28. Straw RC, Powers BE, Withrow SJ, Cooper MF, Turner AS (1992) The effect of intramedullary polymethylmethacrylate on healing of intercalary cortical allografts in a canine model. J Orthop Res 10:434–439
29. Fan TM, Selting KA (2018) Exploring the potential utility of pet dogs with cancer for studying radiation-induced immunogenic cell death strategies. Front Oncol 8:680
30. Coley WB (1907) Sarcoma of the long bones: the diagnosis, treatment and prognosis, with a report of sixty-nine cases. Ann Surg 45:321–368
31. Coley WB (1910) The treatment of inoperable sarcoma by bacterial toxins (the mixed toxins of the Streptococcus erysipelas and the Bacillus prodigiosus). Proc R Soc Med 3:1–48
32. Culp WT, Olea-Popelka F, Sefton J et al (2014) Evaluation of outcome and prognostic factors for dogs living greater than one year after diagnosis of osteosarcoma: 90 cases (1997-2008). J Am Vet Med Assoc 245:1141–1146
33. Lascelles BD, Dernell WS, Correa MT et al (2005) Improved survival associated with postoperative wound infection in dogs treated with limb-salvage surgery for osteosarcoma. Ann Surg Oncol 12:1073–1083
34. Liptak JM, Dernell WS, Ehrhart N, Lafferty MH, Monteith GJ, Withrow SJ (2006) Cortical allograft and endoprosthesis for limb-sparing surgery in dogs with distal radial osteosarcoma: a prospective clinical comparison of two different limb-sparing techniques. Vet Surg 35:518–533
35. Mehl ML, Withrow SJ, Seguin B et al (2001) Spontaneous regression of osteosarcoma in four dogs. J Am Vet Med Assoc 219:614–617
36. Ogihara Y, Takeda K, Yanagawa T, Hirasawa Y (1994) Spontaneous regression of lung metastases from osteosarcoma. Cancer 74:2798–2803
37. Carlsen NL (1999) Spontaneous regression of malignant tumors. Ugeskr Laeger 161:3980
38. Biller BJ, Guth A, Burton JH, Dow SW (2010) Decreased ratio of CD8+ T cells to regulatory T cells associated with decreased survival in dogs with osteosarcoma. J Vet Intern Med 24:1118–1123
39. Troyer RM, Ruby CE, Goodall CP et al (2017) Exosomes from osteosarcoma and normal osteoblast differ in proteomic cargo and immunomodulatory effects on T cells. Exp Cell Res 358: 369–376
40. Wedekind MF, Wagner LM, Cripe TP (2018) Immunotherapy for osteosarcoma: where do we go from here? Pediatr Blood Cancer 65:e27227
41. Dyson KA, Stover BD, Grippin A et al (2019) Emerging trends in immunotherapy for pediatric sarcomas. J Hematol Oncol 12:78
42. Fox MH, Armstrong LW, Withrow SJ et al (1990) Comparison of DNA aneuploidy of primary and metastatic spontaneous canine osteosarcomas. Cancer Res 50:6176–6178
43. LaRue SM, Fox MH, Withrow SJ et al (1994) Impact of heterogeneity in the predictive value of kinetic parameters in canine osteosarcoma. Cancer Res 54:3916–3921
44. Roberts RD, Lizardo MM, Reed DR et al (2019) Provocative questions in osteosarcoma basic and translational biology: a report from the Children's Oncology Group. Cancer 125:3514–3525
45. Khanna C, Fan TM, Gorlick R et al (2014) Toward a drug development path that targets metastatic progression in osteosarcoma. Clin Cancer Res 20:4200–4209

46. Alexandrov LB, Nik-Zainal S, Wedge DC et al (2013) Signatures of mutational processes in human cancer. Nature 500:415–421
47. Angstadt AY, Thayanithy V, Subramanian S, Modiano JF, Breen M (2012) A genome-wide approach to comparative oncology: high-resolution oligonucleotide aCGH of canine and human osteosarcoma pinpoints shared microaberrations. Cancer Genet 205:572–587
48. Schiffman JD, Breen M (2015) Comparative oncology: what dogs and other species can teach us about humans with cancer. Philos Trans R Soc Lond Ser B Biol Sci 370:20140231
49. Gardner HL, Sivaprakasam K, Briones N et al (2019) Canine osteosarcoma genome sequencing identifies recurrent mutations in DMD and the histone methyltransferase gene SETD2. Commun Biol 2:266
50. Kovac M, Blattmann C, Ribi S et al (2015) Exome sequencing of osteosarcoma reveals mutation signatures reminiscent of BRCA deficiency. Nat Commun 6:8940
51. Bousquet M, Noirot C, Accadbled F et al (2016) Whole-exome sequencing in osteosarcoma reveals important heterogeneity of genetic alterations. Ann Oncol 27:738–744
52. Grobner SN, Worst BC, Weischenfeldt J et al (2018) The landscape of genomic alterations across childhood cancers. Nature 555:321–327
53. Chanmee T, Ontong P, Konno K, Itano N (2014) Tumor-associated macrophages as major players in the tumor microenvironment. Cancers (Basel) 6:1670–1690
54. Tsutsui S, Yasuda K, Suzuki K, Tahara K, Higashi H, Era S (2005) Macrophage infiltration and its prognostic implications in breast cancer: the relationship with VEGF expression and microvessel density. Oncol Rep 14:425–431
55. Buddingh EP, Kuijjer ML, Duim RA et al (2011) Tumor-infiltrating macrophages are associated with metastasis suppression in high-grade osteosarcoma: a rationale for treatment with macrophage activating agents. Clin Cancer Res 17:2110–2119
56. Gomez-Brouchet A, Illac C, Gilhodes J et al (2017) CD163-positive tumor-associated macrophages and CD8-positive cytotoxic lymphocytes are powerful diagnostic markers for the therapeutic stratification of osteosarcoma patients: an immunohistochemical analysis of the biopsies fromthe French OS2006 phase 3 trial. Onco Targets Ther 6:e1331193
57. Dhupkar P, Gordon N, Stewart J, Kleinerman ES (2018) Anti-PD-1 therapy redirects macrophages from an M2 to an M1 phenotype inducing regression of OS lung metastases. Cancer Med 7:2654–2664
58. Withers SS, Skorupski KA, York D et al (2019) Association of macrophage and lymphocyte infiltration with outcome in canine osteosarcoma. Vet Comp Oncol 17:49–60
59. Withers SS, York D, Choi JW et al (2019) Metastatic immune infiltrates correlate with those of the primary tumour in canine osteosarcoma. Vet Comp Oncol 17(3):242–252
60. Sottnik JL, Rao S, Lafferty MH et al (2010) Association of blood monocyte and lymphocyte count and disease-free interval in dogs with osteosarcoma. J Vet Intern Med 24:1439–1444
61. Hingorani P, Maas ML, Gustafson MP et al (2015) Increased CTLA-4(+) T cells and an increased ratio of monocytes with loss of class II (CD14(+) HLA-DR(lo/neg)) found in aggressive pediatric sarcoma patients. J Immunother Cancer 3:35
62. Liu T, Fang XC, Ding Z, Sun ZG, Sun LM, Wang YL (2015) Pre-operative lymphocyte-to-monocyte ratio as a predictor of overall survival in patients suffering from osteosarcoma. FEBS Open Bio 5:682–687
63. Selmic LE, Lafferty MH, Kamstock DA et al (2014) Outcome and prognostic factors for osteosarcoma of the maxilla, mandible, or calvarium in dogs: 183 cases (1986-2012). J Am Vet Med Assoc 245:930–938
64. Tuohy JL, Lascelles BD, Griffith EH, Fogle JE (2016) Association of canine osteosarcoma and monocyte phenotype and chemotactic function. J Vet Intern Med 30:1167–1178
65. Wong KL, Yeap WH, Tai JJ, Ong SM, Dang TM, Wong SC (2012) The three human monocyte subsets: implications for health and disease. Immunol Res 53:41–57
66. Zhang W, Dziak R (1996) Tumor necrosis factor alpha stimulates arachidonic acid metabolism in human osteoblastic osteosarcomal cells. Prostaglandins Leukot Essent Fatty Acids 54:427–431
67. Segaliny AI, Mohamadi A, Dizier B et al (2015) Interleukin-34 promotes tumor progression and metastatic process in osteosarcoma through induction of angiogenesis and macrophage recruitment. Int J Cancer 137:73–85
68. Kalinski P (2012) Regulation of immune responses by prostaglandin E2. J Immunol 188:21–28
69. Mohammed SI, Coffman K, Glickman NW et al (2001) Prostaglandin E2 concentrations in naturally occurring canine cancer. Prostaglandins Leukot Essent Fatty Acids 64:1–4
70. Shor S, Fadl-Alla BA, Pondenis HC et al (2015) Expression of nociceptive ligands in canine osteosarcoma. J Vet Intern Med 29:268–275
71. Millanta F, Asproni P, Cancedda S, Vignoli M, Bacci B, Poli A (2012) Immunohistochemical expression of COX-2, mPGES and EP2 receptor in normal and reactive canine bone and in canine osteosarcoma. J Comp Pathol 147:153–160
72. Wasserman J, Diese L, VanGundy Z, London C, Carson WE, Papenfuss TL (2012) Suppression of canine myeloid cells by soluble factors from cultured canine tumor cells. Vet Immunol Immunopathol 145:420–430

73. Dumars C, Ngyuen JM, Gaultier A et al (2016) Dysregulation of macrophage polarization is associated with the metastatic process in osteosarcoma. Oncotarget 7:78343–78354
74. Pahl JH, Kwappenberg KM, Varypataki EM et al (2014) Macrophages inhibit human osteosarcoma cell growth after activation with the bacterial cell wall derivative liposomal muramyl tripeptide in combination with interferon-gamma. J Exp Clin Cancer Res 33:27
75. Zhang X, Zhou H, Wang J et al (2002) Arsenic trioxide, retinoic acid and Ara-c regulated the expression of annexin II on the surface of APL cells, a novel co-receptor for plasminogen/tissue plasminogen activator. Thromb Res 106:63–70
76. Diaz-Montero CM, Finke J, Montero AJ (2014) Myeloid-derived suppressor cells in cancer: therapeutic, predictive, and prognostic implications. Semin Oncol 41:174–184
77. Corzo CA, Condamine T, Lu L et al (2010) HIF-1alpha regulates function and differentiation of myeloid-derived suppressor cells in the tumor microenvironment. J Exp Med 207:2439–2453
78. Zhang H, Maric I, DiPrima MJ et al (2013) Fibrocytes represent a novel MDSC subset circulating in patients with metastatic cancer. Blood 122:1105–1113
79. Goulart MR, Hlavaty SI, Chang YM et al (2019) Phenotypic and transcriptomic characterization of canine myeloid-derived suppressor cells. Sci Rep 9:3574
80. Goulart MR, Pluhar GE, Ohlfest JR (2012) Identification of myeloid derived suppressor cells in dogs with naturally occurring cancer. PLoS One 7:e33274
81. Sherger M, Kisseberth W, London C, Olivo-Marston S, Papenfuss TL (2012) Identification of myeloid derived suppressor cells in the peripheral blood of tumor bearing dogs. BMC Vet Res 8:209
82. Biller BJ, Elmslie RE, Burnett RC, Avery AC, Dow SW (2007) Use of FoxP3 expression to identify regulatory T cells in healthy dogs and dogs with cancer. Vet Immunol Immunopathol 116:69–78
83. O'Neill K, Guth A, Biller B, Elmslie R, Dow S (2009) Changes in regulatory T cells in dogs with cancer and associations with tumor type. J Vet Intern Med 23:875–881
84. Portela RF, Fadl-Alla BA, Pondenis HC et al (2014) Pro-tumorigenic effects of transforming growth factor beta 1 in canine osteosarcoma. J Vet Intern Med 28:894–904
85. Rissetto KC, Rindt H, Selting KA, Villamil JA, Henry CJ, Reinero CR (2010) Cloning and expression of canine CD25 for validation of an anti-human CD25 antibody to compare T regulatory lymphocytes in healthy dogs and dogs with osteosarcoma. Vet Immunol Immunopathol 135:137–145
86. Fritzsching B, Fellenberg J, Moskovszky L et al (2015) CD8(+)/FOXP3(+)-ratio in osteosarcoma microenvironment separates survivors from non-survivors: a multicenter validated retrospective study. Onco Targets Ther 4:e990800
87. Lussier DM, O'Neill L, Nieves LM et al (2015) Enhanced T-cell immunity to osteosarcoma through antibody blockade of PD-1/PD-L1 interactions. J Immunother 38:96–106
88. Modiano JF, Bellgrau D, Cutter GR et al (2012) Inflammation, apoptosis, and necrosis induced by neoadjuvant fas ligand gene therapy improves survival of dogs with spontaneous bone cancer. Mol Ther 20:2234–2243
89. Zhu Z, Jin Z, Zhang M et al (2017) Prognostic value of programmed death-ligand 1 in sarcoma: a meta-analysis. Oncotarget 8:59570–59580
90. Shen JK, Cote GM, Choy E et al (2014) Programmed cell death ligand 1 expression in osteosarcoma. Cancer Immunol Res 2:690–698
91. Merchant MS, Wright M, Baird K et al (2016) Phase I clinical trial of ipilimumab in pediatric patients with advanced solid tumors. Clin Cancer Res 22:1364–1370
92. Le Cesne A, Marec-Berard P, Blay JY et al (2019) Programmed cell death 1 (PD-1) targeting in patients with advanced osteosarcomas: results from the PEMBROSARC study. Eur J Cancer 119:151–157
93. Tawbi HA, Burgess M, Bolejack V et al (2017) Pembrolizumab in advanced soft-tissue sarcoma and bone sarcoma (SARC028): a multicentre, two-cohort, single-arm, open-label, phase 2 trial. Lancet Oncol 18:1493–1501
94. Hartley G, Faulhaber E, Caldwell A et al (2017) Immune regulation of canine tumour and macrophage PD-L1 expression. Vet Comp Oncol 15:534–549
95. Maekawa N, Konnai S, Ikebuchi R et al (2014) Expression of PD-L1 on canine tumor cells and enhancement of IFN-gamma production from tumor-infiltrating cells by PD-L1 blockade. PLoS One 9:e98415
96. Koirala P, Roth ME, Gill J et al (2016) Immune infiltration and PD-L1 expression in the tumor microenvironment are prognostic in osteosarcoma. Sci Rep 6:30093
97. Palmerini E, Agostinelli C, Picci P et al (2017) Tumoral immune-infiltrate (IF), PD-L1 expression and role of CD8/TIA-1 lymphocytes in localized osteosarcoma patients treated within protocol ISG-OS1. Oncotarget 8:111836–111846
98. Coy J, Caldwell A, Chow L, Guth A, Dow S (2017) PD-1 expression by canine T cells and functional effects of PD-1 blockade. Vet Comp Oncol 15:1487–1502
99. Nemoto Y, Shosu K, Okuda M, Noguchi S, Mizuno T (2018) Development and characterization of monoclonal antibodies against canine PD-1 and PD-L1. Vet Immunol Immunopathol 198:19–25
100. Giraldo NA, Becht E, Remark R, Damotte D, Sautes-Fridman C, Fridman WH (2014) The immune contexture of primary and metastatic human tumours. Curr Opin Immunol 27:8–15

101. Sundara YT, Kostine M, Cleven AH, Bovee JV, Schilham MW, Cleton-Jansen AM (2017) Increased PD-L1 and T-cell infiltration in the presence of HLA class I expression in metastatic high-grade osteosarcoma: a rationale for T-cell-based immunotherapy. Cancer Immunol Immunother 66:119–128
102. Tuohy JL, Somarelli JA, Borst LB, Eward WC, Lascelles BDX, Fogle JE (2020) Immune dysregulation and osteosarcoma: Staphylococcus aureus downregulates TGF-beta and heightens the inflammatory signature in human and canine macrophages suppressed by osteosarcoma. Vet Comp Oncol 18(1):64–75
103. Redelman-Sidi G, Glickman MS, Bochner BH (2014) The mechanism of action of BCG therapy for bladder cancer—a current perspective. Nat Rev Urol 11:153–162
104. Bech-Nielsen S, Brodey RS, Fidler IJ, Abt DA, Reif JS (1977) The effect of BCG on in vitro immune reactivity and clinical course in dogs treated surgically for osteosarcoma. Eur J Cancer 13:33–41
105. Owen LN, Bostock DE (1974) Effects of intravenous BCG in normal dogs and in dogs with spontaneous osteosarcoma. Eur J Cancer 10:775–780
106. Betton GR, Gorman NT, Owen LN (1979) Cell mediated cytotoxicity in dogs following systemic or local BCG treatment alone or in combination with allogeneic tumour cell lines. Eur J Cancer 15:745–754
107. Gorman NT (1979) Alveolar macrophage cytotoxicity in dogs following intravenous BCG. Eur J Cancer 15:1051–1059
108. Thrall DE, Withrow SJ, Powers BE et al (1990) Radiotherapy prior to cortical allograft limb sparing in dogs with osteosarcoma: a dose response assay. Int J Radiat Oncol Biol Phys 18:1351–1357
109. Sottnik JL, U'Ren LW, Thamm DH, Withrow SJ, Dow SW (2010) Chronic bacterial osteomyelitis suppression of tumor growth requires innate immune responses. Cancer Immunol Immunother 59:367–378
110. MacEwen EG, Kurzman ID, Rosenthal RC et al (1989) Therapy for osteosarcoma in dogs with intravenous injection of liposome-encapsulated muramyl tripeptide. J Natl Cancer Inst 81:935–938
111. Kurzman ID, MacEwen EG, Rosenthal RC et al (1995) Adjuvant therapy for osteosarcoma in dogs: results of randomized clinical trials using combined liposome-encapsulated muramyl tripeptide and cisplatin. Clin Cancer Res 1:1595–1601
112. Kurzman ID, Shi F, Vail DM, MacEwen EG (1999) In vitro and in vivo enhancement of canine pulmonary alveolar macrophage cytotoxic activity against canine osteosarcoma cells. Cancer Biother Radiopharm 14:121–128
113. Smith BW, Kurzman ID, Schultz KT, Czuprynski CJ, MacEwen EG (1993) Muramyl peptides augment the in vitro and in vivo cytostatic activity of canine plastic-adherent mononuclear cells against canine osteosarcoma cells. Cancer Biother 8:137–144
114. Shi F, MacEwen EG, Kurzman ID (1993) In vitro and in vivo effect of doxorubicin combined with liposome-encapsulated muramyl tripeptide on canine monocyte activation. Cancer Res 53:3986–3991
115. Mitchell DH, Withrow SJ, Johnston MR, Kruse CA (1991) Cytotoxicity against autologous, allogeneic, and xenogeneic tumor targets by human recombinant interleukin-2-activated lymphocytes from healthy dogs and dogs with lung tumors. Am J Vet Res 52:1132–1136
116. Khanna C, Hasz DE, Klausner JS, Anderson PM (1996) Aerosol delivery of interleukin 2 liposomes is nontoxic and biologically effective: canine studies. Clin Cancer Res 2:721–734
117. Khanna C, Anderson PM, Hasz DE, Katsanis E, Neville M, Klausner JS (1997) Interleukin-2 liposome inhalation therapy is safe and effective for dogs with spontaneous pulmonary metastases. Cancer 79:1409–1421
118. Rivoltini L, Viggiano V, Spinazze S et al (1993) In vitro anti-tumor activity of eosinophils from cancer patients treated with subcutaneous administration of interleukin 2. Role of interleukin 5. Int J Cancer 54:8–15
119. Dow S, Elmslie R, Kurzman I, MacEwen G, Pericle F, Liggitt D (2005) Phase I study of liposome-DNA complexes encoding the interleukin-2 gene in dogs with osteosarcoma lung metastases. Hum Gene Ther 16:937–946
120. Regan DP, Coy JW, Chahal KK et al (2019) The angiotensin receptor blocker losartan suppresses growth of pulmonary metastases via AT1R-independent inhibition of CCR2 signaling and monocyte recruitment. J Immunol 202:3087–3102
121. Regan D, Dow S (2015) Manipulation of innate immunity for cancer therapy in dogs. Vet Sci 2:423–439
122. Liu H, Wang SH, Chen SC, Chen CY, Lo JL, Lin TM (2016) Immune modulation of CD4(+)CD25(+) regulatory T cells by zoledronic acid. BMC Immunol 17:45
123. Green J, Clezardin P (2010) The molecular basis of bisphosphonate activity: a preclinical perspective. Semin Oncol 37(Suppl 1):S3–S11
124. Coscia M, Quaglino E, Iezzi M et al (2010) Zoledronic acid repolarizes tumour-associated macrophages and inhibits mammary carcinogenesis by targeting the mevalonate pathway. J Cell Mol Med 14:2803–2815
125. Wolfe TD, Pillai SP, Hildreth BE 3rd et al (2011) Effect of zoledronic acid and amputation on bone invasion and lung metastasis of canine osteosarcoma in nude mice. Clin Exp Metastasis 28:377–389
126. Gorlick R, Huvos AG, Heller G et al (1999) Expression of HER2/erbB-2 correlates with survival in osteosarcoma. J Clin Oncol 17:2781–2788
127. Roth M, Linkowski M, Tarim J et al (2014) Ganglioside GD2 as a therapeutic target for antibody-mediated therapy in patients with osteosarcoma. Cancer 120:548–554

128. Dobrenkov K, Ostrovnaya I, Gu J, Cheung IY, Cheung NK (2016) Oncotargets GD2 and GD3 are highly expressed in sarcomas of children, adolescents, and young adults. Pediatr Blood Cancer 63:1780–1785
129. Maniscalco L, Iussich S, Morello E et al (2015) Increased expression of insulin-like growth factor-1 receptor is correlated with worse survival in canine appendicular osteosarcoma. Vet J 205:272–280
130. Anderson PM, Meyers DE, Hasz DE et al (1995) In vitro and in vivo cytotoxicity of an anti-osteosarcoma immunotoxin containing pokeweed antiviral protein. Cancer Res 55:1321–1327
131. Mostowy S, Cossart P (2012) Virulence factors that modulate the cell biology of Listeria infection and the host response. Adv Immunol 113:19–32
132. Wood LM, Paterson Y (2014) Attenuated Listeria monocytogenes: a powerful and versatile vector for the future of tumor immunotherapy. Front Cell Infect Microbiol 4:51
133. Shahabi V, Seavey MM, Maciag PC, Rivera S, Wallecha A (2011) Development of a live and highly attenuated Listeria monocytogenes-based vaccine for the treatment of Her2/neu-overexpressing cancers in human. Cancer Gene Ther 18: 53–62
134. Ahmed N, Salsman VS, Yvon E et al (2009) Immunotherapy for osteosarcoma: genetic modification of T cells overcomes low levels of tumor antigen expression. Mol Ther 17:1779–1787
135. Flint AF, U'Ren L, Legare ME, Withrow SJ, Dernell W, Hanneman WH (2004) Overexpression of the erbB-2 proto-oncogene in canine osteosarcoma cell lines and tumors. Vet Pathol 41:291–296
136. Mata M, Vera JF, Gerken C et al (2014) Toward immunotherapy with redirected T cells in a large animal model: ex vivo activation, expansion, and genetic modification of canine T cells. J Immunother 37:407–415
137. Mason NJ, Gnanandarajah JS, Engiles JB et al (2016) Immunotherapy with a HER2-targeting Listeria induces HER2-specific immunity and demonstrates potential therapeutic effects in a phase I trial in canine osteosarcoma. Clin Cancer Res 22:4380–4390
138. Al-Ramadi BK, Fernandez-Cabezudo MJ, El-Hasasna H et al (2008) Attenuated bacteria as effectors in cancer immunotherapy. Ann N Y Acad Sci 1138:351–357
139. Thamm DH, Kurzman ID, King I et al (2005) Systemic administration of an attenuated, tumor-targeting Salmonella typhimurium to dogs with spontaneous neoplasia: phase I evaluation. Clin Cancer Res 11:4827–4834
140. Fritz SE, Henson MS, Greengard E et al (2016) A phase I clinical study to evaluate safety of orally administered, genetically engineered Salmonella enterica serovar Typhimurium for canine osteosarcoma. Vet Med Sci 2:179–190
141. Peruzzi D, Mesiti G, Ciliberto G, La Monica N, Aurisicchio L (2010) Telomerase and HER-2/neu as targets of genetic cancer vaccines in dogs. Vaccine 28:1201–1208
142. Marchini A, Daeffler L, Pozdeev VI, Angelova A, Rommelaere J (2019) Immune conversion of tumor microenvironment by oncolytic viruses: the protoparvovirus H-1PV case study. Front Immunol 10:1848
143. Le Boeuf F, Selman M, Son HH et al (2017) Oncolytic Maraba virus MG1 as a treatment for sarcoma. Int J Cancer 141:1257–1264
144. Naik S, Galyon GD, Jenks NJ et al (2018) Comparative oncology evaluation of intravenous recombinant oncolytic vesicular stomatitis virus therapy in spontaneous canine cancer. Mol Cancer Ther 17:316–326
145. Laborda E, Puig-Saus C, Rodriguez-Garcia A et al (2014) A pRb-responsive, RGD-modified, and hyaluronidase-armed canine oncolytic adenovirus for application in veterinary oncology. Mol Ther 22:986–998
146. Trauth BC, Klas C, Peters AM et al (1989) Monoclonal antibody-mediated tumor regression by induction of apoptosis. Science 245:301–305
147. Modiano JF, Bellgrau D (2016) Fas ligand based immunotherapy: a potent and effective neoadjuvant with checkpoint inhibitor properties, or a systemically toxic promoter of tumor growth? Discov Med 21:109–116
148. Rensing-Ehl A, Frei K, Flury R et al (1995) Local Fas/APO-1 (CD95) ligand-mediated tumor cell killing in vivo. Eur J Immunol 25:2253–2258
149. Theoleyre S, Mori K, Cherrier B et al (2005) Phenotypic and functional analysis of lymphocytes infiltrating osteolytic tumors: use as a possible therapeutic approach of osteosarcoma. BMC Cancer 5:123
150. Gasser S, Raulet DH (2006) Activation and self-tolerance of natural killer cells. Immunol Rev 214:130–142
151. Huang YC, Hung SW, Jan TR et al (2008) CD5-low expression lymphocytes in canine peripheral blood show characteristics of natural killer cells. J Leukoc Biol 84:1501–1510
152. Foltz JA, Somanchi SS, Yang Y, Aquino-Lopez A, Bishop EE, Lee DA (2016) NCR1 expression identifies canine natural killer cell subsets with phenotypic similarity to human natural killer cells. Front Immunol 7:521
153. Canter RJ, Grossenbacher SK, Foltz JA et al (2017) Radiotherapy enhances natural killer cell cytotoxicity and localization in pre-clinical canine sarcomas and first-in-dog clinical trial. J Immunother Cancer 5:98
154. Ramirez O 3rd, Dodge RK, Page RL et al (1999) Palliative radiotherapy of appendicular osteosarcoma in 95 dogs. Vet Radiol Ultrasound 40:517–522

155. Green EM, Adams WM, Forrest LJ (2002) Four fraction palliative radiotherapy for osteosarcoma in 24 dogs. J Am Anim Hosp Assoc 38:445–451
156. Knapp-Hoch HM, Fidel JL, Sellon RK, Gavin PR (2009) An expedited palliative radiation protocol for lytic or proliferative lesions of appendicular bone in dogs. J Am Anim Hosp Assoc 45:24–32
157. Panjwani MK, Smith JB, Schutsky K et al (2016) Feasibility and safety of RNA-transfected CD20-specific chimeric antigen receptor T cells in dogs with spontaneous B cell lymphoma. Mol Ther 24:1602–1614
158. Yin Y, Boesteanu AC, Binder ZA et al (2018) Checkpoint blockade reverses anergy in IL-13Ralpha2 humanized scFv-based CAR T cells to treat murine and canine gliomas. Mol Ther Oncolytics 11:20–38
159. Ahmed N, Salsman VS, Yvon E et al (2009) Immunotherapy for osteosarcoma: genetic modification of T cells overcomes low levels of tumor antigen expression. Mol Ther 17:1779–1787
160. Rainusso N, Brawley VS, Ghazi A et al (2012) Immunotherapy targeting HER2 with genetically modified T cells eliminates tumor-initiating cells in osteosarcoma. Cancer Gene Ther 19:212–217
161. DeRenzo C, Gottschalk S (2014) Genetically modified T-cell therapy for osteosarcoma. Adv Exp Med Biol 804:323–340
162. Galluzzi L, Buque A, Kepp O, Zitvogel L, Kroemer G (2017) Immunogenic cell death in cancer and infectious disease. Nat Rev Immunol 17:97–111
163. Gupta A, Probst HC, Vuong V et al (2012) Radiotherapy promotes tumor-specific effector CD8+ T cells via dendritic cell activation. J Immunol 189:558–566
164. Sharma A, Bode B, Studer G et al (2013) Radiotherapy of human sarcoma promotes an intratumoral immune effector signature. Clin Cancer Res 19:4843–4853
165. Dewan MZ, Galloway AE, Kawashima N et al (2009) Fractionated but not single-dose radiotherapy induces an immune-mediated abscopal effect when combined with anti-CTLA-4 antibody. Clin Cancer Res 15:5379–5388
166. Seetharam S, Staba MJ, Schumm LP et al (1999) Enhanced eradication of local and distant tumors by genetically produced interleukin-12 and radiation. Int J Oncol 15:769–773
167. Rech AJ, Dada H, Kotzin JJ et al (2018) Radiotherapy and CD40 activation separately augment immunity to checkpoint blockade in cancer. Cancer Res 78:4282–4291
168. Qu C, Ping NN, Wu Q et al (2018) Radiotherapy priming chimeric antigen receptor T cell therapy is a safe and promising approach in relapsed/refractory diffuse large B cell lymphoma patients with high tumor burden. In: 60th American Society of Hematology Annual Meeting and Exposition, San Diego
169. Coomer A, Farese J, Milner R, Liptak J, Bacon N, Lurie D (2009) Radiation therapy for canine appendicular osteosarcoma. Vet Comp Oncol 7:15–27
170. McEntee MC (1997) Radiation therapy in the management of bone tumors. Vet Clin North Am Small Anim Pract 27:131–138
171. Takahashi Y, Yasui T, Tamari K et al (2017) Radiation enhanced the local and distant anti-tumor efficacy in dual immune checkpoint blockade therapy in osteosarcoma. PLoS One 12:e0189697

Index

E. S. Kleinerman, R. Gorlick (eds.), *Current Advances in the Science of Osteosarcoma*, Advances in Experimental Medicine and Biology 1258, https://doi.org/10.1007/978-3-030-43085-6

D

E

The manufacturer's authorised representative in the EU is Springer Nature Customer Service Centre GmbH, Europaplatz 3, 69115 Heidelberg, Germany. If you have any concerns regarding our products, please contact ProductSafety@springernature.com

Printed and bound by CPI Group (UK) Ltd, Croydon, CR0 4YY
22/09/2026
02221060-0020